Advances in Geographical and Environmental Sciences

Series Editor

R. B. Singh, University of Delhi, Delhi, India

Advances in Geographical and Environmental Sciences synthesizes series diagnostigation and prognostication of earth environment, incorporating challenging interactive areas within ecological envelope of geosphere, biosphere, hydrosphere, atmosphere and cryosphere. It deals with land use land cover change (LUCC), urbanization, energy flux, land-ocean fluxes, climate, food security, ecohydrology, biodiversity, natural hazards and disasters, human health and their mutual interaction and feedback mechanism in order to contribute towards sustainable future. The geosciences methods range from traditional field techniques and conventional data collection, use of remote sensing and geographical information system, computer aided technique to advance geostatistical and dynamic modeling.

The series integrate past, present and future of geospheric attributes incorporating biophysical and human dimensions in spatio-temporal perspectives. The geosciences, encompassing land-ocean-atmosphere interaction is considered as a vital component in the context of environmental issues, especially in observation and prediction of air and water pollution, global warming and urban heat islands. It is important to communicate the advances in geosciences to increase resilience of society through capacity building for mitigating the impact of natural hazards and disasters. Sustainability of human society depends strongly on the earth environment, and thus the development of geosciences is critical for a better understanding of our living environment, and its sustainable development.

Geoscience also has the responsibility to not confine itself to addressing current problems but it is also developing a framework to address future issues. In order to build a 'Future Earth Model' for understanding and predicting the functioning of the whole climatic system, collaboration of experts in the traditional earth disciplines as well as in ecology, information technology, instrumentation and complex system is essential, through initiatives from human geoscientists. Thus human geosceince is emerging as key policy science for contributing towards sustainability/survivality science together with future earth initiative.

Advances in Geographical and Environmental Sciences series publishes books that contain novel approaches in tackling issues of human geoscience in its broadest sense—books in the series should focus on true progress in a particular area or region. The series includes monographs and edited volumes without any limitations in the page numbers.

More information about this series at https://link.springer.com/bookseries/13113

Praveen Kumar Rai · Varun Narayan Mishra ·
Prafull Singh
Editors

Geospatial Technology for Landscape and Environmental Management

Sustainable Assessment and Planning

 Springer

Editors
Praveen Kumar Rai
Department of Geography
Khwaja Moinuddin Chishti Language
University
Lucknow, Uttar Pradesh, India

Varun Narayan Mishra
Centre for Climate Change and Water
Research
Suresh Gyan Vihar University
Jaipur, Rajasthan, India

Prafull Singh ⓘ
Department of Geology,
School of Earth,
Biological and Environmental Sciences
Central University of South Bihar
Gaya, Bihar, India

ISSN 2198-3542   ISSN 2198-3550 (electronic)
Advances in Geographical and Environmental Sciences
ISBN 978-981-16-7375-7   ISBN 978-981-16-7373-3   (eBook)
https://doi.org/10.1007/978-981-16-7373-3

Preface

Landscapes are spatial mosaics of interacting biophysical and socioeconomic components that provide living space for mankind. A landscape can be described by its composition and configuration. The composition refers to the variety and abundance of patch types within a landscape. The configuration refers to the spatial character of patches within a landscape. These two aspects can separately or together affect the ecological processes. The heterogeneity, scale, pattern, hierarchy, disturbance, coupled ecological–social dynamics, and sustainability are the key factors in landscape monitoring. The geospatial distribution and quantification of landscape components is of critical significance in support of sustainable land-use planning, natural resource management, and environmental monitoring. The landscape change can also provide important information to land managers relating to environmental impacts of various types of land use and patterns. The continuous monitoring of landscape and environment at a large scale and at a regular interval can only be possible by using advance technologies.

With the advent of multi-scale geospatial information on composition and configuration of landscapes, the acquisition of spatial information on landscape metrics using geospatial technology has become a common practice. These technologies offer considerable potential for sustainable assessment and planning of landscape and environment. Geospatial technology is a combination of state-of-the-art remote sensing, GIS, and global navigation satellite system (GNSS) technology for the mapping and monitoring of landscapes. The main thrust of using geospatial technology is to understand the causes, mechanisms, and consequences of spatial heterogeneity, while its ultimate objective is to provide a scientific basis for developing and maintaining ecologically, economically, and socially sustainable landscapes.

This book will highlight the state-of-the-art new approaches, various modeling aspects, the role of field and earth observations technologies in landscape, and environmental management practices. It addresses the interests of a wide spectrum of readers with a common interest in geospatial science, geology, water resource management, database management, planning and policy making, and resource management. The chapters in this edited book mostly emphasize the study of landscape and environmental components through advanced computational techniques in conjunction with earth observation data sets and GIS for better assessment, planning, and management of landscape and environment. It also seeks to provide a comprehensive approach for professionals, policy makers, researchers, and academicians across the countries.

In Chap. 1, an attempt is made to analyze the spatiotemporal variability of channel planform dynamics of Ichamati River, West Bengal, India. In this study, the US-Army Toposheet of 1922 and multi-temporal Landsat images of 1976, 1996, and 2016 were used. Besides, the cross-sectional survey of the river was conducted with the help of an echo-sounder and a GPS during 2013–2018. A questionnaire survey was also performed to acquire information regarding the expansion of brick kilns, annual rate of brick production, sediment extraction rate, and land-use activities of brick kilns. This study revealed that the natural shifting and meander mobility of the river have been gradually decreasing with the expansion of brick kilns in the last few decades. The excessive siltation within the channel causes upliftment of the riverbed as measured 9 cm year^{-1} during 2013–2018.

In Chap. 2, a study has been carried out to prepare an integrated water resource action plan for conjunctive use of available water resource in a sub-humid tropical watershed of East India. Delineation of potential zones is done by developing various thematic maps using satellite imagery and associated databases. Overlaying the thematic maps and keeping in view the available groundwater resource, an integrated water resource plan is prepared in ArcGIS software. The net annual groundwater availability is found to be 80.989 ha.m, while the annual draft is 10.18 ha.m. After superimposing all thematic layers, four groundwater potential zones such as poor, moderate, good to moderate, and very good are identified. More than 50% area of the watershed falls in a moderate prospect zone.

In Chap. 3, a hydrologic modeling using the HEC-RAS model in combination with watershed modeling system (WMS) tools compared with the flood hazard index (FHI) method using GIS in the Seyad basin situated in the southwestern region of Morocco. The HEC-RAS approach combines the surface hydrologic model and the digital terrain model data. This combination allows the mapping of the flood zones by using the WMS tools. A weight is calculated from the analytic hierarchy process method and applied to each parameter. HEC-RAS method allows the mapping of a flood with a flood water surface profile that shows the depth of flood for annual exceedance probability (AEP), while FHI permits establishing flood risk level without indicating the depth of water.

In Chap. 4, a geospatial modeling is performed in assessing sustainable water resource management in a Gondia district, India, using remote sensing (RS) and geographical information system (GIS) techniques. The monsoon rains in Gondia district are concentrated in the four months from June to September and receive 90.81% rainfall, post-monsoon 1.86%, pre-monsoon 4.83%, and winter 2.48%. The distribution of annual rainfall in study area is very uneven. Out of the total area in the district, reserved forest area is 56.4%, protected forest area is 26.3%, and unclassified forest area is 17.3%. There are 192 small irrigated ponds below 100 hectares, and its projected irrigation capacity is 10,897 hectares. There are also 294 Kolhapuri-type dams with a projected irrigation capacity of 10,075 hectares and 1559 storage dams with a projected irrigation capacity of 14,817 hectares. The water is neutral to alkaline in nature with pH ranging from 6.6 to 8.92 with high TDS range from 140 ppm to 2184 ppm. The study can be useful for area management, restoration, and conservation of natural resources in the future.

In Chap. 5, groundwater samples were obtained in the 2016 and 2017 pre-monsoon and post-monsoon seasons at 105 locations from dug wells and bore wells along the coast of Andhra Pradesh in the Krishna and Godavari deltas. Groundwater samples are tested for large ions to determine the infiltration of salt water and to classify the salinity sources in the delta zone. Various hydrogeochemical parameters such as pH, electrical conductivity (EC), total dissolved solids (TDS), Ca^{2+}, Mg^{2+}, K^+, Na^+, CO_3, HCO_3-, and Cl^-, and SO_4^{2-} are evaluated for the delineation of the intrusion of saltwater in terms of Ca^{2+}/Mg^{2+}, $Cl^-/(CO^{3+} HCO^{3-})$, Na^+/Cl^- ratios. It is reported that the availability of fresh groundwater is 14 and 62%, respectively, during the pre-monsoon and post-monsoon seasons. The percentage levels of contamination in groundwater for slight, moderate, injuriously, highly, and severely categories are 43, 22, 12, 8, and 1%, respectively, for pre-monsoon season. However, during the post-monsoon season, the levels of contaminations in the above-mentioned categories are 22%, 9%, 4%, 1%, and 3%.

In Chap. 6, an attempt has been made to use the geospatial technology in the land and water resources management of a village enclosed within a micro-watershed. The geospatial database has been generated at the cadastral level or plot level, i.e., at 1:4000 scale. The resource inventory includes land use/land cover, digital elevation model (DEM), slope, geomorphology, ground water prospect, soil, land capability maps. The site suitability analysis technique is used to develop the land and water resources management action plan. In the land resource management plan, seven types of alternative land-use practices are suggested for the study area. The integration of the action plan map on the DEM gives a 3D perspective which will be an add-on in terms of visualization to the administrators and decision makers.

In Chap. 7, the lakes are important ecological units in urban ecosystem which preserve local climate, groundwater, and biodiversity. The unplanned continuous population growth in urban area causes severe destruction to the urban ecosystem across the world. Between 1970–2009, around 108 lakes were lost, and in between 2009–2013, around 230 lakes have been lost and cannot be revived. The present study is an attempt to create an inventory for current status of the lakes and wetlands using remote sensing and GIS techniques and their ecohydrological consequences on the surrounding environment. Time series satellite images from LISS-III and Survey of India maps of 1:50,000 were used for the study. For the study, six lakes have been identified from different parts of Delhi, which are on verge to devastating if not provided with immediate attention.

In Chap. 8, a quantitative assessment of Dhundsir Gad (Gad means stream) watershed has been carried out to study the relationship between hydrological characteristics, lithological, and structural antecedents for management, planning, and conservation activity. The data for morphometric parameters is extracted from remotely sensed images and processed in geographical information system (GIS) platforms for quantitative analysis. Geological field investigation of the watershed is done to investigate inter-relationship between the hydrological characteristics, lithology, and structural attributes. Morphometric characterization was measured from linear, relief, and areal aspects for four sub-watersheds of Dhundsir Gad. The prioritization of the sub-watershed has been done after evaluating and ranking morphometric parameters. Regional patterns of hypsometric integral (HI) and hypsometric curves in the watershed have also been computed to understand the role of tectonics and soil erosion in shaping the relief of the watershed.

In Chap. 9, a study has been carried out to identify the groundwater prolific zones using remote sensing (RS) and geographical information system (GIS) in Kamina sub-watershed of Bhima river basin, Shirur Taluka of Pune District, Maharashtra, India. Several thematic maps were generated to identify the potential zones. Analytical hierarchy process (AHP) is used for the delineation of groundwater potential zones. The AHP proposes a weight for each evaluation criteria according to the decision maker's pairwise comparisons of the criteria. The groundwater potential zone map so generated is divided into four classes (very low, low, moderate, and high) depending on the possibility of groundwater potential. The highest potential area is located toward eastern and southern region of Ghod River and thick soil cover. The findings of the study can help in the formulation of an efficient management plan for sustainable development of the area.

In Chap. 10, a study indicates the effectiveness of remote sensing (RS) and geographic information system (GIS)-based morphometric analysis of a sub-watershed of Damodar river basin in Ramgarh district, Jharkhand, India. The sub-watershed and drainage texture of the study area are extracted by ASTER DEM and topographical map in the GIS environment. Morphometric parameters such as stream order, stream length, bifurcation ratio, drainage density, stream frequency, form factor, and circulatory ratio are calculated. The sub-watershed's total drainage area is 46.71 km^2 and shows a dendritic drainage pattern that designates homogeneous lithology, gentle regional slope, and lack of structural control. The study area is designated as the fifth-order basin with a drainage density (D_d) value range 9.91 km/km^2. Results of the study have immense significance for engineers, managers, and planners for management of soil and water and provide watershed prioritizing management activities in the area.

In Chap. 11, the use of emerging technologies, early warnings, immediate incidence response, and post-recovery activities can well be employed to mitigate the losses that the disaster would lead to. One such emerging technology is the Internet of Things (IoT). This can help in monitoring for the purpose of early warning of disasters. The potential of IoT is described in order to provide rescue, response, mitigation, and preparedness to manage a disaster. This chapter explains the role of IoT in disaster management along with proposing a generic model having distinct layers through well-defined functionality. The chapter also explains the integration of cloud and IoT that could improve the efficiency of IoT applications in disaster management.

In Chap. 12, a work is carried out for developing a framework of systematic analysis to explain e-participation of citizens in India. The paper tries to identify and measure key indicators of e-participation in India. Survey of 200 respondents is analyzed from four smart cities across India. The work applies regression and concludes that all the indicators have a significant impact on e-participation of citizens in smart cities. The study finds that though the government is investing a huge amount of funds in smart cities development but still a lot is to be done. Also, the concept of participatory approach is currently not a prominent research theme among scholars. So, the current work will address this gap to unravel the conceptual framework of e-participation of citizens in smart cities in India.

In Chap. 13, a present study is performed to understand the spatial-temporal variability of LULC of Nagpur city, Maharashtra, from 2000 to 2020. The LULC classification is performed considering four different classes, i.e., barren land, built-up, agriculture, and water bodies. The LULC results show that the built-up area is increased by 26.62% from 2000 (41.24%) to 2020 (67.86%), and with a slight increase in water bodies, 0.19% is also evident. On the other hand, the area covered with vegetation is decreased by 15.93 % from 2000 (30.17%) to 2020 (14.24%), and barren land is reduced by 10.88%. The present study also predicted the LULC scenario using artificial neural network (ANN)-based cellular automata (CA) model with the help of different driving parameters. The prediction model showed an overall accuracy of 81.23% in predicting the 2025 LULC maps with the help of 2015 and 2020 LULC data. The results of the prediction model evident a maximum growth of 30.88% in the built-up area as compared to year 2020.

In Chap. 14, the research attempts to categorize the slums based on living standards, which will help to formulate the sustainable development techniques for better implementation of slum improvement projects. Data about the socioeconomic and physical condition of the slums has been collected using field surveys. For clustering slums in different categories, a $2 \times 2 \times 2$ matrix is formed. For creating an indicative matrix, essential inputs were identified, and an overall matrix table for all the slums with their scores was prepared. A georeferenced very high-resolution satellite imagery with a ward boundary map was used to create a base map. Different maps were generated showing current slum distribution and also the spatial distribution of varying slum categories. Maps were validated with field survey and with field photographs.

In Chap. 15, the study aims to analyze urban growth and sprawl in Dehradun city of Uttarakhand, which is one of the cities in government's smart city project list. The urban sprawl of Dehradun can be analyzed by using the Shannon's entropy approach. As per the result, the entropy value obtained for the year 2008 is 0.877 and 2016 is 1.598, in which the value of 2016 is near to the value of upper limit of log n (i.e., 1.591) which depicts more urban sprawl in 2016 than in 2008. The present study effectively used the Landsat TM data of year 2008 and 2016. This study can help for better planning and sustainable management of resources of a certain region and can help government officials and planners to monitor and analyze current urbanization and plan for future growth and requirements. This particular study can be helpful to understand the urbanization pattern in the city.

In Chap. 16, a study aims to map and analyze the dynamics of land-use/land cover changes using IRS LISS-III data for the years 2011–2012 and 2015–2016 of Andhra Pradesh state, India. On-screen visual interpretation techniques have been used to delineate the land-use/land cover classes in ArcGIS environment and cross-tabulation used for quantifying the changes in land-use pattern. The study reveals that built-up area, agriculture land, and water bodies have been increased about 0.21% (343.06 km^2), 0.11% (176.21 km^2), and 0.02% (32.08 km^2), respectively, while area under other land categories such as forest area, wastelands, and wetlands have decreased about 0.02% (107.44 km^2), 0.20% (333.38 km^2), and 0.07% (110.53 km^2), respectively. The results of this study would be helpful for planners, decision makers, and administrators to plan and implement appropriate decisions in order to sustainable resource utilization.

In Chap. 17, the current work tried to envisage the study on urbanization transitions due to launch of metro rail network in the NCR region. The metro network influenced the periphery and rural areas of Delhi and also contributing toward linear development or development of parallel infrastructures along the metro network. Cities of nearby states have become major intersections of metro network. Thereby, the current work tries to intend to establish the urbanization along with the social scenario, as it has a massive network of nearly 288 km length comprising of six different lines. This study proposes to identify the push and pull to further analyze the urbanization pattern due to the expansion of the network in the region, as the growing population in region requires further expansion. Hence, metro projects are expanding continuously and providing a new stimulus toward the increasing urbanization.

In Chap. 18, the spatiotemporal effects of UHI in Rajkot city has been assessed using Landsat 5 TM and Landsat 8 OLI remote sensing data. The land-use/land cover (LULC) classification is performed using maximum livelihood method on Landsat images for the year 2009 and 2017. Normalized difference vegetation index (NDVI) and land surface temperature (LST) were derived using mono-window algorithm. Subsequently, ambient air temperature was scrutinized and isotherm was derived for three locations in Rajkot city. On the basis of various results derived and analysis of temperature trend of past 60 years, it was determined that UHI effect was more prominent in the Central Business District (CBD) area of the selected regions. The results also revealed that the study region has experienced an increase of 0.3 °C in ambient air temperature in past 60 years. The built-up area and LST for LULC classes have also increased by 8.42% between 2009 and 2017 in Rajkot.

In Chap. 19, various rural towns are developing into urban towns, and hence, for a balanced and a proper development, a planning is required. Remote sensing is the acquiring of the data about the item without contacting it or without physically being present there. Due to advanced technology and new innovations, satellite imaging has enabled to collect and interpret various data which earlier was done physically and consumed a lot of time. A surface analysis is conducted with the help of remote sensing which gives a lot of information regarding various aspects, whereas it also interprets the physical data with other socioeconomic data. This interpretation helps in getting a link to the planning process. The information collected through satellites helps planning in various formats such as time, efficiency, and other ways.

In Chap. 20, a study was carried out to assess green spaces in Vijayawada's Urban Local Body. Because of better economic opportunities, the city has seen a surge in population inflow. As a result, there has been a decrease in urban green spaces from 2012 to 2020. The results show that there exists a negative correlation of -0.46 between per capita green and population; therefore, with every 1 unit increase of population, the demand for built up and urban amenities will increase, thereby impacting the per capita green and overall greenness index of the city negatively. In this study, transformed difference vegetation index (TDVI) has proven to be superior to normalized difference vegetation index (NDVI) for urban green analysis. NDVI shows vegetation of $21.25 \, km^2$, whereas TDVI shows vegetation of $16 \, km^2$. There has been an increase of merely 2% of vegetation in past 8 years span in the Vijayawada city.

In Chap. 21, magnetic susceptibility measurements were carried out of agricultural soil which was collected from 23 locations from Kopargaon area of Ahmadnagar district, Maharashtra State of India, using AGICO-MFK1-FA Multifunction Frequency Kappabridge KLY4S with low-frequency susceptibility (F1) 976 Hz and high-frequency susceptibility (F2) 15616 Hz. The magnetic susceptibility values were observed at low and high frequencies. This significant magnetic enhancement is an indication of the presence of ferromagnetic minerals in agricultural soil from the studied area. Heavy metals in soil samples were analyzed by using double-beam atomic absorption spectrophotometer. The evaluation of anthropogenic influence and contamination with trace elements in soil from study area was carried out using geoaccumulation index. The interpretation of the obtained field measurements and the laboratory analyses indicate that Cd, Pb, and Ni provide the potential risk, while the other heavy metals are in the safe limits.

This edited book entitled *Geospatial Technology for Landscape and Environmental Management—Sustainable Assessment and Planning* comprises the chapters written by scholarly academicians, well-known researchers, and experts. The primary motivation of this book is to fill the gap in the available literature on the subject by bringing together the concepts, theories, and experiences of the specialists and professionals in this field.

Lucknow, India   Praveen Kumar Rai
Jaipur, India   Varun Narayan Mishra
Gaya, India   Prafull Singh

Acknowledgements

The completion of this edited book entitled *Geospatial Technology for Landscape and Environmental Management—Sustainable Assessment and Planning* could not have been possible without the grace of Almighty God.

The editors would like to express sincere thankfulness to all the members of editorial advisory board for their boundless support and valuable instructions at all stages of the preparation of this edited book. We also express our gratitude to all the reviewers for their kind and timely support during the review process. We humbly extend our sincere thanks to all concerned person for their relentless and moral support. The editors are also very much thankful to Springer Nature for giving the opportunity to publish with them.

July 2021
Praveen Kumar Rai
Varun Narayan Mishra
Prafull Singh

Contents

Chapter 1
Spatio-Temporal Variability of Channel Planform Dynamics in Response to Spatial Expansion of Brick Kilns: A Case Study of the Downstream Course of Ichamati River, West Bengal, India

Soumen Ghosh⃝ⓘ **and Souvik Biswas**⃝ⓘ

Abstract Channel planform reflects the quasi-natural equilibrium in response to energy distribution and carrying capacity of the river. If the river is unable to carry its sediment load, then accretion on the river bed and the formation of several channel bars are inevitable consequences for alluvial rivers. The Ichamati River is a distributary channel of the Mathabhanga River, disconnected from its parent source at Majdia in Nadia district of West Bengal and hardly received any water from the Mathabhanga River except monsoon season. The downstream part of the river is chiefly maintained by groundwater and tidal activity. However, the shortage of water supply from the upstream and unwise downstream human activities has been decreasing the flow velocity and transportation capacity of the river and causing excessive siltation on the river bed. This abundant source of clay-rich fine sediment and perennial source of river water boost the rapid growth of brick kilns along the riverbank over the years. The incursion of sediment-rich tidal water by adjacent brick kilns of the Ichamati River has altered the sediment–water budget of the river. Most of the brick kilns were established within 200 m peripheries of the riverbank violating the guidelines of the Pollution Control Act of 1986 which have directly or indirectly affected different aspects of the channel planform like channel width, depth, meander geometry, cut-off formation process and natural mobility of the river. To study this spatio-temporal planimetric variability of the river, the US Army Toposheet of 1922 and multi-temporal Landsat images of 1976, 1996 and 2016 were used in this study. Besides, the cross-sectional survey of the river was conducted with the help of an echo-sounder and a GPS during 2013–2018. A questionnaire survey was also performed to acquire information regarding the expansion of brick kilns, annual rate of brick production, sediment extraction rate and land-use activities of brick kilns. This study

S. Ghosh (✉)
Department of Geography, The University of Burdwan, Bardhaman, West Bengal 713104, India

S. Biswas
Forest Survey of India, Ministry of Environment Forest and Climate Change, Dehradun 248195, India

© The Author(s), under exclusive license to Springer Nature Singapore Pte Ltd. 2022   1
P. K. Rai et al. (eds.), *Geospatial Technology for Landscape and Environmental Management*, Advances in Geographical and Environmental Sciences,
https://doi.org/10.1007/978-981-16-7373-3_1

revealed that the natural shifting and meander mobility of the river have been gradually decreasing with the expansion of brick kilns in the last few decades. The channel width was drastically reduced, and the temporal change rate of the channel width was also decreasing due to the control movement of bank lines during 1976–2016. The excessive siltation within the channel causes upliftment of the river bed as measured 9 cm year^{-1} during 2013–2018. This river is slowly decaying with time and may disappear in future if the proper restoration planning will not implement to revive the river.

Keywords Ichamati River · Channel planform · Channel shifting · Brick kilns · River decaying · GIS

1.1 Introduction

The channel planform change is a dynamic process and evolving a distinctive channel characteristic with time depending upon flow–sediment interaction. Understanding the controlling factors of the channel planform dynamics is quite difficult and complex as they vary with time and scale (Brierley and Fryirs 2013). Historically, the adjustment of the channel form based on landscape characteristics of youth, mature and old stage was first qualitatively conceptualized by Davis (1899, 1909) through his concept 'cycle of erosion', and later, Gilbert (1914) was elaborated using different quantitative techniques to measure channel forms and processes (Fuller et al. 2013). In the 1950s and 1960s, some pioneering research studies were conducted by Leoplold and Wolman (1957), Schumm (1960), Dury (1964) and Brice (1964) in the field of process geomorphology based on observations of channel forms in USA, and later, these studies were paved the way for further research on the process-based study of channel forms. Recent progress in remote sensing and GIS technology and publicly accessible high-resolution satellite data make current studies technically more accurate and help to overcome the economic hurdle for studying the channel planform at different spatial extent (Fisher et al. 2012). In south-east Asian countries, a few studies were conducted on the pattern of channel planforms such as Gupta (2012) on the Ganga–Padma river system, Deb and Ferreira (2014) on Manu River flowing within the territory of North East India and Bangladesh, Midha and Mathur (2014) on Sharda River which acts as an international border between India and Nepal and notable work of Sinha et al. (2014) on Koshi River to assess avulsion threshold and planform dynamics using quantitative techniques. These previous studies led us to conduct the present study, on a meandering river of South Bengal, namely the Ichamati River, which forms the international boundary between India and Bangladesh along its downstream course, and the characteristics of the channel planform significantly changed over time at the human–nature interface.

The Ichamati River is flowing on the lower Gangetic Plain of West Bengal and carrying millions of tons of sediments (Mondal and Satpati 2017). The huge demand for bricks in the adjacent densely populated regions, transportation facility and avail-ability of raw materials (sediment and water) are primarily responsible for the massive expansion of brick kilns along this riverbank (Biswas et al. 2020). Previously, some researchers have focused on the morphological and hydrological characteristics of the Ichamati River (Mondal and Satpati 2012, 2019; Mondal, and Bandyopadhyay 2014; Mondal et al. 2016). The recent research works conducted by Biswas et al. (2020) and Mondal et al. (2020) have highlighted the impact of human intervention on the downstream course of the Ichamati River. However, the channel planform vari-ability with relation to the growth of brick kilns along the riverbank has been received the least attention from the earlier researchers. Therefore, the present research work is mainly aiming to fulfil this research gap following the three major objectives. Firstly, this study shows the spatial and temporal outgrowth of brick kilns along the downstream course of the river. Secondly, to analyse the sediment incursion rate from the river to estimate the potential loss in discharge capacity of the river, and thirdly, this study highlights the spatio-temporal relationship between the spatial expansion of brick kilns and channel planform variability considering channel meandering and morphological attributes of the river. The quantitative assessment of the channel plan-form may provide relevant information about the system dynamicity of the riverine processes and may also help to understand the changing morphometric attributes of the river for better management of the river basin (Hooke 2007; Brierley and Fryirs 2000).

1.2   Study Area

The Ichamati River is considered a lifeline of the densely inhabited North 24 Parganas district of South Bengal (Biswas et al. 2020). The entire South Bengal is a part of the Bengal basin developed by alluvium deposition of the Ganga–Brahmaputra river system, and the delta took its present configuration during the recent Holocene period (Goodbred and Kuehl 2000). This river is a spill channel of Mathabhanga River originated at Majdia in Krishnaganj Block of Nadia district and flows eastward about 20 km and entered into Bangladesh near Mubarakpur. Then, the river was turned towards India at Duttafulia in Nadia district after flowing 35 km path in Bangladesh. The river forms an international boundary between India and Bangladesh in North 24 Parganas district from Angail to Berigopalpur on a path of almost 21 km, before it debouches into the Bay of Bengal through Raimangal and Kalindi River (Rudra 2008). The 248 km length of the river is divided into two stretches based on the homogeneity of the river morphology. The upper stretch extends from the off-take point of the river to Swarupnagar Block of 24 Parganas (N) district and the lower stretch lies within Swarupnagar to the Bay of Bengal (Mondal et al. 2016; Mondal and Bandyopadhyay 2014). The channel width from the source to the estuary is gradually prograding due to the effect of semi-diurnal tidal activity. The present study is mainly focusing on the lower course of the river covering an area of 270 km^2. The present study area lies within 88°52′42″E to 88°54′47″E longitude and

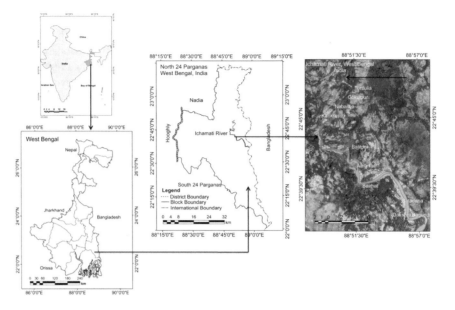

Fig. 1.1 Location map of the study area

22°36′45″N to 22°48′51″N latitude (Fig. 1.1). This study area is characterized by the tropical monsoon climate with an average temperature of 25.5 °C and rainfall of 1579 mm (Mondal et al. 2020).

1.3 Methods and Methodology

To study various aspects of the channel planform, different multi-temporal Landsat images were captured using Multispectral Scanner System (MSS), Thematic Mapper (TM) and Operational Land Imager (OLI) sensor for the year 1976, 1996 and 2016, respectively, and incorporated in this study downloaded from the United States Geological Survey (USGS). US Army Toposheet was obtained from the University of Texas library, published in 1955. The bilinear resample technique was applied for the Landsat MSS data in the ArcGIS environment to merge spatial resolution with other Landsat data. The detailed description of the datasets used in this study is shown in Table 1.1 (Fig. 1.2).

Different morphological attributes mainly mobility of meander, channel shifting, changes in channel width and depth were analysed to understand the variability of the channel planform pattern over the years. The bank lines of the river were precisely digitized using the polyline feature from the temporal datasets in the GIS platform. The lower course of the river was classified into four equal segments, and equal numbers of cross sections were placed across the river to assess the rate of temporal

Table 1.1 Descriptions of data sources

Datasets	Description	Resolution		Date of acquisition	Path and row	Source
		Pixel size (m)				
Satellite Images	Landsat MSS	60 m (30 m*)		10 January 1976	148 and 44	https://earthexplorer.usgs.gov
	Landsat TM	30 m		30 November 1996	138 and 44	
	Landsat 8 OLI	30 m		06 January 2016	138 and 44	
	Toposheet No	Scale		Year of Surveying	Year of Compilation	Source
US Army Toposheet	NF-45-08	1:250,000		1920s	1955	www.lib.utexas.edu

*Resampled resolution of Landsat MSS data for the year 1976

Fig. 1.2 Cross
section-based delineation of
channel shifting

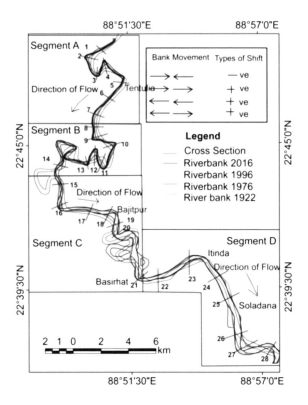

shifting of the river. The positions of cross sections were chosen based on visual interpretation of overlaying vector layers showing maximum channel shifting. If the riverbanks of the successive two years shifted towards each other, this change treated as negative, whereas the riverbanks move the opposite direction, and the shifting was considered as positive (Fig. 1.2). The erosion rate was calculated by the average distance of bank line shifting divided by the numbers of years. The method followed to delineate channel shifting was identical to the previous research works of Hazarika et al. (2015); Chakraborty and Mukhopadhyay (2015); Debnath et al. (2017) and Biswas et al. (2020). A total of eight meanders were chosen to estimate the temporal change of the meander planform particularly, area, length, sinuosity and the average width of meanders using spatial analysis tool, raster calculator and ArcGIS measurement tool. The method was comparable to the previous work of Deb and Ferreira (2014). The lateral channel stability index (LCSI) was calculated for each meander to show the temporal change of the channel mobility using Eq. 1.1, and the method of LCSI is shown in Fig. 1.3. The value ranges from 0 to 1. The value close to 0 denotes high mobility, whereas the value close to 1 specifies the stability of the channel. The entire method of LCSI was similar to the works of Richard et al. (2005) and Esfandiary and Rahimi (2019).

Fig. 1.3 Delineation method of meander mobility

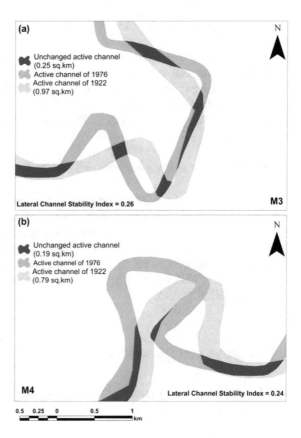

$$\text{LCSI} = \frac{\text{Unchanged area of active channel between two successive years}}{\text{the previous active channel area}} \quad (1.1)$$

To observe the spatio-temporal response of the channel behaviour with relation to the expansion of brick kilns, the Pearson correlation method was adopted using these two variables of this study. The channel width was measured using the ArcGIS measurement tool, and the temporal change rate was delineated for each segment with relation to the segment-wise spatial growth of brick kilns.

The area of brick kilns was measured from Landsat data using polygon feature in the GIS environment, and the area was verified from the high-resolution Google Earth image for better accuracy. The total area of brick kilns was measured for each segment, selected meander bends and also for the entire lower reach of the river for spatial and temporal correlation to different variables of the channel planform. The number of brick kilns was counted from Google Earth image and also verified during field visits. To collect data regarding the sediment extraction rate of brick kilns, an intensive field survey was conducted with the help of a structured questionnaire during 2017–2018. The sediment extraction rate of brick kilns was calculated by multiplying the volume of a mud-brick with the average brick production rate per year, and it helps to estimate the overall sediment incursion rate from the river along its downstream course.

The velocity of the river was measured using the Price AA current metre at different depth and calculated the average velocity to estimate the discharge capacity of that respective place. The cup of the current metre revolved for 30 s and delineated velocity (v) by using the formula $v = 0.0069 +$ Revolution Per Second (RPS) 0.0838. The river discharge of a place was calculated using Eq. (1.2)

$$\text{River discharge} = \text{Cross-sectional area of flow} \times \text{average flow velocity} \quad (1.2)$$

The cross-sectional survey was conducted at an irregular distance using the specified local benchmark, and each benchmark was shifted to the riverbank by profile survey. The locations of the cross-sectional survey are shown in Fig. 1.4. The depth of the river for every cross section was surveyed with the help of the modern echo-sounder along with a GPS receiver for positional accuracy. The cross-sectional survey was conducted in 2013 and 2018 to understand the temporal change of the river depth during this period.

1.4 Results and Discussion

1.4.1 Expansion of Brick Kilns Along the Riverbank of Ichamati River

Around 210 brick kilns can be found along the downstream riverbank of the Ichamati, depending on sediment availability, water availability, transport facility through road and waterways and the huge demand for bricks in adjacent urban and sub-urban areas. The area of brick kilns was measured at almost 11.09 km^2 in 2016 as compared to 5.45 km^2 in 1996 and 0.18 km^2 in 1976 (Fig. 1.5a). The area of brick kilns has been gradually increasing towards the downstream course of the river. Therefore, segments C and D of the river face huge pressure of intensive brick kilns growth and consequent land-use activities as compared to segments A and B of the river. Satellite data shows that the expansions of brick kilns were started after the 1980s. However, this rate of expansion has been drastically increasing since 2000. Most of these brick kilns established within 200 m peripheries of the riverbank which is considered as an absolute violation of the Pollution Control Act of 1986 (Fig. 1.5b). According to the field survey, the annual average production rate of bricks estimated at approximately 55.5 lakh per year. However, the temporal production rate may vary due to climatic factors and the availability of labours. The survey revealed that the production rate of bricks in 173 brick kilns was recorded less than the average rate of production whereas 73 brick kilns were produced above the average rate of brick production during the survey period (2017–2018).

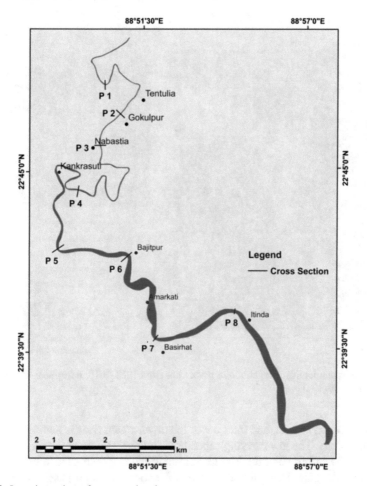

Fig. 1.4 Location points of cross-sectional survey

1.4.2 Sediment–Water Budget

To estimate the sediment–water budget of Ichamati River in consideration of the water incursion rate of brick kilns, the discharge capacity of the Ichamati River was measured at Basirhat in different seasons during 2017–2018 as it varies with seasonal change. The average discharge capacity of the river was comparatively higher in the monsoon season as measured at $3540.0 \, \mathrm{m}^3\mathrm{s}^{-1}$. However, the discharge rate of the river was recorded comparatively low in pre-monsoon ($1010.0 \, \mathrm{m}^3\mathrm{s}^{-1}$) and post-monsoon season ($1038 \, \mathrm{m}^3\mathrm{s}^{-1}$). The average annual discharge of the river was estimated at 1862 $\mathrm{m}^3\mathrm{s}^{-1}$ in Basirhat. The river has carried a huge amount of sediment, and local brick kilns are continuously extracting this sediment-rich tidal water from the river for bricks production. The river has hardly received any water from its upstream course

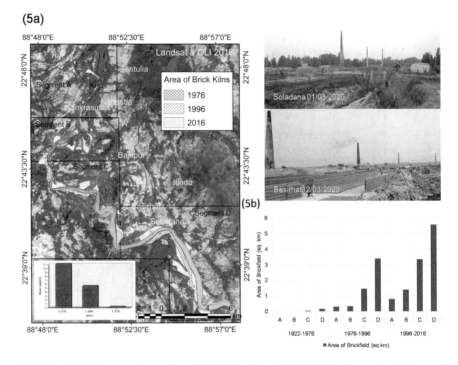

Fig. 1.5 a Spatio-temporal expansion of brick kilns from 1922–2016, **b** Segment-wise temporal expansion of brick kilns

except monsoon season, and the flow of the downstream river is mainly influenced by tidal activity. The low carrying capacity of the river causes excessive siltation on the river bed during ebb tide. The local brick kilns have extracted silted clay from sand bars emerging during low tide mainly for commercial purposes. Based on the analysis of the field survey data, it was estimated that $469,224 \times 10^5$ L of water incurred every day from the river which is 5.82% of the total discharge of the river. The mathematically derived sediment–water budget of the river is presented in Table 1.2.

1.4.3 Variability of Channel Planform Dynamics with Relation to the Expansion of Brick Kilns

1.4.3.1 Impact on Meander Planform

The changes in meander planform have been observed during pre-brick kilns time (1922–1976) and post-brick kilns time (1976–2016) using US Army Toposheet and Landsat images. The quantitative assessment was done to compute the variations

Table 1.2 Sediment–water budget of Ichamati River

Rate of average discharge	1862 m³/s or 1862,000 L
Tidal hour	7 h
Total incursion of tidal water	$1862 \times 3600 = 6{,}703{,}200{,}000$
	$6{,}703{,}200{,}000 \times 7\,h = 46{,}922{,}400{,}000$
The volume of a brick	$2.34375 \times 10^{-03}\ m^3$
Average production of brick kilns	555×10^4
Total brick kilns in the study area	210
Total production of bricks	$11{,}655 \times 10^5$
The volume of sediment used	$(11{,}655 \times 10^5) \times (2.34375 \times 10^{-03}\ m^3) = 2{,}731{,}640.6$ $m^3/2{,}731{,}641 \times 10^3\ L$
Total discharge (%)	5.82%

in the meander planform pattern over the years as shown in Table 1.3. This study revealed that the average width of meanders was increasing in the case of M5, M6, M7 and M8 whereas the average width of M1, M2, M3 and M4 was decreasing during 1922–1976. The rate of change of meander width during this time was comparatively

Table 1.3 Characteristics of meander planform of the selected reach of Ichamati River

Meander	Meander width (m)				Meander length (km)			
	1922	1976	1996	2016	1922	1976	1996	2016
1	199.80	140.55	62.38	44.28	3.3	3.9	4.7	5
2	201.88	159.48	70.05	36.68	3.1	2.8	2.8	2.9
3	192.14	157.03	85.76	71.50	2.2	2.7	3.4	3.7
4	273.39	225.74	137.62	93.90	3.6	5.2	5.9	6.7
5	257.55	258.01	185.70	143.04	2.4	2.8	2.9	2.9
6	225.28	319.92	231.48	275.10	2.7	2.7	2.9	3.1
7	253.34	399.65	239.24	233.64	2.1	2.3	2.3	2.4
8	301.42	311.66	288.79	291.42	4.5	4.5	4.7	4.7
Meander	Sinuosity				Area (km²)			
	1922	1976	1996	2016	1922	1976	1996	2016
1	2.00	2.31	2.66	2.82	0.66	0.57	0.28	0.21
2	1.54	1.35	1.39	1.42	0.63	0.42	0.21	0.1
3	1.26	1.54	1.9	2.06	0.42	0.43	0.3	0.23
4	2.06	2.83	3.08	3.22	0.92	1.02	0.76	0.6
5	1.17	1.39	1.46	1.55	0.61	0.67	0.51	0.45
6	1.54	1.22	1.36	1.41	0.55	0.86	0.76	0.82
7	1.16	1.62	1.36	1.35	0.53	0.82	0.55	0.54
8	1.20	1.21	1.26	1.26	1.35	1.44	1.39	1.37

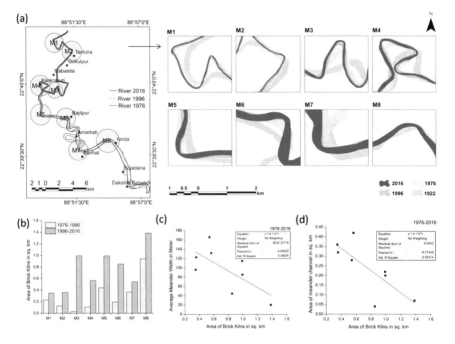

Fig. 1.6 **a** Location of meanders and temporal changes in channel width, **b** spatial expansion of brick kilns within meander bends, **c** correlation between area of brick kilns and average meander width, **d** correlation between area of brick kilns and area of meander channel

less than the post-brick kilns period. However, the meander width was abruptly decreasing during 1976–2016 (Fig. 1.6a). The spatial expansion of brick kilns has been increasing within the meander bends towards the downstream course of the river (Fig. 1.6b). There was a negative correlation ($r^2 = 0.476$, $r = -0.690$) found between the expansion of brick kilns within mender bends and the decrease of average channel width for the period of 1976–2016 (Fig. 1.6c). Similarly, the area of meander bends was estimated to increase except M1 and M2 in the absence of brick kilns, but this trend was reversed for all meanders in the presence of brick kilns. The correlation value between the spatial expansion of brick kilns and the meander area was found negative ($r^2 = 0.59$, $r = -0.77$) during 1976–2016 (Fig. 1.6d). Contrarily, meander length and meander sinuosity were gradually increasing during the entire study period. The concave bank of the meander bend is continuously eroded as the channel thalweg remains close to the concave bank, and the maximum flow velocity coupled with the formation of numbers of eddies causes continuous erosion through bank scour and mass failure processes. The eroded materials deposited inside the opposite convex bank and resulted in continuous shifting of the respective meander. As the apex of the meander bends shifted, the meander sinuosity was also increasing with time consequent upon increasing meander length. Therefore, there was a positive correlation ($r^2 = 0.653$, $r = 0.808$) found between meander length and sinuosity (Fig. 1.7).

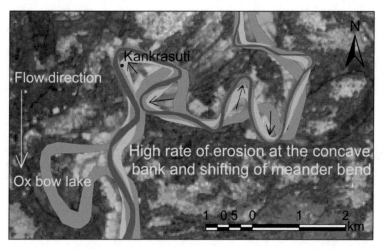

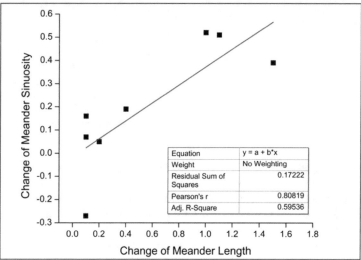

Fig. 1.7 Correlation between meander length and sinuosity

1.4.3.2   Impact on Meander Mobility

The meander mobility was measured using channel stability index with relation to the temporal expansion of brick kilns from 1922 to 1976 and 1976–2016. The post-brick kilns period was again classified into 1976–1996 and 1996–2016 to show the spatial and temporal correlation of meander mobility and expansion rate of brick kilns within meander bends. This study revealed that meanders were highly unstable due to high channel mobility during pre-brick kilns time. In the absence of any human imposed obstruction, the channel was shifting more freely on the alluvial flood plain during 1922–1976. However, the channel movement was controlled in

Table 1.4 Temporal mobility of studied meanders (1922–2016)

Meander stability index (MSI)	M1	M2	M3	M4	M5	M6	M7	M8
1922–1976	0.19	0.32	0.26	0.22	0.39	0.20	0.57	0.62
1976–2016	0.21	0.21	0.19	0.39	0.60	0.70	0.54	0.81
1976–1996	0.33	0.48	0.39	0.56	0.73	0.71	0.55	0.83
1996–2016	0.50	0.33	0.50	0.60	0.78	0.96	0.92	0.96

Fig. 1.8 Correlation between channel stability index (CSI) and expansion of brick kilns (1976–2016)

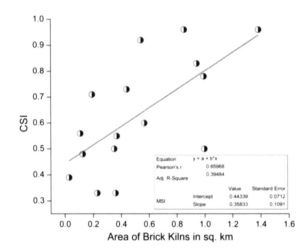

the later phase due to human inventions through the spatial expansion of brick kilns and resultant land-use activities along the riverbank. The calculation of the channel stability index revealed that meanders were comparatively more stable during the post-brick kilns period (1976–2016). The satellite images of this period proved the mushrooming expansion of brick kilns along the river. As a result, the meander stability was increasing for almost all meanders in this period (Table 1.4). There was a positive correlation observed between the expansion of brick kilns within meander bends and consequent meander stability during 1976–2016 (Fig. 1.8).

1.4.3.3 Impact on Channel Width and Depth

The segment-wise average channel width of the river was measured for the years 1922, 1976, 1996 and 2016. The river was comparatively more dynamic and active during 1922–1976 and continued to widen its channel through riverine processes. The channel widening was more prominent towards the downstream course mainly because of strong diurnal tidal activity throughout the year. The average width of the channel was estimated to increase from 264.40 m to 271.42 m (+7 m) during pre-brick kilns time. However, the channel width was abruptly reduced after 1976 due to massive human intervention on the river. As mentioned previously, the expansion of brick kilns which is one of the major anthropogenic factors is responsible for channel narrowing in recent decades. This study revealed that the average channel width was decreasing from 271.42 m in 1976 to 177.59 m in 2016. The average width was reduced by almost 94 m in the last 40 years. Similar to the pre-brick kilns time, the channel narrowing was more intensive in upstream segments (Segments A & B) as compared to the downstream segments primarily due to insignificant channel activity in the upstream course of the river (Table 1.5). Segment-wise change rate of channel width was assessed concerning the growth of brick kilns in each segment of the river. The channel movement was controlled at the human–nature interface, and shifting of bank lines was restricted by imprudent land-use activities of brick kilns. Therefore, the change rate of channel width was shown an inverse correlation with the spatial expansion of brick kilns and directly correlated to the shifting of bank lines during the study period (Fig. 1.9).

Table 1.5 Temporal change of channel width (1922–2016)

Channel Width (m)	1922	1976	1922–1976	1996	2016	1976–2016	1922–2016
Segment A	217.01	156.59	−60.42	69.99	50.32	−19.67	−166.69
Segment B	241.67	186.57	−55.10	35.45	80.81	45.36	−160.86
Segment C	264.75	320.24	55.49	236.36	232.94	−3.42	−31.81
Segment D	337.30	422.29	84.99	391.00	346.33	−44.67	9.03
Average	264.40	271.42	7.02	198.20	177.59	−20.61	−89.81

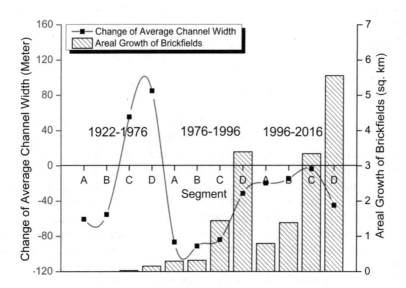

Fig. 1.9 Segment-wise change rate of channel width with relation to expansion of brick kilns (1922–2016)

Table 1.6 Cross section-wise change of channel width and depth of Ichamati River (2013–2018)

CS	Location	Average depth (m)			Cross-sectional area (m^2)		
		2013	2018	2013–2018	2013	2018	2013–2018
P 1	Ramchandrapur	2.5	2.3	−0.2	135	124	−11
P 2	Gokulpur	2.6	2.2	−0.4	122	103	−19
P 3	Nabastia	2.7	2.3	−0.4	162	138	−24
P 4	Kankrasuti	2.9	2.4	−0.5	184	158	−26
P 5	Baduria	3.7	3.2	−0.5	427	390	−37
P 6	Bajitpur	4.8	4.2	−0.6	850	714	−136
P 7	Basirhat	6.5	5.9	−0.6	967	920	−47
P 8	Itinda	5.8	5.1	−0.7	1044	918	−126

A cross-sectional survey was conducted on the Ichamati River in 2013 and 2018 to measure the changes in channel depth and cross-sectional area (CA) of the river. As observed during the field survey, several sandbars appear during low tide because of excessive siltation on the river bed. The increasing human intervention on and along the Ichamati River, not only because of brick kilns but also for the construction of bridge pillars, embankment and sand mining activities, is equally responsible for the rapid decaying of the river over the years. The cross-sectional survey was conducted in eight places from upstream to downstream course of the river. According to this survey, CA of the river was reduced with time consequent upon decrease of channel depth. The average depth was measured to decline from 3.9 m to 3.45 m from 2013 to 2018, and similarly, CA was estimated to reduce from 486.53 m in 2013 to 433.35 m in 2018 (Table 1.6; Fig. 1.10). The survey revealed that the river bed is gradually uplifted with time and this rate was measured 9 cm per year during 2013–2018.

1.4.3.4   Impact on Channel Shifting

The shifting and migration of an alluvial channel is a quasi-natural phenomenon due to human intervention in the riverine process. Like a meandering river, channel avulsion is a common phenomenon for the Ichamati River. The rate of shifting and erosion activity of the river is increasing during the monsoon season as a result of high flow velocity consequent upon the high discharge of the Ichamati River. However, the natural mobility of the river has been restricted by several human imposed obstructions. The mobility of the bank lines has been controlled by unsystematic land-use practices of brick kilns and also mobilized the channel flow to reduce the flow velocity and erosional capacity of the river in some places. It was estimated that the average channel shifting was measured 268.19 m during 1922–1976 which was reduced to 31.82 m during 1996–2016. Similarly, the erosion rate was estimated a sudden decrease over time as measured 13.40 m year^{-1} during 1922–1976, 6.63 m year^{-1} during 1976–1996 and 3.18 m year^{-1} during 1996–2016 (Table 1.7). To understand

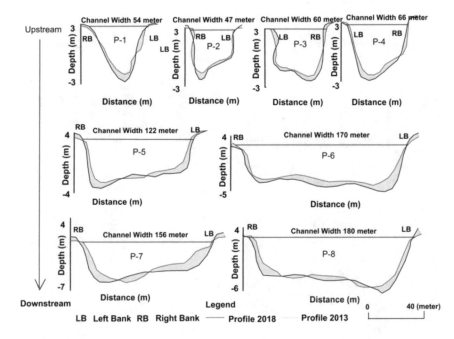

Fig. 1.10 Spatio-temporal variations of cross-profiles of the Ichamati River (2013–2018)

Table 1.7 Temporal change of erosion rate (1922–2016)

Segment	1922–1976	1976–1996	1996–2016
	Rate of erosion (m year^{-1})	Rate of erosion (m year^{-1})	Rate of erosion (m year^{-1})
A	10.54	5.75	2.07
B	14.54	8.46	4.24
C	16.62	4.90	1.58
D	11.91	7.39	4.82
Average	13.40	6.63	3.18

the impact of brick kilns expansion on channel shifting, the segment-wise assessment was done. The result revealed that there was a negative correlation found between segment-wise channel shifting and spatial expansion of brick kilns as the growth of brick kilns is directly or indirectly affected the natural mobility of the river (Fig. 1.11). The rate of erosion was measured comparatively high towards the downstream part of the river primarily because of the diurnal infiltration of tidal water through the downstream course. The river is gradually decaying with time as a result of the high accretion rate on the river bed.

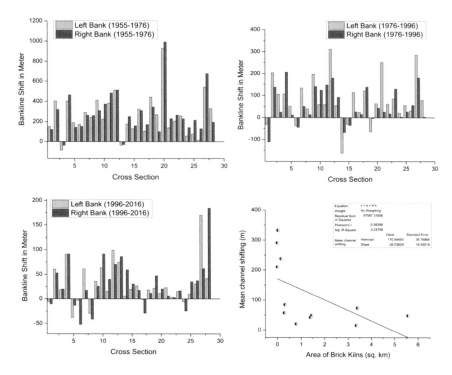

Fig. 1.11 Cross section-wise channels of Ichamati River shifting with relation to growth of brick kilns (1922–2016)

1.5   Conclusion

The channel planform of the Ichamati River has undergone a massive change in the last few decades due to human-induced activities along the riverbank. The area of brick kilns was increasing from 0.18 km² in 1976 to 11.09 km² in 2016. Near about 210 brick kilns were established along the riverbank, and most of them lie within 200 m peripheries of the riverbank. A huge amount of sediment-rich water which was estimated at almost 5.28% of the total discharge of the river extracted for brick production. The hydrological and morphological characteristics of the Ichamati River were abruptly changed during the post-brick kilns period. The meander mobility was reduced in the last four decades (1976–2016) as compared to the pre-brick kilns period. The average width of the meanders was remarkably reduced in response to the expansion of brick kilns within the meander loops during 1976–2016. The average channel width was measured to increase during pre-brick kilns time because of the natural dynamicity of the river, and this trend was reversed in the later phase (1976–2016). However, the change rate of the channel width with relation to the growth of brick kilns was inversely correlated due to the controlled mobility of the channel at the human–nature interface. The river is gradually decaying with time consequent upon excessive siltation on the river bed. The depth of the river was significantly decreasing as estimated at 9 cm per year during 2013–2018. The channel shifting is controlled by human-induced obstructions such as constructions of embankments and various anti-erosion activities. The erosion rate of the river was estimated to decrease from 13.40 m year^{-1} during 1922–1976 to 3.18 m year^{-1} during 1996–2016. The aforesaid discussions evident the profound impact of the mushrooming expansion of brick kilns on the channel planform dynamics. However, it is also noteworthy to mention that the change of the channel planform pattern with time is a complex process and depends on multiple natural and anthropogenic factors, act combinedly to change the system. Therefore, considering any specific factor for assessing the change of channel planform may be treated as a limitation of this study. However, the present work may pave the way for a more comprehensive study of the channel planform dynamics by incorporating other relevant variables. This study will also encourage policymakers to implement proper land-use planning to control the unsystematic outgrowth of brick kilns along the riverbank.

References

Biswas S, Ghosh S, Halder R (2020) Impact of human intervention on assessing downstream channel behaviour of Ichamati River on the lower Gangetic Plain of West Bengal, India. Modeling Earth Syst Environ 1–15. https://doi.org/10.1007/s40808-020-00895-7

Brierley GJ, Fryirs K (2000) River styles, a geomorphic approach to catchment characterization: implications for river rehabilitation in Bega catchment, New South Wales, Australia. Environ Manage 25(6):661–679

Brierley GJ, Fryirs KA (2013) Geomorphology and river management: applications of the river styles framework. Wiley

Chakraborty S, Mukhopadhyay S (2015) An assessment on the nature of channel migration of River Diana of the sub-Himalayan West Bengal using field and GIS techniques. Arab J Geosci 8(8), 5649–5661. https://doi.org/10.1007/s12517-014-1594-5

Davis WM (1899) The geographical cycle. Geogr J 14(5):481–504

Davis WM (1909) The systematic description of land forms. Geogr J 34(3):300–318

Deb M, Ferreira C (2014) Planform channel dynamics and bank migration hazard assessment of a highly sinuous river in the north-eastern zone of Bangladesh. Environ Earth Sci 73(10):6613–6623. https://doi.org/10.1007/s12665-014-3884-3

Debnath J, Pan ND, Ahmed I, Bhowmik M (2017) Channel migration and its impact on land use/land cover using RS and GIS: a study on Khowai River of Tripura, North-East India. Egyptian J Remote Sens Space Sci 20(2):197–210. https://doi.org/10.1016/j.ejrs.2017.01.009

Dury GH (1964) Principles of underfit streams, vol 452. US Government Printing Office

Esfandiary F, Rahimi M (2019) Analysis of river lateral channel movement using quantitative geomorphometric indicators: Qara-Sou River, Iran. Environ Earth Sci 78(15):469. https://doi.org/10.1007/s12665-019-8478-7

Fisher GB, Amos CB, Bookhagen B, Burbank DW, Godard V (2012) Channel widths, landslides, faults, and beyond: the new world order of high-spatial resolution Google Earth imagery in the study of earth surface processes. Geol Soc Am Special Papers 492(01), 1–22. https://doi.org/10.1130/2012.2492(01)

Gilbert GK (1914) The transportation of debris by running water (No. 86). US Government Printing Office

Goodbred SL Jr, Kuehl SA (2000) The significance of large sediment supply, active tectonism, and eustasy on margin sequence development: late quaternary stratigraphy and evolution of the Ganges-Brahmaputra delta. Sed Geol 133(3–4):227–248

Gupta N (2012) Channel planform dynamics of the Ganga-Padma system, India (Doctoral dissertation, University of Southampton)

Hazarika N, Das AK, Borah SB (2015) Assessing land-use changes driven by river dynamics in chronically flood affected Upper Brahmaputra plains, India, using RS-GIS techniques. Egyptian J Remote Sens Space Sci 18(1):107–118. https://doi.org/10.1016/j.ejrs.2015.02.001

Hooke JM (2007) Spatial variability, mechanisms and propagation of change in an active meandering river. Geomorphology 84(3–4):277–296

Leopold LB, Wolman MG (1957) River channel patterns: braided, meandering, and straight. US Government Printing Office

Midha N, Mathur PK (2014) Channel characteristics and planform dynamics in the Indian Terai, Sharda River. Environ Manage 53(1):120–134

Mondal I, Bandyopadhyay J (2014) Environmental change of trans international boundary Indo-Bangladesh border of Sundarban Ichamati River catchment area using geoinformatics techniques, West Bengal, India. Univers J Environ Res Technol 4(3):143–154

Mondal M, Satpati LN (2012) Morphodynamic setting and nature of bank erosion of the Ichamati River in Swarupnagar and Baduria Blocks, 24 Parganas (N), West Bengal. Indian J Spatial Sci 3(2):35–43

Mondal M, Satpati LN (2017) Hydrodynamic character of Ichamati: impact of human activities and tidal management (TRM), WB, India. Indian J Power River Valley Dev 67(3–4):50–62

Mondal M, Satpati L (2019) Human intervention on river system: a control system—a case study in Ichamati River, India. Environ Dev Sustain 1–27. https://doi.org/10.1007/s10668-019-00423-3

Mondal I, Bandyopadhyay J, Paul AK (2016) Estimation of hydrodynamic pattern change of Ichamati River using HEC RAS model, West Bengal, India. Modeling Earth Syst Environ 2(3):125 https://doi.org/10.1007/s40808-016-0138-2

Mondal I, Thakur S, Bandyopadhyay J (2020) Delineating lateral channel migration and risk zones of Ichamati River, West Bengal, India. J Cleaner Prod 244: 118740

Rudra K (2008) Banglar nadikatha. Sahitya Samsad, Kolkata, 11–19

Schumm SA (1960) The shape of alluvial channels in relation to sediment type. US Geol Survey Prof Paper, B 352:17–30

Sinha R, Sripriyanka K, Jain V, Mukul M (2014) Avulsion threshold and planform dynamics of the Kosi River in north Bihar (India) and Nepal: A GIS framework. Geomorphology 216:157–170

Chapter 2
Assessment of Replenishable Groundwater Resource and Integrated Water Resource Planning for Sustainable Agriculture

P. K. Paramaguru, J. C. Paul, B. Panigrahi, and K. C. Panda

Abstract The aim of this study is to prepare an integrated water resource action plan for conjunctive use of available water resource in a sub-humid tropical watershed of East India. This paper describes the quantification and delineation of potential zones of groundwater of a micro-watershed Ghumuda of Odisha along with an integrated plan for sustainable development of water resource. Applying simplistic water balance approach with groundwater controlling parameters estimation of annual dynamic groundwater resource is done. Delineation of potential zones is done by developing various thematic maps using satellite imagery and associated databases. Overlaying the thematic maps and keeping in view the available groundwater resource, an integrated water resource plan is prepared in Arc-GIS software. From the study, the net annual groundwater availability is found to be 80.989 ha m, while the annual draft is 10.18 ha m. After superimposing thematic layers like slope, hydro-geomorphology, line magnets, and land use in the GIS environment, four groundwater potential zones are identified such as poor, moderate, good to moderate, very good. More than 50% area of the watershed falls in a moderate prospect zone. The final action plan map is developed with an aim to utilize the maximum potential of groundwater as well as surface water for sustainable development of agriculture and local residents of the watershed. Different water conservation structures like Nala bunds, percolation tanks, and check dams are recommended along with a proposed groundwater utilization point map for recharge and sustainable discharge of groundwater at convenient sites. The results and the action plan obtained from this study can be helpful for future agricultural growth with sustainable groundwater development. The action plan can be applied to similar hydrologic characteristic areas for overall water resource augmentation.

P. K. Paramaguru (✉)
ICAR-IINRG, Ranchi, Jharkhand, India

J. C. Paul · B. Panigrahi
Department of Soil and Water Conservation Engineering, CAET, OUAT, Bhubaneswar, Odisha, India

K. C. Panda
Department of Farm Engineering, BHU, Varanasi, UP, India

© The Author(s), under exclusive license to Springer Nature Singapore Pte Ltd. 2022   21
P. K. Rai et al. (eds.), *Geospatial Technology for Landscape and Environmental Management*, Advances in Geographical and Environmental Sciences,
https://doi.org/10.1007/978-981-16-7373-3_2

Keywords Dynamic groundwater resource · Draft · GEC · Sustainable development · GIS · Hydro-geomorphology · Groundwater potential zone

2.1 Introduction

The most considerable but vulnerable natural resource is groundwater which is rapidly declining and deteriorating day by day. About 60% of the world's freshwater resource is groundwater (EPA 2009). In this present world, it is the least contaminated water source available for human beings, but due to lack of sustainable foresight and indiscriminate use, we are hampering this sparse resource. Depending on land use, hydro-geomorphology, regional climate, and drainage pattern, groundwater resource varies significantly. In developing countries like India, it is underutilized in some locations and over-utilized in some area. Some regions like North-east region have abundant groundwater resources which remained unexploited till now (Chatterjee and Purohit 2009). For a country like India, groundwater is utilized in almost all area starting from drinking water source to industry, agriculture, and energy productions (Nag and Lahiri 2011; U.S. Geological Survey 2013). The annual replenishable groundwater resource of India is 447 BCM but net annual groundwater availability is only 411 BCM (Anonymous 2011). There is tremendous scope for development of groundwater resource in India as the stage of groundwater development is only 61%.

For sustainable planning of any natural resources in a region, it is inevitable to quantify the resources available in that location. In real world, surface water and groundwater are closely linked with each other hydrologically (Paul et al. 2016). So, both should be planned and utilized in conjunction with each other. With the rapid advent of remote sensing and GIS, quantification of surface water resource is quicker nowadays. But groundwater quantification is most difficult due to the dependence of its availability on lithological layers, subsurface deformation, geomorphic history, and various structural features beneath the soil layers. For sustainable planning of water resource of a location precise and reliable quantification of both surface and groundwater reserves within the boundary of an area is essential (Sahu and Nandi 2016; Rai et al. 2017a, b, 2018).

The main objective of the estimation of groundwater resource is to create a sustainable balance between its availability and its utilization at a regional scale. If the recharge of groundwater cannot meet its regional human pressure then it will lead to an ecological imbalance (Sameena et al. 2000) and groundwater mining situations. So, for maintaining a balance between recharge and discharge of groundwater regionally, we have to estimate the annual recharge of groundwater and to do a micro-regional level planning for its utilization. Numerous techniques are used by researchers nowadays to quantify the groundwater recharge, and hydrological water balance approach is the most common one. It is a widely used method for estimation of annual groundwater recharge over an area due its simplicity and reliability.

The conceptual basis of this water balance approach is on spatio-temporal fluctuation of water level over a particular region (Healy 2010). Other practiced methods to estimate groundwater recharge include Chaturvedi formula (Chaturvedi 1973), recession curve displacement method (Rorabaugh and Simons 1966), and soil water balance method (Rushton and Ward 1979).

Simply quantifying the resource in an area is inadequate for the management of groundwater. Most of the region of India is not suitable for groundwater development and management as 65% of the land area of our country is occupied by hard rock terrains having relatively impermeable subsurface formation (Saraf and Choudhary 1998). So, it is essential to locate potential zones for groundwater recharge and utilization. Groundwater potential zones can provide an alternative to surface water resource that is mostly insufficient in harsh climatic situations like drought (Manap et al. 2013). So, both quantifying and locating potential sites of groundwater should be the primary steps for integrated management plans. Zoning of potential recharge sites should be applied to regional or local scale like in watershed basis (Page et al. 2012) as potential zones of groundwater depend on local parameters like surface water sources, river gradient, aquifer thickness, and its areal extent (Dinesh Kumar et al. 2007; Prasad et al. 2008).

Nowadays, use of geo-informatics and remote sensing in potential zoning has gained widespread attention. Zonation of potential sites has done by different techniques like weighted index overlay method (Muthikrishnan and Manjunatha 2008), analytical hierarchy method (Pradhan 2009), weighted linear combination methods (Vijith 2007; Thomas et al. 2009). Along with groundwater resource, planning of surface water resource should be crucial. In a watershed scale, conjunctive use of water resource can act as a cushion for the period of dry spell. Management of both surface and groundwater can provide a buffer condition to mitigate the risk involved in extreme climatic events (FOSTER et al. 2010). Planning of water resources on a watershed basis is efficient as this is the unit which inherently connects natural resources like land and water with environmental consequences and externalities associated with those resources (Akhouri 1996; Darghouth et al. 2008). Thus, an integrated water resource action plan of a watershed can serve various purposes like groundwater augmentation, soil conservation, regeneration of forest cover, improvement of soil fertility and nutrient status and overall development of local residents (Rao 2005; Chimdesa 2016; FAO 2017).

Most of the watershed management plans only focus on surface water resource and work accordingly on a regional scale. This planning can lead to groundwater recharge in selected sites irrespective of the potential of the subsurface layer to draw the water and to recharge the underground groundwater reserve. So, an attempt has been made to quantify the dynamic groundwater resource and then to locate groundwater potential zones and finally to develop an integrated water resource action plan at the micro-watershed level for sustainable development.

Fig. 2.1 Location map of
the study area

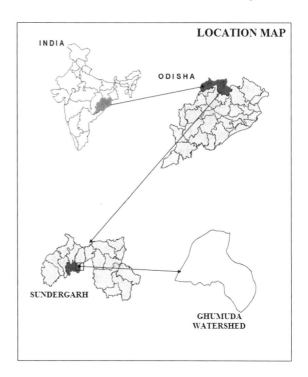

Fig. 2.1 Location map of the study area

2.2 Materials and Methods

2.2.1 Study Area Description

The study was conducted in the Ghumuda micro-watershed, a non-treated watershed present in the north-west part of Odisha and lies between 22° 4′ 0″ N to 22° 5′ 42″ N latitude and 84° 13′ 06″ E to 84° 14′ 46″ E longitude (Fig. 2.1). The elevation of the study area varies from 198 to 301 m above mean sea level having a drainage area of about 851.95 ha. The watershed come under Eastern plateau agro-ecological region with sub-humid climate.

2.2.2 Topography

The study area is in north-western plateau agro-climatic zone and is diversified by undulating tracts with ridges and valleys. The study area also comprises of isolated patches of hills and hill slopes with upland, medium, and low lands which are responsible for creating good drainage network toward the middle of the watershed. A large portion of micro-watershed is surrounded by hillocks and forest area, which contributes major discharge to the small streams and drainage network of the watershed.

2.2.3 Climate and Rainfall

Having sub-tropical humid climate, due to the high elevation of about 300 m from the mean sea level, the study area receives a good rainfall of 1261.63 mm per year. Its temperature varies from 4 °C in winter to a maximum of 46 °C in summer. The entire crop production of the study area depends on monsoon rain which starts from July to October and cool winter from November to March. Remaining months of the year are relatively dry without crop production. The study watershed is severely degraded due to extensive deforestation and erosive rainfall in the dry month.

2.2.4 Soil and Subsurface Formation

Mainly red-lateritic and acidic soil is prominent in the study area. A small portion of erosion slope is occupied by shallow soils, upland by the moderate depth and low land by deep alluvial soil layers. The agriculture system of the watershed is characterized by paddy mono-cropping system with small patches of land under short duration maize, mustard, and cowpea. Presence of high undulation is due to bisected uplands and ridges. The sedentary landscape is predominant in the study area with two distinguished stages namely depositional and erosional. Valley bottoms and some portions of the upland are under erosional phases.

2.2.5 Norms Used for Groundwater Resource Estimation

For the country like India, the dynamic groundwater resource is estimated using the GEC-1997 (Groundwater Estimation Committee) norms. These norms have been formulated by the various expert committee of state and central agencies, academic and scientific institutions which are revised in 2011 and 2015. These norms are basically a simplistic water balance approach, applied to a small assessment unit like blocks and states in the country. In our study, this water balance approach is applied to the micro-watershed level for conjunctive water application and integrated planning purpose.

2.3 Database Used for Potential Zoning and Watershed Planning

2.3.1 Primary Data

Survey of India (SOI) toposheet 73 B4 on a scale of 1:50,000 having contour interval of 20 m was digitized and used to get topographical features of the study area. The toposheet also used to prepare the base map and to locate road network, settlements and to delineate the watershed boundary. The daily rainfall data for 27 years (1991–2017) were obtained from the Ground Water Survey and Investigation Department, Odisha. Groundwater level data, groundwater quality, and vertical electrical sounding (VES) test data of the observation wells were collected from Ground Water Survey and Investigation Office, Sambalpur. Ground truthing and some information like industrial draft, irrigation draft, population data and data regarding water harvesting structures are obtained from local field visit and survey to lessen the erroneousness which may arise during visual interpretation, groundwater prospecting, and watershed planning process.

2.3.2 Secondary Data

Secondary data used in this study mostly includes satellite imagery and digital elevation model (DEM). IRS LISS-III FCC (False Colour Composite) acquired on 3 February 2013 from Bhuvan portal (ISRO geoportal) has been used for the preparation of thematic maps like drainage, land use/land cover, lineament, and geomorphology maps. The satellite imagery thus used was of good quality with less than 10% visible cloud. DEM was downloaded from USGS Earth Explorer: ASTER Global DEM with resolution 30 m × 30 m which acts as an input for the preparation of slope map and assists in study, interpretation and planning of study area providing topographic features.

2.3.3 Software

For the preparation of thematic maps of the study area, Arc-GIS version 10.1 has been used. Then superimposing these prepared maps in ARC-GIS, groundwater potential zone map and final integrated water resource action plan map was prepared.

2.4 Methodology for Estimation of Dynamic Groundwater Resource

2.4.1 Draft Calculations

A simplified unit draft method where the draft or discharge rate of groundwater extraction structures already in use, in the study area was multiplied with their numbers to get the total draft imposed by them on local groundwater resource. Due to unavailability of domestic use rate, the stand norms recommended by GEC-1997 (Anonymous 1997b) with projected per capita demand, 60 L/day/person, was used to calculate domestic draft. The industrial draft was kept as 0 as there was no industry during the considered study period (Paramaguru et al. 2019). The total draft on groundwater was obtained by adding the drafts from individual sources.

2.4.2 Recharge Calculations

Recharge estimation in our assessment unit includes the sum of recharges from precipitation, irrigation, and from the canal and other water storage structures. As our study area is an agricultural non-treated watershed, its major recharge contribution is from precipitation. The recharge from irrigation and water storage structures forms a small portion of total recharge amount. Recharge from rainfall was estimated by dividing it into two parts, monsoon recharge and non-monsoon recharge. Monsoon recharge was estimated as per GEC-1997 using both water level fluctuation method (WLF) and rainfall infiltration factor method (RIF) and non-monsoon season by only RIF method. Then normal rainfall recharge was obtained by adding both the recharge and normalizing the final value. Unit draft method and recharge factor norms of CGWB were used to get recharge from irrigation source and water storage tanks and pond in the study area. The ancillary data like design discharge rate, period of run, water spread area of storage structures were used as input in these recharge calculations. The governing equations used in the study is presented in Table 2.1 and the flowchart of detail methodology followed is presented in Fig. 2.2.

(All the norms or factors used from GEC-1997)

Table 2.1 Governing equations used in recharge estimation

Period of recharge	Approach	Equation	Major input parameters	Norms or factors used	Obtained parameter
Monsoon	WLF	$R = S + DG = h \times S_y \times A + DG$	• Total draft • Groundwater fluctuation • Normal annual rainfall	$S_y = 0.015$, seepage rate for ponds = 0.00144 m/day/ha	Monsoon recharge
		$R_{rfi} = R - R_c - R_{sw} - R_{gw} - R_{wc} - R_t$,			
		$R_{rfi}(\text{normal}) = \frac{R_{i \times r(\text{normal})}}{ri}$			
		$R\text{monsoon} = R_{rfi} + R_c + R_{sw} + R_{gw} + R{wc} + R_t$			
	RIF	$R_{rfi} = F \times A \times$ normal rainfall during monsoon season	• Assessment area • Normal annual rainfall	$F = 0.08$	Monsoon recharge
		$R\text{monsoon} = R_{rfi} + R_c + R_{sw} + R_{gw} + R_{wc} + R_t$			
Non-monsoon	RIF	$R_{rf} = F \times A \times$ normal rainfall during monsoon season	• Assessment area • Normal annual rainfall	$F = 0.08$	Non-monsoon recharge
		$R\text{non-monsoon} = R_{rf} + R_c + R_{sw} + R_{gw} + R_{wc} + R_t$			
Net annual ground water availability		Rmonsoon + Rnon-monsoon − Natural discharge		10% as RIF method is used in calculating rainfall recharge in monsoon season	Annual replenishable groundwater resource

Where R is the recharge during monsoon, S the change in storage ($S = h \times S_y \times A$), DG the gross draft during monsoon season, h the water-level fluctuation between pre-monsoon and post-monsoon, S_y the specific yield, A the area of assessment, R_{rfi} is the rainfall recharge during monsoon season for ith particular year, R_c the recharge from canal seepage during monsoon season (in command areas) for ith particular year, R_{sw} the recharge from surface water irrigation during monsoon season (in command areas) for ith particular year, R_{gw} the recharge from groundwater irrigation during monsoon season for ith particular year, and R_{wc} the recharge from water conservation structures during monsoon season for ith particular year and R_t the recharge from tanks during monsoon season for ith particular year, R_{rf} is the rainfall recharge during monsoon, F the rainfall infiltration factor, and A the area of assessment unit

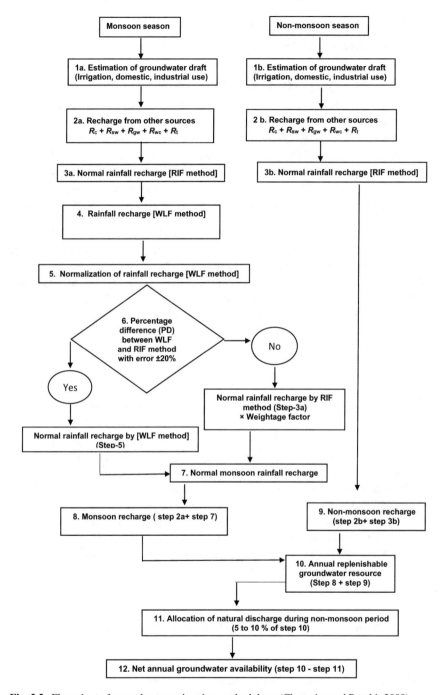

Fig. 2.2 Flow chart of groundwater estimation methodology (Chatterjee and Purohit 2009)

2.5 Methodology Used for Groundwater Prospecting and Watershed Planning

2.5.1 Generation of Thematic Maps

Delineation of the watershed boundary consists of five major steps. These steps are DEM set up, stream definition, outlet and inlet definition, watershed outlet selection, and then the calculation of sub-basin parameters. After the setup of DEM flow direction and watershed boundary was delineated by the model itself. Correction and modification of the delineated boundary were then done by overlaying toposheet and satellite imagery. The slope map was then prepared using a special analyst tool of Arc-GIS. LISS-III image of the study area was used with supervised classification to get land use/land cover map. The visual interpretation technique was followed by overlaying the downloaded satellite imagery on the prepared base map. The map was then clarified and modified by comparing with the geomorphology map prepared by NRSC (National Remote Sensing Centre) in Bhuvan portal. Using the DEM and satellite imagery, demarcation of lineaments was done by the special enhancement technique in 3-6-3 high-pass directional convolution kernels filter. After preliminary interpretation, images were correlated with ground truth data and necessary modification was done.

2.5.2 Preparation of Groundwater Potential Zone Map

Superimposing the thematic maps like slope, land use/land cover, lineament, and geomorphology in Arc info grid environment groundwater potential zone map was prepared (Hutti and Nijagunappa 2011). Intersecting polygons were demarcated by integrating all maps to get the final composite map. Overlaying was done by weighted index overlay method where individual weights were assigned to all thematic maps, and in those maps, various parameters were given distinct ranks depending on their groundwater potentiality. Considering the similar works carried out by various researchers, weights and ranks were assigned in our study (Asadi et al. 2007; Ramamoorthy and Rammohan 2015).

2.5.3 Preparation of Water Resource Action Plan Map

The study aimed to develop an integrated water resource action plan to conserve surface water, recharge groundwater and use it in a conjunctive way for sustainable agricultural development. For integrated planning, water balance study was done to check the stage of groundwater development in the study area. Terrain parameters, subsurface geologic features, land use type, drainage networks, and all necessary data

Table 2.2 Common logic of providing different water conservation structures for water resource development

S. No	Water action plan units	Logic to allocate the site
1	Nalabund	In lower-order stream line(1st, 2nd order) and nearly level to gently sloping land (0–3% slope)
2	Percolation tank	Along or at the intersection of fracture/lineaments with nearly level to gently sloping land (0–3% slope)
3	Check dam	Lower-order streams (1st order) and gently to moderately sloping land (3–10%)
4	Water harvesting structure	Comparatively higher order (up to 3rd order), command area up to 50 ha and nearly level to gently sloping land (0–5%)

were collected or developed in the study. The areas having good groundwater prospect were also selected for further development and to increase artificial recharge. Integrating all the ancillary data, annual recharge amounts, thematic maps, and developed potential zone map, an integrated water resource development map was prepared in the GIS environment. Areas suitable for construction of water harvesting structures were chosen by overlaying prepared thematic maps and the appropriate structure to be made on that location was determined using the guidelines proposed by Integrated Mission for Sustainable Development (Paul et al. 2008) (Table 2.2). The flowchart of preparation of integrated water resource action plan is given in Fig. 2.3.

2.6 Results and Discussion

2.6.1 Computation of Different Parameters of Groundwater Resource

2.6.1.1 Computation of Gross Groundwater Draft

The total of groundwater draft which is the sum of irrigation, industrial and domestic draft of Ghumuda watershed, was found as 10.1807 ha-m. As there was no industry or any production unit, the industrial draft is taken as 0. As per the population of the study area, the domestic draft is calculated as 4.18 ha-m. Out of the existing 30 wells in the study area, 6 wells are used for groundwater irrigation. So, groundwater draft is calculated as 6 ha-m as per the norms recommended by GEC (Kumar et al. 2017).

2.6.1.2 Computation of Recharge Potential

Different recharge components are calculated for both monsoon and non-monsoon period separately using the water balance equation, taking into consideration return

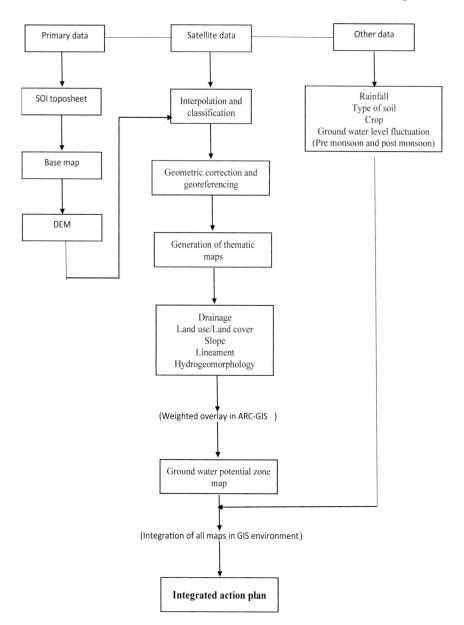

Fig. 2.3 Flowchart of water resource action plan preparation

Table 2.3 Detail recharge parameters of Ghumuda micro-watershed

S. No	Resource attributes	Volume (ha m)
1	Recharge from rainfall	72.584
2	Recharge from canal irrigation	0
3	Recharge from surface water irrigation	11.18
4	Recharge from groundwater irrigation	1.5
5	Recharge from tanks and ponds	4.722
6	Recharge from water harvesting structures	0
7	Natural discharge	8.998
8	Net annual groundwater availability	80.989

flow from irrigation, constant seepage rate from water storage structures and recharge from rainfall depending on the subsurface geologic features of the study area. The total recharge attribute including both monsoon and non-monsoon period are presented in Table 2.3.

2.6.2 Computation of Dynamic Groundwater Resource and Stage of Development

During the calculation of monsoon recharge, the percentage difference (PD) is calculated between the WLF and RIF method and is found less than −20%. So, as per the GEC norms, corrected rainfall recharge is obtained multiplying correction factor with recharge amount obtained from RIF method (Anonymous 2011). The total monsoon recharge is calculated as 64.90 ha m and in non-monsoon period as 25.08 ha m. So total recharge is calculated as 89.98 ha m. As during monsoon recharge estimation, RIF method is used, the natural discharge was taken as 10% of total recharge during that period as per the standard norm. After deducting natural discharge from total recharge, the net annual groundwater availability is obtained as 80.989 ha m. Stage of groundwater development is basically the percentage of the existing use of groundwater in a particular area. So, the ratio of gross annual groundwater draft and net annual groundwater availability is the stage of groundwater development which is only 12. 57% in our study area. So, there is tremendous scope for further utilization and development of groundwater resource in this watershed.

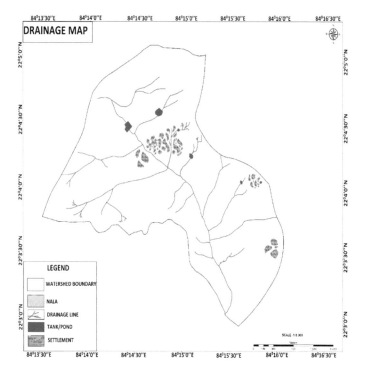

Fig. 2.4 Drainage map

2.6.3 Thematic Layers of Study Area

2.6.3.1 Drainage Map

Drainage line study is a form of indirect analysis of lithological and runoff producing characteristics of landform. The basis of both surface and groundwater development depends on the available drainage network in a watershed. Our study watershed is comprised of a dendritic drainage pattern with isolated patches of undulating hilly tracts. The drainage pattern lies toward the center of the watershed and from west to east direction. The drainage map of the Ghumuda watershed is presented in Fig. 2.4.

2.6.3.2 Slope Map

The slope or degree of steepness of land is an important attribute in groundwater potential zoning as it is the dominating factor in deciding the runoff volume. It is also dominant in integrated water resource planning also as slope forms a basis for choosing suitable sites to reduce erosion and conserve both soil and water. The slope of the study area (Fig. 2.5) is developed by Arc-GIS using DEM file as an input layer.

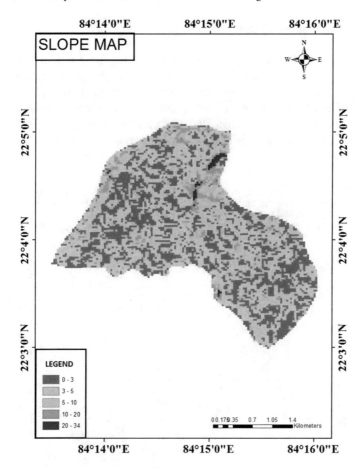

Fig. 2.5 Slope map

The slope map is divided into 5 slope classes (Anonymous 1994) such as <3, 3–5, 5–10, 10–20, 20–34%. Different slope units and their percentage area are given in Table 2.4.

Table 2.4 Slope attributes of Ghumuda micro-watershed

Map unit	Slope unit (%)	Area (ha)	% of total area
1	0–3	267.85	31.44
2	3–5	248.428	29.16
3	5–10	299.204	35.12
4	10–20	26.24	3.08
5	20–34	10.228	1.2

2.6.3.3 Land Use/Land Cover Map

The land use map of the study area is prepared by using IRS-LISS-III FCC satellite image in the GIS environment. The current land use in an area determines the geo-hydrological processes and micro-catchment water flow. So, it as a critical attribute for both surface and groundwater planning. Our study area is basically an agricultural watershed comprised of more than 90% of agricultural land (Fig. 2.6). The information on the existing land use/land cover and its types present in the watershed is given in Table 2.5.

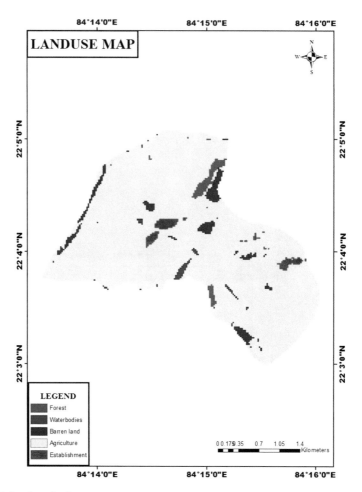

Fig. 2.6 Land use/land cover map

Map unit	Land use/land cover class	Area (ha)	% of the total area
I	Agriculture	776.98	91.2
II	Barren land	40.55	4.76
III	Establishment	14.95	1.755
IV	Forest	13.26	1.557
V	Water bodies	6.21	0.728

Table 2.5 Land use/land cover classification of Ghumuda micro-watershed

2.6.3.4 Hydro-geomorphology Map

Hydro-geomorphology is a distinct feature of the landform which connect hydro-logic phenomena with geomorphic characteristics. Primarily this feature controls both surface and subsurface flow network. In groundwater potential zoning, hydro-geomorphology has got the highest weightage due to its importance in groundwater availability (Anonymous 1995). Five types of hydro-geomorphic units are present in the study area (Fig. 2.7) namely valley fills, moderately buried pediments (BPM), shallow buried pediments (BPS), pediment/valley floor, and inselberg. BPS is the

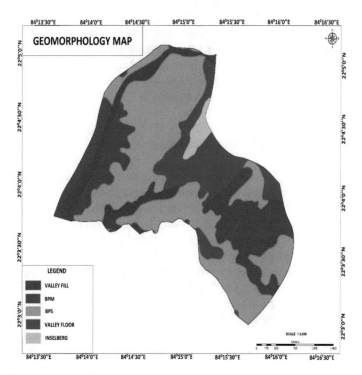

Fig. 2.7 Hydro-geomorphology map

Table 2.6
Hydrogeomorphological parameters of Ghumuda micro-watershed

Map unit	Geomorphic unit	Area (ha)	% of total area
1	Inselberg	36.207	4.25
2	BPM	165.704	19.45
3	BPS	459.286	53.91
4	Valley fill	39.189	4.6
5	Pediment/valley floor	102.489	12.03

predominant unit occupying more than 50% of the study area. Different hydrogeomorphological units and their occupied percentage area are presented in Table 2.6.

2.6.3.5 Lineament Map

Lineaments are nothing but cracks or faults of the landscape developed due to long-term geomorphic and hydrologic processes (Clark and Wilson 1994). Lineaments can be characterized by distinct features like abrupt truncation in hills, linear ridgelines, and offsetting of stream courses. By image processing in a GIS environment, lineament map is developed for Ghumuda watershed. In our study area, 10 lineaments are identified and most of them are present toward the east, nearer to the outlet of watershed. These lineaments assist in better groundwater movement, thus represent potential zones for future development and recharge. The information of the lineaments of the micro-watershed is presented in Fig. 2.8.

2.6.3.6 Weighted Index Overlay Analysis

Weighted index overlay is a simple and straightforward technique where influencing thematic maps like land use, slope, hydro-geomorphology, and lineaments were analyzed and given weight as per their impact on groundwater availability. The effect of each thematic map is different and independent of each other. Each group of main four thematic layers are placed into one of the following class as (i) Very good, (ii) Good to moderate, (iii) Moderate, (iv) Poor. Favorable rank on a scale of '1–4' has been given to each type of a particular thematic map. Geomorphology and lineament maps got more weights as compared to land use and slope map. The geomorphic features such as pediments got the lowest rank in geomorphology map where valley fill got the highest rank. Similarly, in the land use map, water body got the highest rank and establishment got the lowest. The detailed weights and ranks given to all features are given in Table 2.7. After assigning weights, raster calculated superimposed the thematic maps to get final groundwater potential zone map. The potential index was estimated by using the following relation.

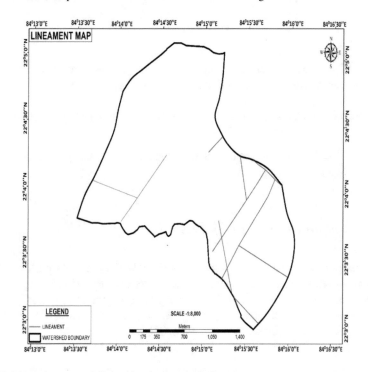

Fig. 2.8 Lineament map of Ghumuda micro-watershed

Table 2.7 Assigning ranking and weightages to various parameters

S. No	Thematic layer	Individual feature	Map weight	Ranks
1	Geomorphology	Valley fill	30	4
		BPM		3
		BPS		2
		Pediment/valley floor		1
		Inselberg		1
2	Lineament	Present	30	4
		Absent		1
3	Slope (%)	0–3	20	4
		3–5		3
		5–10		2
		10–34		1
4	Land use	Waterbody	20	4
		Forest		3
		Agriculture		2
		Barren land		1
		Establishment		1

$$\text{GWP} = \Sigma \, W \times R$$

where GWP = Groundwater potential, W = Weightage, R = Rank. The groundwater potential index value was identified using the following formula as;

$$\text{GWPI} = (\text{GW} \times \text{GR} + \text{LW} \times \text{LR} + \text{SW} \times \text{SR} + \text{LUW} \times \text{LUR})/\text{Total weightage}$$

where GWPI = Groundwater potential index value, G = Geomorphology, S = Slope, L = Lineament, LU = Land use.

2.6.4 Groundwater Potential Zone Map

For delineation of potential zones superimposition in Arc-GIS was performed after assigning weights and ranks to all thematic layers and presented in Fig. 2.9. The groundwater potential zone map is divided into four potential zones, i.e., very good, good to moderate, moderate, and poor. Percentage and areal distribution of groundwater potential zones are presented in Table 2.8.

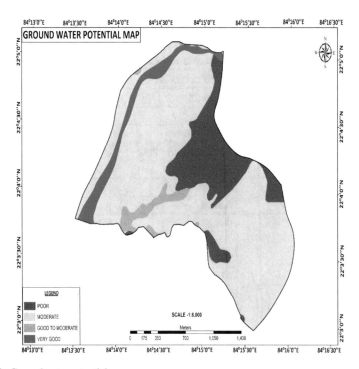

Fig. 2.9 Groundwater potential zone map

Table 2.8 Percentage and areal distribution of groundwater potential zones

Map unit	Ground water potential zone	Area (ha)	% of total area
1	Very good	49.159	5.77
2	Good to moderate	17.89	2.1
3	Moderate	659.92	77.46
4	Poor	124.981	14.67

2.6.5   Very Good

This potential zone class includes mostly hydro-geomorphic units like valley fills and intersection points of lineaments. This zone possesses abundant groundwater resource due to the subsurface behavior and presence of aqueducts due to lineaments.

2.6.6   Good to Moderate

This zone includes some isolated patches of valley fills, water bodies, and gentle slope areas. It mostly consists of alluvial plains with unconsolidated sandy layers. This zones also holds a fairly good amount of groundwater recharge capability due to its subsurface homogeneity.

2.6.7   Moderate

Both BPM and BPS are the moderate groundwater prospect units. Homogeneity of subsurface layers in this zone is less as compared to valley fills but due to presence of weathered and fractured rocks it can also hold a good quantity of groundwater. Top confining moderate or shallow unconsolidated layers also makes it a suitable zone for recharge. Our study area mostly consists of this moderate zone having more than 77% occupied area.

2.6.8   Poor

This unit is mostly dominated by inselberg, valley floor, and residual hills. This zone has negligible groundwater storage, but due to its steep slope and subsurface condition, it is an ideal zone for runoff generation. So, not directly but indirectly it can help in groundwater recharge and development. Its aquifer is mostly granite or quartzite formation with inadequate recharge capacity.

2.6.9 Proposed Groundwater Utilization Points

The annual dynamic groundwater resource of the Ghumuda watershed is 80.98 ha m but the annual draft is only 10.18 ha m. So, there is a huge gap in the availability and use of groundwater in our study area. So, we proposed some groundwater utilization points for further groundwater extraction to supply year-round irrigation facilities to the surrounding agricultural lands. Lineaments are the geologic features responsible for natural groundwater movement and its intersection point is the best site for groundwater extraction (Paul et al. 2016). In proposed groundwater extraction points (Fig. 2.10), seven tubewells are recommended at the intersection points of lineaments in the study area. The proposed tubewells can increase the draft of groundwater to 70% of net annual groundwater availability which is within the safe range of groundwater development (Anonymous 1997b). The recommended extraction points with their latitude and longitude are presented in Table 2.9. The proposed tubewells can supply water for irrigation during the dry seasons which can lead to the addition of 10 ha per tubewell cultivable area in the watershed (Anonymous 1997a, b).

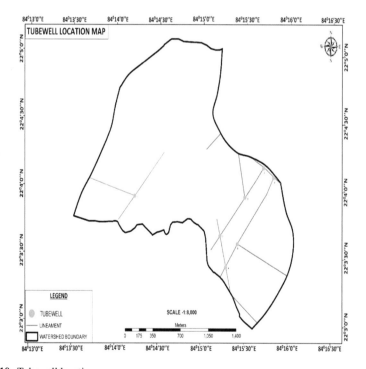

Fig. 2.10 Tubewell location map

Table 2.9 Suggested tubewell location sites of the study area

S. No	Village	Latitude	Longitude
1	Bijadihi	84° 14′ 13″	22° 3′ 57″
2	Bauridihi	84° 15′ 30″	22° 3′ 57″
3	Bauridihi	84° 15′ 42″	22° 4′ 10″
4	Bauridihi	84° 15′ 49″	22° 4′ 6″
'5	Bauridihi	84° 15′ 24″	22° 3′ 36″
6	Kulta	84° 15′ 16″	22° 3′ 26″
7	Kulta	84° 13′ 13″	22° 3′ 39″

Table 2.10 Vertical electrical sounding test data of the study area

Sundergarh Well code—29M16A18 Well type—Borewell	Depth below ground level (in meter)	Lithological layer
	0	Sandy clay with mica
	4	Weathered quartz, mica, and schist
	12	Quartz, mica, and schist

2.6.10 Ground Truthing

To validate the potential zone map and proposed groundwater development sites, vertical electrical sounding (VES) test data were obtained from Ground Water Survey and Investigation Department of Government of Odisha. The test data revealed that the planned location for groundwater extraction by tubewells are feasible points due to having good subsurface layers. The VES test data are given in Table 2.10.

2.6.11 Water Resources Action Plan

The aim of integrated water resource action plan is to conserve surface runoff as much as possible to utilize it in agriculture along within groundwater recharge to attain an overall sustainable development of watershed and its local residents. With this aim, an integrated water resource action plan (Fig. 2.11) is prepared by overlaying all the thematic maps produced along with groundwater potential zone map for the study area. The location of water harvesting structures is selected about 200–300 m upstream of the problematic area but suitable for augmenting the groundwater level. They have located mostly on 1st to 3rd stream with good geomorphic layer and optimum land slope. The number and type of proposed engineering structures are given in Table 2.11.

Fig. 2.11 Integrated water resource action plan map

Table 2.11 Number of proposed Structures for water resources development in Ghumuda micro-watershed

S. No	Engineering structures	Numbers
1	Renovation of water bodies	2
2	Nala bund	3
3	Percolation tank	9
4	Check dam	1
5	Water harvesting structure	2

2.7 Conclusions

Assessment of groundwater resource is an inevitable subject in the context of water resource planning. In this study, both quantification and zonation of groundwater resource are done to assist the sustainable planning of water resource. The quantification of replenishable groundwater resources shows that the study area has enormous scope for groundwater development as the stage of development is only 12.57% with 80.98 ha m net annual groundwater resource. Development of potential zone map can be useful in the selection of suitable location for groundwater recharge and extraction points. The potential zone map shows that more than 50% area of watershed comes under moderate groundwater prospect with high potential to accommodate more groundwater recharge. Final integrated water resource action plan is the summation of all the maps developed which aim to provide a conjunctive application plan for the available water resource. It includes various recharge structure like Nala bunds, percolation tanks, water harvesting structures, with the renovation of existing but unused water bodies. This plan will not only help in groundwater recharge but also helps in a further increase in agricultural land by supplying sufficient water for irrigation during dry seasons of the year.

Acknowledgements The authors would like to express their gratitude to the Groundwater Survey and Investigation Office, Sundergarh, Odisha for providing essential information related to lithological layer and VES data and sharing their experience about groundwater investigations in the study area. Also, sincere thanks go to Dr. S.K. Khatua, P.D. Sundergarh, Directorate of Soil Conservation and Watershed Mission, Odisha, for providing data and help in interpreting the results. We also acknowledge anonymous referees who directly or indirectly contributed to the preparation and modification of the manuscript.

References

Abd M, Sulaiman WNA, Ramli MF, Pradhan B, Surip N (2013) A knowledge-driven GIS modeling technique for groundwater potential mapping at the Upper Langat Basin, Malaysia. Arab J Geosci 6:1621–1637

Anonymous (1994) Manual of land use/land cover mapping. National Remote Sensing Agency, Dept. of Space, Govt. of India.

Anonymous (1995) Technical guidelines. Integrated mission for sustainable development. National Remote Sensing Agency, Department of Space, Govt. of India

Anonymous (1997a) Groundwater resources of Odisha, Central Ground Water Board, Ministry of Water Resources, Govt. of India, Bhubaneswar

Anonymous (1997b) Report of the ground water resource estimation committee, Central Ground Water Board, Ministry of Water Resources, Govt. of India, New Delhi

Anonymous (2011) Dynamic ground water resources of India, Central Ground Water Board, Ministry of Water Resources, Govt. of India, New Delhi.

Asadi SS, Vuppala P, Reddy MA (2007) Remote sensing and GIS techniques for evaluation of groundwater quality in municipal corporation of Hyderabad (Zone-V), India. Int J Environ Res Public Health 4:45–52

Basavaraj H, Nijagunappa R (2011) Identification of groundwater potential zone using geoinformatics in Ghataprabha basin, North Karnataka, India . Int. J. Geomat. Geosc.2:91–109

Chatterjee R, Purohit RR (2009) Estimation of replenishable groundwater resources of India and their status of utilization. Current Sci. 1581–1591

Chaturvedi RS (1973) A note on the investigation of ground water resources in western districts of Uttar Pradesh. Ann Rep UP Irrigation Res Inst 1973:86–122

Chimdesa G (2016) Historical perspectives and present scenarios of watershed management in Ethiopia. Int J Nat Res Ecol Manage 1:115–127

Clark CD, Wilson C (1994) Spatial analysis of lineaments. Comput Geosci 20:1237–1258

Darghouth S, Ward C, Gambarelli G, Styger E, Roux J (2008) Watershed management approaches, policies, and operations: lessons for scaling up

Dinesh Kumar PK, Gopinath G, Seralathan P (2007) Application of remote sensing and GIS for the demarcation of groundwater potential zones of a river basin in Kerala, southwest coast of India. Int J Remote Sens 28:5583–5601

EPA (2009) United States Environmental Protection Agency

FAO (2017) Watershed management in action: lessons learned from FAO field projects. Mt Res Dev 39

Foster S, van Steenbergen F, Zuleta J, Garduño H (2010) Conjunctive use of groundwater and surface water. GW-Mate, Strateg Overview Ser Num 2:26

Healy RW (2010) Estimating groundwater recharge. Cambridge University Press

Krishna AP (1996) Remote sensing approach for watershed based resources management in the Sikkim Himalaya: a case study. J Ind Soc Rem Sens 24:71–83

Kumar A, Anand S, Kumar M, Chandra R (2017) Groundwater assessment: a case study in Patna and Gaya District of Bihar, India. Int J Current Microbiol Appl Sci 6:184–195

Le Page M, Berjamy B, Fakir Y, Bourgin F, Jarlan L, Abourida A, Benrhanem M, Jacob G, Huber M, Sghrer F, Simonneaux V (2012) An integrated DSS for groundwater management based on remote sensing. The case of a semi-arid aquifer in Morocco. Water Resour Manage 26:3209–3230

Muthukrishnan A, Manjunatha V (2008) Role of remote sensing and GIS in artificial recharge of the ground water aquifer in the shanmuganadi sub watershed in the Cauvery River basin, Tiruchirappalli District, Tamil Nadu

Nag SK, Lahiri A (2011) Integrated approach using remote sensing and GIS techniques for delineating groundwater potential zones in Dwarakeswar watershed, Bankura distict, West Bengal. Int J Geomat Geosc 2:430–442

Paramaguru PK, Paul JC, Panigrahi B (2019) Estimation of replenishable groundwater resource for sustainable development: a case study for Ghumuda watershed of Odisha. J Soil Water Conserv 18(1):76–84

Paul JC, Mishra JN, Pradhan PL, Sharma SD (2008) Remote sensing and GIS aided land and water management plan preparation of watershed–a case study. J Agric Eng 45:27–33

Paul JC, Panigrahi B, Padhi GC, Mishra P (2016) Geo-informatics based groundwater plan preparation of Kichna nala watershed of Odisha. J Soil Water Conserv 15:325–331

Pradhan B (2009) Groundwater potential zonation for basaltic watersheds using satellite remote sensing data and GIS techniques. Central Eur J Geosc 1:120–129

Prasad RK, Mondal NC, Banerjee P, Nandakumar MV, Singh VS (2008) Deciphering potential groundwater zone in hard rock through the application of GIS. Environ Geol 55:467–475

Rai PK, Mishra VN, Mohan K (2017a) A study of morphometric evaluation of the Son Basin India using geospatial approach. Rem Sens Appl: Soc Environ 7:9–20

Rai PK, Mishra VN, Singh P (2018) Hydrological inferences through morphometric analysis of lower Kosi River Basin of India for water resource management based on remote sensing data. Appl Water Sci (springer) 8(15):1–16. https://doi.org/10.1007/s13201-018-0660-7

Rai PK, Chaubey PK, Mohan K, Singh P (2017b) Geoinformatics for assessing the inferences of quantitative drainage morphometry of the Narmada Basin in India Applied Geomatics 1–23https://doi.org/10.1007/s12518-017-0191-1

Ramamoorthy P, Rammohan V (2015) Assessment of groundwater potential zone using remote sensing and GIS in Varahanadhi watershed, Tamil Nadu, India. Int J Res Appl Sci Eng Technol 3:695–702

Rao RJ (2005) Participatory watershed management (PWM): an approach for integrated development of rural India: a case study from Karnataka, southern India. Int J Environ Technol Manage 5:107–115

Rorabaugh MI, Simons WD (1966) Exploration of methods of relating ground water to surface water. Columbia River basin-Second phase. Tacoma, WA: U.S. Geological Survey

Rushton KR, Ward C (1979) The estimation of groundwater recharge. J Hydrol 41:345–361

Sahu PC, Nandi D (2016) Groundwater resource estimation and budgeting for sustainable growth in agriculture in a part of drought prone Sundergarh district, Odisha, India. Int Res J Earth Sci 4:9–14

Sameena M, Ranganna G, Krishnamurtty J, Rao M, Jayaraman V (2000) Targeting ground-water zones and artificial recharge sets using remote sensing and GIS techniques. In: Abs 12th Convention IGC, pp 113–114

Saraf AK, Choudhury PR (1998) Integrated remote sensing and GIS for groundwater exploration and identification of artificial recharge sites. Int J Remote Sens 19:1825–1841

Thomas BC, Kuriakose SL, Jayadev SK (2009) A method for groundwater prospect zonation in data poor areas using remote sensing and GIS: a case study in Kalikavu Panchayath of Malappuram district, Kerala, India. Int J Digital Earth 2:155–170

United States Geological Survey (USGS) 2013. "Groundwater."

Vijith H (2007) Groundwater potential in the hard rock terrain of Western Ghats: a case study from Kottayam district, Kerala using Resourcesat (IRS-P6) data and GIS techniques. J Ind Soc Rem Sens 35:163

Chapter 3
Spatial Prediction of Flood Frequency Analysis in a Semi-Arid Zone: A Case Study from the Seyad Basin (Guelmim Region, Morocco)

Fatima Zahra Echogdali, Rosine Basseu Kpan, Mohammed Ouchchen, Mouna Id-Belqas, Bouchra Dadi, Mustapha Ikirri, Mohamed Abioui, and Said Boutaleb

Abstract Flood, a constant phenomenon especially in the semi-arid areas and flood plain regions, can be seen as one of the most destructive natural hazards jeopardizing the life of a population, their property, and their physical and economic environment. This paper focus on hydrologic modeling using the HEC-RAS model in combination with Watershed Modeling System (WMS) tools compares to the Flood Hazard Index (FHI) method using GIS in the Seyad basin situated in the southwestern region of Morocco with an area of 1512.85 km². The goal sought in this study is to evaluate flood risk in the Seyad basin that covers the cities of Taghjijt, Aday, Amtoudi, Tagri-ante, and Timoulayn'Ouamalougt that are areas with important agricultural lands. The HEC-RAS approach combines the surface hydrologic model and the digital terrain model data. This combination allows the mapping of the flood zones by using the WMS tools. This approach predicts flood occurrence probability for different times and determines the intensity of the flood (depth and velocity of floodwater) by using the existing hydrological data. On the other hand, The Flood Hazard Index method presents a multi-criteria index to assess flood risk areas, using six physical parameters namely: Permeability, slope, distance from the drainage network, land use, drainage network, and flow accumulation. A weight is calculated from the analytic hierarchy process method and applies to each parameter. HEC-RAS method allows the mapping of a flood with a flood water surface profile that shows the depth of flood for Annual Exceedance Probability (AEP) while FHI permits establishing flood risk level without indicating the depth of water. In both approaches, six types of simulations were performed with the return periods of 10, 20, 50, 100, 200, and 500 years and the simulation revealed that the most susceptible areas to flooding are the area along the Seyad River.

F. Z. Echogdali · M. Ouchchen · M. Id-Belqas · B. Dadi · M. Ikirri · M. Abioui (✉) · S. Boutaleb
Department of Earth Sciences, Faculty of Sciences, Ibn Zohr University, Agadir, Morocco
e-mail: m.abioui@uiz.ac.ma

R. B. Kpan
Regional Water and Environmental Sanitation Centre Kumasi, Kumasi, Ghana

© The Author(s), under exclusive license to Springer Nature Singapore Pte Ltd. 2022   49
P. K. Rai et al. (eds.), *Geospatial Technology for Landscape and Environmental Management*, Advances in Geographical and Environmental Sciences,
https://doi.org/10.1007/978-981-16-7373-3_3

Keywords HEC-RAS · WMS · Flood Hazard Index · Seyad basin · Flood risk

3.1 Introduction

Floods are among the most frequent and devastating types of disasters in the world. Studies show that economic damages and loss of human lives caused by floods are always increasing and present a very high level of risk (Creach et al. 2016; Di Salvo et al. 2017; Abdessamed and Abderrazak 2019; Khalfallah and Saidi 2018; Nkwunonwo et al. 2019; Abu El-Magd et al. 2020; Aitali et al. 2020; Haque et al. 2021). The overflow, the break, and the sagging (subsidence) of banks, and the ebb flow of the river by the conversion due to variations in water level and velocity are the main reason for flood risk (Ghosh and Kar 2018). The kingdom of Morocco is one of the most vulnerable countries to flood risk due to climate change manifested by intense violence and frequent precipitations.

In semi-arid regions, precipitation is characterized by very short durations and high intensities often leading to localized flash floods (Al-Zahrani et al. 2017). Rapid improvement in the natural science domain has led to additional studies in numerous countries located in these semi-arid regions such as Morocco (e.g. Echogdali et al. 2018a, b; Theilen-Willige et al. 2015), Tunisia (e.g. Souissi et al. 2020; Khalfallah and Saidi 2018), Israel and USA (e.g. Metzger et al. 2020), Namibia (Morin et al. 2009a, b) and Algeria (e.g. Abdessamed and Abderrazak 2019) suffering from repetitive and devastating events. It is, therefore, necessary to approach and study the risks caused by inundations to ensure a healthy and sustainable environment.

The floods have killed more than 1036 people and caused an economic loss of 267,009,000 US$ between 1980 and 2010 in the south of Morocco, the impact of these events is due to the socio-economic vulnerability of that region. In 1985, the province of Guelmim was affected by disastrous floods caused by the overflow of the Oum Laachar River in the centre of Guelmim and Bouizakarne with a discharge of 1000m3/s (EVICC 2011). The damages caused by these natural disasters keep increasing due to the demographic explosion raising the impervious zone area further exacerbating floods.

A comparative analysis of the flood hazard of the Seyad basin with other research studies in the same region in southern Morocco shows that the flooding has caused enormous and catastrophic damage. Theilen-Willige et al. (2015) studied flooding that particularly affects the region of Guelmim located downstream of the Seyad basin. They demonstrated that severe storms in November 2014 led to flash floods and rivers flooding, produced enormous damages, and the Guelmim region was declared an area of catastrophe. After the outbreak, when the river banks submerged, several areas were wholly flooded, several roads were dysfunctional and the electricity networks were destroyed. Hundreds of houses were completely or partially damaged, and dozens of roads were blocked. Heavy rains produced flooding in the Wadi Boussafen, around 32 km south of Guelmim and in the wadi Oum el Aachar, west of Guelmim, flooding trees, cars, roads, and bridges. These floods led to the loss of life in Guelmim, with more than 32 people killed (Zurich and Targa-AIDE 2015).

During the last decades, spatial mapping of flood risk zones has improved thanks to Geographic Information System (GIS) technologies, remote sensing, and hydraulics studies (Ikirri et al. 2021; Souissi et al. 2020; Lyu et al. 2020; Ghosh and Kar 2018; Echogdali et al. 2018b; Guerriero et al. 2018; Kabenge et al. 2017; Theilen-Willige et al. 2015; Das 2019, 2020). Research done in the world has considerably improved the simulation capacity of floods. Among the most used model, we have HEC-RAS and WMS characterized by the fast simulation and the ease of application and thereby allow developing different scenarios of simulation for different return periods. Al-Zahrani et al. (2017) combined these two models to estimate flood potential in an arid watershed in Saudi Arabia. Khalfallah and Saidi (2018) mapped floodplains using HEC-RAS and GIS tools in the Mejerda basin, Tunisia. Khattak et al. (2016) coupled HEC-RAS and ArcGIS for the mapping of floodplains in the Kaboul basin, Pakistan. The establishment of new approaches is more important to mitigate and prepare for the destructive impacts from floods that are sudden and in a short duration. The multi-criteria approach is part of the tools used to identify areas of high flood risk (Ajjur and Mogheir 2020; Hosseini et al. 2020). A great number of recent studies used that approach and showed its efficiency. This approach proposes several factors controlling floods and is the subject of many studies. We can consider flow accumulation, distance from drainage, land use, slope, geology, and density of drainage. For instance, Kazakis et al. (2015) proposed other factors such as elevation and precipitation for flood mapping in the Rhodope-Evros region, Greece. El-Haddad et al. (2020) considered TWI, curvature, and slope aspects for the delineation of the most vulnerable areas to flooding in the Qena Wadi basin, Egypt. The determination of those factors represents a crucial step in allowing the generation of a vulnerability model to flood risks. The accuracy of risk mapping models has become more promising.

In this chapter, a study was carried out in the Seyad basin in Guelmim, southeast Morocco. The objective of this study is to assess the extent of flooding to create a flood risk management plan for the Guelmim region to ensure the protection of areas at high risk of flooding using hydraulic and hierarchical process analysis (HPA) approaches within a GIS environment. One of the principal keys to natural risk management is the comprehension of the evaluation and the identification of flood risks. The elaborated results permit decision-making regarding flood prevention and flood control.

3.2   Study Area

Seyad River is situated in the Guelmim region in the south of Morocco, between $X = 97{,}550.890$ to $X = 147{,}779.799$ and $Y = 224{,}373.266$ to $Y = 280{,}605.698$ (latitudes $29°0'0''$N to $29°30'0''$N and longitude $9°32'0''$W to $9°04'0''$W) (Fig. 3.1). The basin covers an area of 1512.85 km^2. The main river length is 63.73 km (Fig. 3.2a), with an average slope of 10%. The low slopes (<14%) cover vast plains that are more prone to

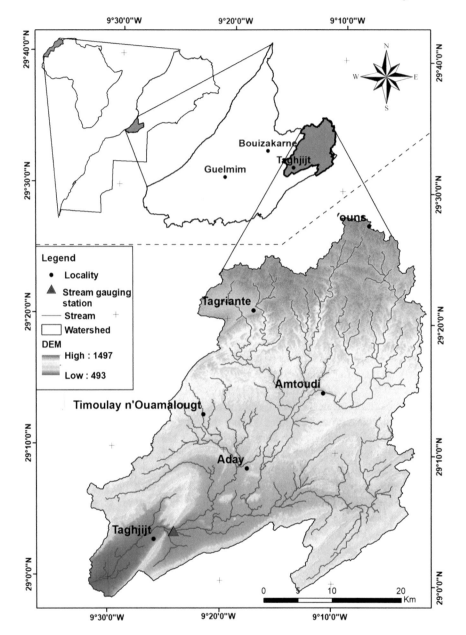

Fig. 3.1 **a** Location map of Seyad basin, **b** Seyad river and stream gauging stations with digital elevation model

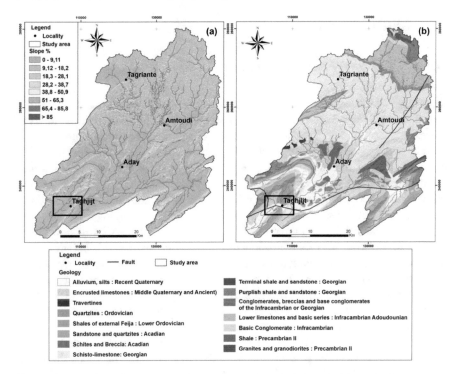

Fig. 3.2 Parameters of Seyad basin: **a** Slope map and **b** geological map of Seyad basin

flooding (Fig. 3.2b). The elevation varies from 493 to 1497 m with an average annual rainfall of approximately 120 mm, with high variability between 15 and 300 mm, and an average annual temperature around 21 °C, with a maximum monthly average exceeding 38 °C. The study area was selected for having been frequently subject to flooding which causes serious damages.

Geologically, the Oued Seyad basin located on the western Anti-Atlas extends on the southern edge of the Kerdous Massif, bounded to the south by Jbel Tayert, to the east by Adoudounien formations of the Kerdous Massif, to the west by the Guelmim plain, and to the northwest by the Lakhssas plateau (Choubert 1963).

The basin is represented by upper Acadian outcrops generally formed of shales with the terminal part consisting of quartzite bars that appear in the centre of the plain at the Jbel Tayert. The upper Acadian shales overlie Georgian limestones, which in turn lie on ancient Precambrian terrains (granite, granodiorite, and shales), outcropping over a small area in the northeast of the basin (Choubert and Faure-Muret 1970). The shales that constitute the Guelmim plain appear also on the oldest terrains. The quaternary generally outcrops downstream of the basin in the wadi, in the form of terraced alluvium or veneer on ancient rocks, in the form of mantle alteration and incrusted limestones overlying the Acadian schists of the substratum. Although mountainous in appearance, the basin is mainly made up of soft and rolling hills (Yazidi 1976; Atbir 2014).

3.3 Materials and Methods

3.3.1 Hydrology and Climatology

Hydrological data are the basic input parameters required for hydraulic modelling. For that, we require maximum annual discharges to estimate magnitude floods of different year return periods (RP) measured at Taghjijt hydrological station located closest to the inundation areas (Table 3.1). The annual maximum method takes into account only one maximum flow in the year discarding all other historical flooding events for that same year (Karim et al. 2017). The full record of 27-years of peak discharges (1986–2013) of data was used for this study, supplied by the Hydraulic Basin Agency of Souss-Massa.

Seyad River originates on the southern slopes of the Anti-Atlas, it flows in an east–west direction over its entire length and receives numerous tributaries on its right bank, the most important of which are: Kelmt, Tanzirt, Taouimarht, Ifrane, Ben Rhezrou, and Oum Aachar. On a large scale, the branching of the network is more pronounced upstream than downstream, given the very abundant geological accidents in the medium and high altitudes. The Oued is bounded by relatively strong banks that become a stream in the rainy season.

The monthly evolution of precipitation shows the existence of two distinct rainy seasons: The wet season, going from November to March, during which the region receives about 70% of the annual rain. The dry season, from April to October, during which the region receives only 30% of the annual rain. Precipitation over the study area shows great spatial and temporal variability. The maximum is recorded

Table 3.1 Maximum annual instantaneous flow rate of Seyad River

Year	Q max (m³/s)	Year	Q max (m³/s)
1985/1986	44	1999–2000	0.023
1986/1987	357	2000–2001	0.018
1987–1988	288	2001–2002	1068.837
1988–1989	639	2002–2003	381.942
1989–1990	52.7	2003–2004	337.901
1990–1991	40.4	2004–2005	21.802
1991–1992	169	2005–2006	32.968
1992–1993	0.026	2006–2007	141.532
1993–1994	0.038	2007–2008	0.005
1994–1995	353	2008–2009	261.136
1995–1996	136	2009–2010	150.453
1996–1997	3.032	2010–2011	324.96
1997–1998	52.915	2011–2012	1.069
1998–1999	47.803	2012–2013	155.953

in February with 114 mm of rain. The annual average is around 88.9 mm in Taghjijt but with a very significant irregularity between the years.

The Guelmim region is characterized by a semi-Saharan climate but much less hot and much less dry than typically Saharan regions, due to moderating ocean influences, except during the periods of the chergui (the desert wind) during which the mercury climbs even higher than inside the Sahara. The average annual flow of the Wadi Seyad at the Taghjijt station is 24.75 m³/s. The strongest floods reach a flow of 120m³/s recorded during the years 1988 and 2002.

3.3.2 Flood Modelling Using HEC-RAS and WMS Method

3.3.2.1 Hydraulic and Flood Plain Modelling

Using numerical complexes in solving the conservation equations for free surface flow is essential in the hydrodynamic modelling of a river with floodplains (Pinos and Timbe 2019). As numerical models represent reality, a key characteristic of hydrodynamic modelling adequately represents the topography of the river channel and the neighbouring floodplains (Casas et al. 2006). Many numerical tools enable the modelling of rivers and floodplains with one-dimensional (1D), two-dimensional (2D), or three-dimensional (3D) approaches (Bladé et al. 2014; Anees et al. 2016; Amellah et al. 2020; Naiji et al. 2021). Those models change considerably in terms of capabilities and data needs (Singh and Frevet 2006). Based on the available meteorological and spatial data of the study area, two models were selected: HEC-RAS and WMS models.

HEC-RAS is a windows-based hydraulic model built by the U.S. Army Corps of Engineers constructed to execute one-dimensional hydraulic computation for a full network of natural and constructed channels. The HEC-RAS model calculates the heights of water surface at all locations of interest for given values (Rivera et al. 2007). It employs the equation of Bernoulli (1) for subcritical flow at each cross-section (Bedient and Huber 2002):

$$Z_2 + Y_2 + \frac{\alpha_2 V_2^2}{2g} = Z_1 + Y_1 + \frac{\alpha_1 V_1^2}{2g} + h_e \qquad (3.1)$$

where

Z_1 and Z_2 represent the elevation of the principal channel inverts; Y_1, Y_2 = depth of water at cross-sections; V_1, V_2 = mean flow speed (total discharges/total surface flow); α_1, α_2 = velocity coefficients; g = gravitational acceleration; and h_e = energy head loss.

To define the width of the floodplain area for various years of return periods, HEC-RAS is not adequate to generate a flood inundation map. For this purpose, the Watershed Modelling System (WMS) was applied for developing model input and output of HEC-RAS to generate flood hazard maps. WMS is software encoded to promote hydrologic and hydraulic modelling of a catchment area (WMS 2018). The Environmental Modelling Research Laboratory, Brigham Young University, in collaboration with the U.S. Army Corps of Engineers Waterways Experiment Station created this software (Brigham Young University 2002). WMS provides useful tools to carry out operations such as automated delineation of basins, computation of geometric parameters, calculations of GIS overlays, cross-section extraction from terrain data, floodplain delineation, map design, tempest drain analysis, and floodplain boundaries can be drawn automatically in the generated hydrological model.

3.3.2.2 Methods for Simulated Floodplain Mapping

Estimation of flooding required various spatial databases including the Digital Elevation Model (DEM), land use, and flow data of Seyad River that were collected from diverse sources. For creating a flood simulation of the Seyad basin, there are four important basic steps:

1. Preparation of a Triangular Irregular Network (TIN) that accounts for the study area topography, created from digital contour data of digital elevation model (DEM) with a 12×12 m spatial resolution, sourced from https://asf.alaska.edu, Alaska Satellite Facility, UAF (June 2007).
2. Preparing geometric data, this stage consists of the creation of a stream centre-line, main channel banks, material zones which are called "channel geometry", and cross-section lines were chosen perpendicular to the river and used to extract elevation values from the terrain model (Echogdali et al. 2018a). A total of 31 sections across the river were considered for the data that is extracted to be more precise as seen in Fig. 3.3. Reducing that number results in poorer quality inundation maps (Aaron and Venkatesh 2009), and for this reason, the number of sections must be adequate and greater to obtain more details on the flood maps. A cross-section water surface elevation is shown in Fig. 3.3b. Manning's n-values were applied in the model to determine the roughness of the various cross-sections (Goodell and Warren 2006). The n-values were assigned after defining land use characteristics for the study area extracted using Sentinel-2 satellite images by supervised classification in Envi 5.3 software. Manning's n-values were attributed to areas as followed: 0.03 for the river, 0.08for the residential area, 0.10 for agricultural lands, and 0.04 for bare soils.
3. Mapping of flood areas: After the simulation has been made in HEC-RAS which allowed calculating water surface elevation along the Seyad River, the project is exported to the WMS software which uses the flood analysis function to the mapping of flood inundation areas. This function is based on an interpolation of water surface elevation on the cross-sectional area over 9500 m. Figure 3.4 illustrates the methodology of the flood modelling process using WMS and HEC-RAS methodology.

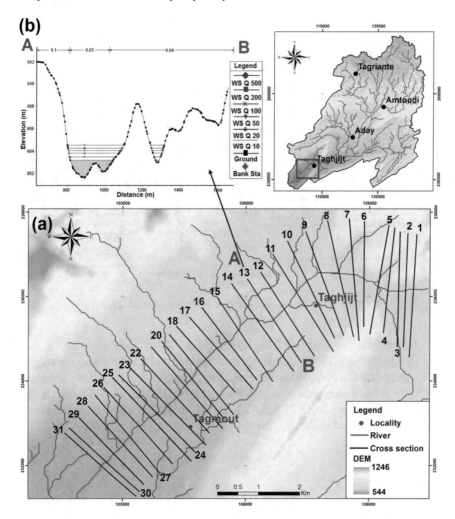

Fig. 3.3 a Cross-section along the Seyad River; **b** water surface elevations at cross-section profile for six scenarios (10, 20, 50, 100, 200 and 500 years)

4. Frequency analysis of the maximum annual discharges: the analysis of frequencies is a statistical approach popularly utilized in hydrology to report the extent of extreme events (e.g. floods or low flows) to the probability of occurrence (Tramblay et al. 2008). After that, the flood amplitude for a given return period can be evaluated based on the design of flood sluicing buildings, bridges, culverts, and other flood control projects (Cong and Hu 1989). Several formulas exist to calculate the frequency distribution of the maximum annual series of flood discharge. In this study, five methods of allotment types that are frequently applied for hydrologic frequency analysis were selected. To determine the best distributions. El Adlouni et al. (2007) proposed two selection criteria: the Akaike

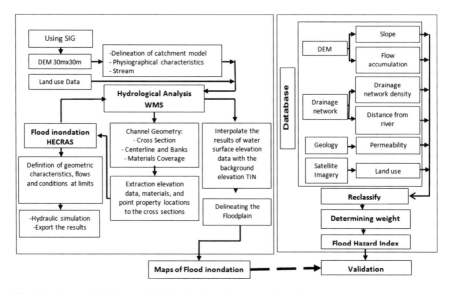

Fig. 3.4 The methodology applied for the Seyad basin (Echogdali et al. 2018b)

Information Criterion (AIC) (Akaike 1973) and Bayesian Information Criterion (BIC) (Schwarz 1978). Criteria that are founded on the differences between the appropriate distribution and the empirical probability with a penalty depending on the distribution parameter numbers and sample size. The distribution associated with the smallest BIC and AIC values is that better match (Rao and Hamed 2001) as defined in Eqs. (3.2) and (3.3):

$$\text{AIC} = -2\log(L) + 2k \tag{3.2}$$

$$\text{BIC} = -2\log(L) + 2k\,\log(N) \tag{3.3}$$

L likelihood function; k number of parameters; N sample size.

3.3.3 Flood Modelling Using FHI Method

Input data information acquired from various data sources is processed, analysed, and combined in the GIS (Fig. 3.4). The model was developed to carry out a multi-criteria analysis including Flood Risk Index (FHI). Six FHI factors have been selected which include slope, flow accumulation, distance from the river, drainage network density, land use, and geology (Elkhrachy 2015; Kazakis et al. 2015) (Fig. 3.5a–f, Table 3.2). The choice of those factors was theoretically founded on their pertinence to flood hazards (Haan et al. 1994).

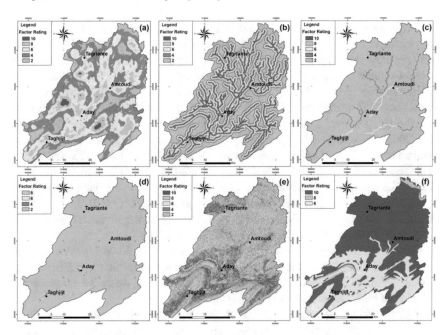

Fig. 3.5 Factors using in FHI method for Seyad basin: **a** Flow accumulation; **b** distance from river; **c** drainage network density; **d** land use; **e** slope; **f** geology

Table 3.2 Classes of the flood control factors

Factors	Class	Rating	Weight	Factors	Class	Rating	Weight
Flow accumulation (pixels)	1,178,608.7–1,902,185	10	3.31	Slope (%)	0–12.9	10	0.71
	499,789.8–1,178,608.7	8			13–25.8	8	
	231,246–499,789.8	6			25.9–42.5	6	
	52,216.8–231,246	4			42.6–65.3	4	
	0–52,216.8	2			65.4 <	2	
Distance from river (m)	0–200	10	3.08	Drainage network density (m/km²)	1.123–1.779	10	1.34
	200–400	8			0.858–1.123	8	
	400–700	6			0.628–0.858	6	
	700–1000	4			0.377–0.628	4	
	1000–2000	2			0–0.377	2	
Land use	Stream	8	1.08	Geology	Low permeability	10	0.49
	Residence	6			Medium permeability	8	
	Agricultural land	4			High permeability	6	
	Soil	2					

The parameter weight is established using the Analytical Hierarchy Process (AHP) to integrate and evaluate the impacts of different factors (Saaty 1990a, b). The AHP was originally designed by Saaty (1980) to supply a scheme to solve various kinds of multi-criteria decision problems based on the relative priorities attributed to each criterion's role in the achievement of the mentioned goal. AHP is an advantage measurement (scoring) model that is based on multiple criteria; these inputs are transformed into marks employed to evaluate each of the possibilities (Handfield et al. 2002). To assess each factor's weight, we allocate values on a marking scheme that varies from 2 to 10 as proposed by Kazakis et al. (2015) (Table 3.2). The classes of the flow accumulation and drainage network density were determined using Jenks (1967), the slope class was determined by Demek (1972), while the class assigned to distance from the river was determine by Kazakis et al. (2015). Land use and geological formation have been classified according to the characteristics of the study area (Tehrany et al. 2013; Ouma and Tateishi 2014).

Flow accumulation: The sum amount of water flowing downslope is correlated with higher discharges and thereby proportionate to the relative flood hazard potential (Kazakis et al. 2015). To extract the higher flood hazard area, we assigned a rate equal to 10 to the high values of flow accumulation (Fig. 3.5a).

Distance from the river: The risk of flooding is higher in areas closest to the river. As the distance from the river increases, the risk of riverbank flooding reduces. In the Seyad basin, areas within 400 m to the river and especially at the confluence points of the rivers are at high risk of flooding, while the effect of this parameter decreases over distances of more than 700 m (Fig. 3.5b).

Drainage network density: Network drainage density in an area affects the time of concentration of runoff and therefore is correlated to flow accumulation tracks and the probability of flooding (Schmitt et al. 2004). A detailed drainage network density map of the whole basin was set which shows that the intensity of the risks is very high around the Seyad River, the density ranged between 6 and 8 in this part (Fig. 3.5c).

Land use: The land use map is generated from the LANDSAT image of the Seyad basin. The land use of the Taghjijt basin is very contrasted, in terms of class size, their properties (density, area, location …), and temporal variability, but roughly it is mainly formed by agriculture, this activity is widespread along the beds of the river in the form of furnished terraces, bare soil, this class is the most dominant and residential area. The type of land use determines the rate of infiltration of precipitation in the soil. Therefore, residential is the most exposed to flood risk than other lands; the study of this parameter is important which allows attributing rates equal to 6 for residential areas, respectively (Fig. 3.5d).

Slope: Areas with lower slopes endure extended flooding whilst modest to raised slopes allow quicker floodwater drainage (Ghosh and Kar 2018). The slope map of the area was generated from ALSA Satellite which shows that the downstream part of the basin is characterized by a very shallow slope, which makes it more susceptible to flooding because runoff would flow much more slowly and accumulate (Fig. 3.5e).

Geology: The analysis of the geological map of the Seyad basin shows a dominance of Palaeozoic formations (limestones and shales) which represent more than

70% of the basin surface followed by quaternary formations (20%) and then the Infra-Cambrian formations (10%). The lithology of the Seyad basin is dominated by low to medium permeability formations. Crystalline and carbonate formations (low to medium permeability) account for 80% of the extent, while high permeability formations (alluvium) account for about 20% of the extent of the Seyad basin. The schistose limestone, crusted limestone, and crystalline formations constitute a major risk by amplifying the destructive power of floods in the Seyad basin, for this reason, we assigned a rate equal to 10. Whereas a lower rating was given to alluvial and continental deposits, thanks to their greater capacity for infiltration, promoting infiltration causing groundwater flow and reducing the intensity of floods (Fig. 3.5f).

The combination of six different factors were performed using the FHI method. The FHI is calculated using Eq. (3.4).

$$\text{FHI} = \sum_{i=1}^{n} W_i . r_i \tag{3.4}$$

where W_i represents the weight of the factors causing flooding, r_i the score rating of the factor causing flooding and n is the number of factors causing flooding.

In this study, a matrix of pairwise comparisons of the criteria for the AHP was defined. Each factor was evaluated mutually with other factors by attributing a relative importance value scale from 1 to 9 (Table 3.3), where 1 implicates that factors have equal value and 9 indicating that a given factor is highly valuable compare to another (Saaty 2008).

For the verification of the consistency between the pairwise comparisons and the accuracy of the obtained weights, the consistency ratio (CR) can be calculated. The value of the CR must be less than or equal to 0.10, which indicates an acceptable matrix (Saaty 2012). The CR is calculated according to Eq. (3.5).

$$\text{CR} \frac{\text{CI}}{\text{RI}} \tag{3.5}$$

Table 3.3 Pairwise comparison matrix using for FHI

Factor	Flow accumulation	Distance	Drainage	Land use	Slope	Geology
Flow accumulation	1	2	3	3	5	4
Distance from river	1/2	1	6	3	4	6
Drainage network density	1/3	1/6	1	2	3	3
Land use	1/3	1/3	1/2	1	3	2
Slope	1/5	1/4	1/3	1/3	1	3
Geology	1/4	1/6	1/3	1/2	1/3	1

With CI representing the evenness of index calculated following Eq. (3.6):

$$CI = \frac{\lambda_{max} - n}{n - 1} \tag{3.6}$$

where λ_{max} is the greatest Eigenvalue of the matrix of comparison and n the number of criteria. RI is the random index calculated from the mean evenness index of as ample and indiscriminately provided of a 500 pairwise comparison matrix as shown in Table 3.4 (Saaty 1977).

In this study, based on the values of Table 3.5, CI = 0.108 (λ_{max} = 6.54, n = 6), RI = 1.24 and CR = 0.08, which conforms to the standard of CR < 0.1.

After determining weights (Wi) derived using AHP (Table 3.5), the maps for six factors (Fig. 3.5a–f) were integrated to prepare the Flood Hazard Index map using Eq. (3.7):

$$\begin{aligned} FHI = {} & \text{Flow accumulation} \times 3.31 + \text{Distance from river} \\ & \times 3.08 + \text{Drainage network density} \times 1.34 + \text{Land use} \\ & \times 1.08 + \text{Slope} \times 0.71 + \text{Geology} \times 0.49 \end{aligned} \tag{3.7}$$

Table 3.4 Random indices used to calculate the consistency ratio

n	1	2	3	4	5	6	7	8	9	10
RI	0	0	0.58	0.9	1.12	1.24	1.32	1.41	1.45	1.49

Table 3.5 Normalized flood hazard for different factors

Factor	Flow accumulation	Distance	Drainage	Land use	Slope	Geology	Mean	Wi
Flow accumulation	0.38	0.51	0.27	0.31	0.31	0.21	0.33	3.31
Distance from river	0.19	0.26	0.54	0.31	0.24	0.32	0.31	3.08
Drainage network density	0.13	0.04	0.09	0.20	0.18	0.16	0.13	1.34
Land use	0.13	0.09	0.04	0.10	0.18	0.11	0.11	1.08
Slope	0.08	0.06	0.03	0.03	0.06	0.16	0.07	0.71
Geology	0.10	0.04	0.03	0.05	0.02	0.05	0.05	0.49

3.4 Results and Discussion

3.4.1 Flood Hazard Mapping Using HEC-RAS and WMS Method

The maximum annual discharge data of the Seyad River was used for the analysis of flood frequency (Table 3.1). For this study, Gumbel, Exponential, Gamma, Weibull, and GEV distributions were used in the flood frequency analysis. Based on the values of AIC and BIC criteria, the Gamma distribution proved to be the most applicable and adequate and was thus used to estimate flood peaks for Seyad River for different return periods (RP) (Table 3.6; Fig. 3.6). The peak flood estimated by Gamma distribution for return periods of 5, 10, 20, 50, 100, 200, and 500 years were 281 m^3/s, 529 m^3/s, 810 m^3/s, 1210 m^3/s, 1530 m^3/s, 1860 m^3/s, 2310 m^3/s, respectively (Table 3.7).

P (Mi) a priori probability, P (Mi/x) a posterior probability (Method of Schwarz), *AIC* akaike information criterion, *BIC* Bayesian information criterion.

The floodwater level and flood amplitude are acquired by the use of HEC-RAS and WMS software and the floodplain maps for the different return periods; the 50 and 500 years floodplain maps are illustrated in Fig. 3.7. Flood mapping in the study area shows that the levels of actual hazard vary from 0.43 to 8 m and shows that flooded areas are essentially situated along the Seyad River. The floodwater level in the river banks is higher and can reach depths of 6.45 m causing severe damage; even floods with depths of less than 0.8 m can cause severe damage. The higher values were due to geomorphological characteristics of the terrain, characterized by lower depressions, causing significant lateral extensions of the floods. The depressions located in height are between 493 and 725 m. These regions in southern Morocco are characterized by prolonged hills and crests developing obstructions to surface water runoff and flow into the Atlantic. This natural obstacle leads to the retention of water in the case of heavy rainfall and therefore is one of the factors supporting flash flood risk (Theilen-Willige et al. 2015). Approximately 16.04 and 25.85% of the whole area of building residences and agriculture area was inundated. However, such flood visualization maps are important tools for real-time flood alerts in case of emergency and to decrease the rate of loss in human lives (Mai and De Smedt 2017). Comparing the simulated floodplain depths of different return periods showed that

Table 3.6 Results of the best-fitted distributions using AIC and BIC criteria of Seyad River

Return period: $T = 100$	Number of observation: 27				
Model	XT	P(Mi)	P(Mi|x)	BIC	AIC
Gamma (Maximum Likelihood)	2	1534.039	100	312.733	310.068
Weibull (Maximum Likelihood)	2	1130.097	0	332.98	330.315
Exponential (Maximum Likelihood)	2	856.587	0	355.744	353.079
Gumbel (Maximum Likelihood)	2	724.032	0	375.204	372.54
GEV (Maximum Likelihood)	2	1023.158	0	376.997	373.001

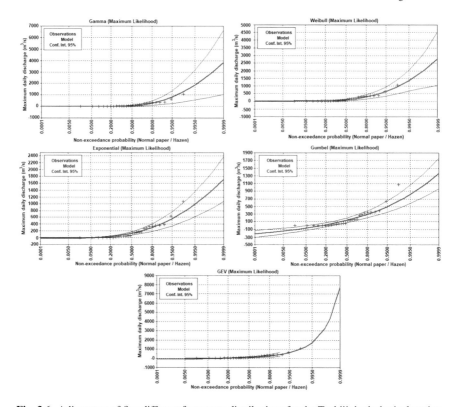

Fig. 3.6 Adjustment of five different frequency distributions for the Taghjijt hydrological station

Table 3.7 Discharges for different scenarios (10, 20, 50, 100, 200, and 500 years) of Seyad River

Return periods (years)	Discharges of Seyad river (m³/s)
500	2310
200	1860
100	1530
50	1210
20	810
10	529
5	281

important overflows from the two riverbanks, as shown in the eastern part of the city of Seyad, make these areas particularly vulnerable to flood inundation and increases the risk of losses.

To present the floodplain areas, the entire Seyad River was separated into upper and lower zones, and distinct floodplain maps were presented for each zone (Fig. 3.8). The water levels exceeded Seyad River bank heights during 500 years return periods with a discharge rate of 2310 m³/s, showing significant flood depths on both sides

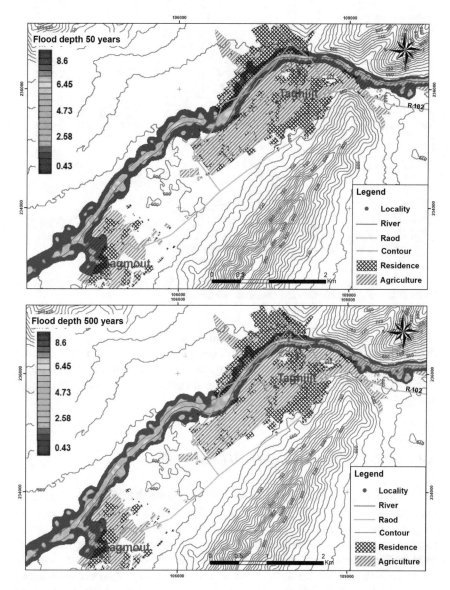

Fig. 3.7 Flood water depths map of Seyad River simulated by the HEC-RAS and WMS for 50 (upper) and 500 (lower) years return periods

of the river, and the most seriously damaged villages are Taghjijt and Tagmout. As we can see on the floodplain map, floodwater was nearly 8 m deep at these critical regions, indicating the severity of the flood. A comparison of water surface elevations in cross-sections for 10, 100, and 500 years return period floods with a discharge rate of 529, 1530, and 2310 m³/s, respectively, showed that the greater overflow occurs

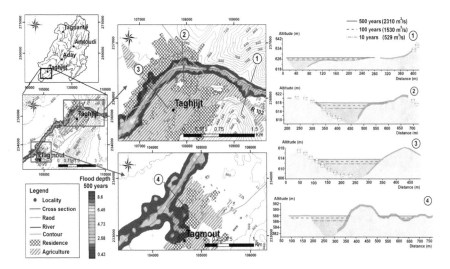

Fig. 3.8 Flood water depths map of Seyad River for 500 years and water surface elevations of floods of 10, 100, and 500 years at different cross-sections

along the Seyad River, causing extended flooding around it. The maximum affected flood area lies in the agricultural area, residences, and roads close to the Seyad River. The river continues to present the risk of dangerous conditions during severe flooding. This will permit better planning of stream restoration to reduce losses and damage during floods.

3.4.2   Flood Hazard Index (FHI) Approach

The maps in Fig. 3.9 represent flood risk maps using the FHI at vulnerable areas in the study area. The identification of factors were used to assess the most sensitive factors to the flood risk. This assessment revealed that the most susceptible areas to flooding are concentrated around the Seyad River due to the flow accumulation, distance from the river, and density of the drainage network. These are the factors that influence FHI, which is similarly found by other studies under the same conditions by Echogdali et al. (2018b). Land use, slope, and geology are considered the least important factor in the study area. The Seyad basin is entirely drained by rivers, which makes these areas more susceptible to flooding.

FHI permits establishing flood risk levels without indicating the depth of water. Accordingly, the Seyad basin has been classified into five levels of risk groups arranged from very minimal risk to very high risk (Fig. 3.9). Floodplain areas with very low to low, medium, and high to very high risk represent, respectively, 47.69%, 21.18%, and 31.12% of the Seyad basin. The flooding risk severity is really important in the localities of Tagmout and Taghjijt (Fig. 3.9a), which are characterized by a

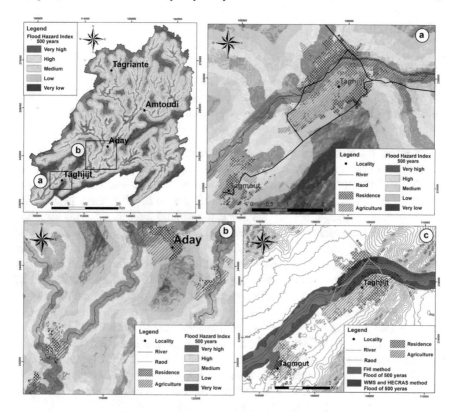

Fig. 3.9 Flood hazard map using FHI in the **a** downstream and **b** upstream part of the basin, and **c** comparison between the simulated flood using HEC-RAS-WMS and FHI process in the study area of Seyad River for 500 years return periods

plain with low altitude and a low slope. The spatial distribution of risk indicates that a large number of residential and agricultural areas are at very high risk and also the infrastructures, including roads, are affected by massive damage. The risk scenario is more similar to that of the Aday region (Fig. 3.9b), the areas in proximity to the river are highly exposed to the risk, the magnitude of the risk ranges from high to very high.

A comparative study of the results of the HEC-RAS-WMS approach and the FHI scenario for a return period of 500 years is presented in Fig. 3.8c. The result shows that the FHI model tends to modestly overestimate the magnitude of the inundation area. This overestimation can be explained by the presence of the areas of confluence between the Seyad River and other affluent that feed this river. Consequently, the results of the FHI model agree closely with the results of the HEC-RAS-WMS model, demonstrating the ability of the FHI model to map flood areas in other regions without the presence of hydrological data from a measuring station. However, the FHI model generates and estimates flood risk for the entire Seyad basin.

3.5 Conclusions

In this research, the conclusion is based on the assumption that two methods were used to assess flood hazards in the Seyad basin. The two methods which are different in their methodology allowed us to see that each one is useful and important in the resolution of the problems related to the management of floods. This study highlighted several important points:

1. For the first method using the combination of the HEC-RAS model and WMS model, water surface profile was generated and flood depths were estimated for 20, 50, 100, 200, and 500 years return period. The simulation and mapping of flood hazard using this model show that already at 20 years return period with 810 m^3/s discharge, houses, and agricultural land are flooded. This situation worsens as the return period increases to 500 years where the discharge is up to 2310 m^3/s. This method estimates the distance and the actual depth of the floodwaters over several periods based on current and prior data. However, it cannot be applied to the whole basin unless it is divided into several small portions and modelled individually. Also, flood hazards for tributaries can be assessed based on the flood height on the main channel.
2. The Flood Hazard Index method used specific factors to assess flood hazards in the study area. With this method, the whole basin is modelled and the results were similar to the results found in using HEC-RAS and WMS methods. The flood risk is higher at the bank of the river and decreases as we go away from the bank. It also gives a larger hazard area and then can be considered as the most suitable method in a development project.
3. Finally, with the two methods, we observed that the high risk areas are mostly along the river. However, the flood depth given by HEC-RAS and WMS model shows that the low-risk areas have a flood depth between 0.43 and 2.58 m, and can be damaging for people and property. By comparing the two methods, we can conclude that they are complementary in the sense that one gives a better horizontal view and the other a better vertical view. The developed map can be used as a decision-making material in building road infrastructure or incorporated into land use planning decisions for the towns within the catchment.

References

Aaron C, Venkatesh M (2009) Effect of topographic data, geometric configuration and modeling approach on flood inundation mapping. J Hydrol 377(1–2):131–142

Abdessamed D, Abderrazak B (2019) Coupling HEC-RAS and HEC-HMS in rainfall-runoff modeling and evaluating floodplain inundation maps in arid environments: case study of Ain Sefra city, Ksour Mountain, SW of Algeria. Environ Earth Sci 78(19):586

Abu El-Magd SA, Amer RA, Embaby A (2020) Multi-criteria decision-making for the analysis of flash floods: A case study of Awlad Toq-Sherq, Southeast Sohag, Egypt. J Afr Earth Sci 162:103709

Aitali R, Snoussi M, Kasmi S (2020) Coastal development and risks of flooding in Morocco: the cases of Tahaddart and Saidia coasts. J Afr Earth Sci 164:103771

Ajjur SB, Mogheir YK (2020) Flood hazard mapping using a multi-criteria decision analysis and GIS (case study Gaza Governorate, Palestine). Arab J Geosci 13(2):44

Akaike H (1973) Information theory and an extension of the maximum likelihood principle. In: Petrovand BN, Csaki F (eds) Proceedings of the 2nd international symposium on information theory. Akademiai Kiado, Budapest, pp. 267–281

Al-Zahrani M, Al-Areeq A, Sharif HO (2017) Estimating urban flooding potential near the outlet of an arid catchment in Saudi Arabia. Geomat Nat Haz Risk 8(2):672–688

Amellah O, El Morabiti K, Ouchar Al-djazouli M (2020) Spatialization and assessment of flood hazard using 1D numerical simulation in the plain of Oued Laou (north Morocco). Arab J Geosci 13(14):635

Anees MT, Abdullah K, Nawawi MNM, Ab Rahman NNN, Piah ARM, Zakaria NA, Syakir MI, Omar AKM (2016) Numerical modeling techniques for flood analysis. J Afr Earth Sci 124:478–486

Atbir H (2014) La Feija de Bouizakarne-Guelmim: Géomorphologie et perspectives environnementales. Ph.D. dissertation, Ibn Zohr University (In French)

Bedient PB, Huber WC (2002) Hydrology and floodplain analysis, 3rd edn. Prentice Hall, New Jersey

Bladé E, Cea L, Corestein G (2014) Numerical modelling of river inundations. Ingeniería Del Agua 18(1):71–82

Brigham Young University (2002) WMS v7.0 Help. Environmental Modelling Research Laboratory, Provo, UT

Casas A, Benito G, Thorndycraft VR, Rico M (2006) The topographic data source of digital terrain models as a key element in the accuracy of hydraulic flood modelling. Earth Surf Proc Land 31(4):444–456

Choubert G (1963) Histoire géologique du Précambrien de l'Anti-Atlas. Notes Mém Serv Géol Maroc 162, 352 p

Choubert G, Faure-Muret A (1970) Livret-guide de l'excursion: Anti-Atlas occidental et central. Notes Mém Serv Géol Maroc 299, 259 p

Cong SZ, Hu SY (1989) Some problems on flood-frequency analysis. Chin J Appl Probab Statist 5:358–368

Creach A, Chevillot-Miot E, Mercier D, Pourinet L (2016) Vulnerability to coastal flood hazard of residential buildings on Noirmoutier Island (France). J Maps 12(2):371–381

Das S (2020) Flood susceptibility mapping of the Western Ghat coastal belt using multi-source geospatial data and analytical hierarchy process (AHP). Remote Sens Appl Soc Environ 20:100379

Das S (2019) Geospatial mapping of flood susceptibility and hydro-geomorphic response to the floods in Ulhas basin, India. Remote Sens Appl Soc Environ 14:60–74

Demek J (1972) Manual of detailed geomorphological mapping. Academia, Prague

Di Salvo C, Ciotoli G, Pennica F, Cavinato GP (2017) Pluvial flood hazard in the city of Rome (Italy). J Maps 13(2):545–553

Echogdali FZ, Boutaleb S, Jauregui J, Elmouden A (2018) Cartography of flooding Hazard in semi-arid climate: the case of Tata Valley (South-East of Morocco). J Geograph Nat Disasters 8(1):1–11

Echogdali FZ, Boutaleb S, Jauregui J, Elmouden A, Ouchchen M (2018) Assessing flood hazard at river basin scale with comparison between HECRAS-WMS and Flood Hazard Index (FHI) methods: the case of El Maleh Basin, Morocco. J Water Resour Prot 10(9):957–977

El Adlouni S, Ouarda TBMJ, Zhang X, Roy R, Bobée B (2007) Generalized maximum likelihood estimators for the nonstationary generalized extreme value model. Water Resour Res 43(3):W03410

El-Haddad BA, Youssef AM, Pourghasemi HR, Pradhan B, El-Shater AH, El-Khashab MH (2020) Flood susceptibility prediction using four machine learning techniques and comparison of their performance at Wadi Qena Basin Egypt. Nat Hazards 105(1):83–114

Elkhrachy I (2015) Flash flood hazard mapping using satellite images and GIS tools: a case study of Najran City, Kingdom of Saudi Arabia (KSA). Egypt J Remote Sens Space Sci 18(2):261–278

EVICC (2011) Evaluation de la vulnérabilité et des impacts du changement climatique dans les oasis du Maroc et structuration de stratégies territoriales d'adaptation. Mission 1.1 : Bilan-Diagnostic des vulnérabilités climatiques et des capacités d'adaptation en situation actuelle. Supplementary report, October 2011 (In French)

Ghosh A, Kar SK (2018) Application of analytical hierarchy process (AHP) for flood risk assessment: a case study in Malda district of West Bengal India. Nat Hazards 94(1):349–368

Goodell C, Warren C (2006) Flood inundation mapping using HEC-RAS. Obras y Proyectos 2:18–23

Guerriero L, Focareta M, Fusco G, Rabuano R, Guadagno FM, Revellino P (2018) Flood hazard of major river segments, Benevento Province Southern Italy. J Maps 14(2):597–606

Haan CT, Barfield BJ, Hayes JC (1994) Design hydrology and sedimentology for small catchments. Elsevier, New York

Handfield R, Walton SV, Sroufe R, Melnyk SA (2002) Applying environmental criteria to supplier assessment: a study in the application of the analytical hierarchy process. Eur J Oper Res 141(1):70–87

Haque MM, Seidou O, Mohammadian A, Ba K (2021) Effect of rating curve hysteresis on flood extent simulation with a 2D hydrodynamic model: a case study of the Inner Niger Delta, Mali, West Africa. J Afr Earth Sci 178:104187

Hosseini FS, Choubin B, Mosavi A, Nabipour N, Shamshirband S, Darabi H, Haghighi AT (2020) Flash-flood hazard assessment using ensembles and Bayesian-based machine learning models: application of the simulated annealing features election method. Sci Total Environ 711:135161

Ikirri M, Faik F, Boutaleb S, Echogdali FZ, Abioui M, Al-Ansari N (2021) Application of HEC-RAS/WMS and FHI models for the extreme hydrological events under climate change in the Ifni River arid watershed from Morocco. In: Nistor MM (ed) Climate and land use impacts on natural and artificial systems: mitigation and adaptation. Elsevier, Amsterdam, pp 251–270

Jenks GF (1967) The data model concept in statistical mapping. Int Yearb Cartograph 7:186–190

Kabenge M, Elaru J, Wang H, Li F (2017) Characterizing flood hazard risk in data-scarce areas, using a remote sensing and GIS-based flood hazard index. Nat Hazards 89(3):1369–1387

Karim F, Hasan M, Marvanek S (2017) Evaluating annual maximum and partial duration series for estimating frequency of small magnitude floods. Water 9(7):481

Kazakis N, Kougias I, Patsialis T (2015) Assessment of flood hazard areas at a regional scale using an index based approach and analytical hierarchy process: application in Rhodope-Evros Region, Greece. Sci Total Environ 538:555–563

Khalfallah CB, Saidi S (2018) Spatiotemporal floodplain mapping and prediction using HEC-RAS-GIS tools: case of the Mejerda River, Tunisia. J Afr Earth Sci 142:44–51

Khattak MS, Anwar F, Saeed TU, Sharif M, Sheraz K, Ahmed A (2016) Floodplain mapping using HEC-RAS and ArcGIS: a case study of Kabul River. Arab J Sci Eng 41(4):1375–1390

Lyu HM, Zhou WH, Shen SL, Zhou AN (2020) Inundation risk assessment of metro system using AHP and TFN-AHP in Shenzhen. Sustain Cities Soc 56:102103

Mai DT, De Smedt F (2017) A combined hydrological and hydraulic model for flood prediction in Vietnam applied to the Huong river basin as a test case study. Water 9(11):879

Metzger A, Marra F, Smith JA, Morin E (2020) Flood frequency estimation and uncertainty in arid/semi-arid regions. J Hydrol 590:125254

Morin E, Grodek T, Dahan O, Benito G, Kulls C, Jacoby Y, Langenhove GV, Seely M, Enzel Y (2009) Flood routing and alluvial aquifer recharge along the ephemeral arid Kuiseb River, Namibia. J Hydrol 368(1–4):262–275

Morin E, Jacoby Y, Navon S, Bet-Halachmi E (2009) Towards flash-flood prediction in the dry Dead Sea region utilizing radar rainfall information. Adv Water Resour 32(7):1066–1076

Naiji Z, Mostafa O, Amarjouf N, Rezqi H (2021) Application of two-dimensional hydraulic modelling in flood risk mapping. A case of the urban area of Zaio, Morocco. Geocarto Int 36(2):180–196

Nkwunonwo UC, Whitworth M, Baily B (2019) Urban flood modelling combining cellular automata framework with semi-implicit finite difference numerical formulation. J Afr Earth Sci 150:272–281

Ouma YO, Tateishi R (2014) Urban flood vulnerability and risk mapping using integrated multi-parametric AHP and GIS: methodological overview and case study assessment. Water 6(6):1515–1545

Pinos J, Timbe L (2019) Performance assessment of two-dimensional hydraulic models for generation of flood inundation maps in mountain river basins. Water Sci Eng 12(1):11–18

Rao AR, Hamed KH (2001) Flood frequency analysis. CRC Press, New York

Rivera S, Hernandez AJ, Ramsey RD, Suarez G (2007) Predicting flood hazard areas: a SWAT and HEC-RAS simulations conducted in Aguan river basin of Honduras, central America. In: ASPRS 2007 annual conference, Tampa, Florida

Saaty TL (2012) Decision making for leaders: the analytic hierarchy process for decisions in a complex world, 3rd edn. RWS Publications, Pittsburgh

Saaty TL (2008) Decision making with the analytic hierarchy process. Int J Serv Sci 1(1):83–98

Saaty TL (1990) How to make a decision: the analytic hierarchy process. Eur J Oper Res 48(1):9–26

Saaty TL (1990) An exposition of the AHP in reply to the paper remarks on the analytic hierarchy process. Manage Sci 36(3):259–268

Saaty TL (1980) The analytical hierarchy process. McGraw-Hill, New York

Saaty TL (1977) A Scaling method for priorities in hierarchical structures. J Math Psychol 15(3):234–281

Schmitt TG, Thomas M, Ettrich N (2004) Analysis and modeling of flooding in urban drainage systems. J Hydrol 299(3–4):300–311

Schwarz G (1978) Estimating the dimension of a model. Ann Stat 6(2):461–464

Singh V, Frevert D (2006) Watershed models. Taylor & Francis, London

Souissi D, Zouhri L, Hammami S, Msaddek MH, Zghibi A, Dlala M (2020) GIS-based MCDM–AHP modeling for flood susceptibility mapping of arid areas, southeastern Tunisia. Geocarto Int 35(9):991–1017

Tehrany MS, Pradhan B, Jebur MN (2013) Spatial prediction of flood susceptible areas using rule based decision tree (DT) and a novel ensemble bivariate and multivariate statistical models in GIS. J Hydrol 504:69–79

Theilen-Willige B, Charif A, Ouahidi AE, Chaibi M, Ougougdal MA, AitMalek H (2015) Flash floods in the Guelmim area/southwest Morocco–use of remote sensing and GIS-tools for the detection of flooding-prone areas. Geosciences 5(2):203–221

Tramblay Y, St-Hilaire A, Ouarda TB (2008) Frequency analysis of maximum annual suspended sediment concentrations in North America. Hydrol Sci J 53(1):236–252

WMS (Watershed Modelling System) (2018) Reference manual, user manual (v10.1). Environmental Modeling Research Laboratory of Brigham Young University, Provo, UT

Yazidi A (1976) Les formations sédimentaires et volcaniques de la boutonnière d'Ifni (Maroc): Lithostratigraphie et chronologie du Précambrien supérieur. Ph.D dissertation, Grenoble Alpes University (In French)

Zurich and Targa-AIDE (2015) Inondations au Maroc en 2014: Quels enseignements tirer de Guelmim et Sidi Ifni? Report, 40p

Chapter 4
Geospatial Modeling in the Assessment of Environmental Resources for Sustainable Water Resource Management in a Gondia District, India

Nanabhau Santujee Kudnar

Abstract The present study is geospatial modeling in the assessment of environmental resources for sustainable water resource management in a Gondia District, India, using geographical information system (GIS) and remote sensing (RS) techniques. The monsoon rains in Gondia District are concentrated in the four months from June to September and receive 90.81% rainfall, post-monsoon 1.86%, pre-monsoon 4.83%, and winter 2.48%. The distribution of annual rainfall in Gondia is very uneven. The major river is Wainganga tributaries are Bagh, Pangoli, Gadhvi, Chor, Chandan, and Bawanthadi. Out of the total area received in the district, reserved forest area is 56.4%, protected forest area is 26.3%, and unclassified forest area is 17.3%. There are 192 small irrigated ponds below 100 ha, and its projected irrigation capacity is 10,897 ha. There are also 294 Kolhapuri type dams with a projected irrigation capacity of 10,075 ha and 1559 storage dams with a projected irrigation capacity of 14,817 ha. The water is neutral to alkaline in nature with pH ranging from 6.6 to 8.92 with high TDS range from 140 to 2184 ppm. The aim of this present study was to evaluate environmental resource units that have been delineated based on the geospatial modeling of environment parameters with appropriate weights in GIS and RS techniques. The data can be used for area management, utilized in restoration and conservation of natural resources studies in the future.

Keywords Geospatial modeling · Environmental resources · Water resource management

Abbreviations

APML	Adani Power Maharashtra Limited
IMD	Indian Meteorological Department
SOI	Survey of India

N. S. Kudnar (✉)
C. J. Patel College Tirora, Gondia, Maharashtra 441911, India

WGS   World Geodetic System
WQI   Water quality index

4.1   Introduction

Resources have a very important place in the economic development of a nation. Resources are considered the foundation of economic life. The country's economic and social development depends on wealth. The explosion of the world's population, the ever-increasing proportion of land and human resources, the growing stress on resource wealth are causing many problems today (Aslam et al. 2020; Tsihrintzis et al. 2017; Akhtar et al. 2020; Azhar et al. 2015; Chamine et al. 2019; Soutter et al. 2009). Along with the economic development of the nation, the use of resources has also become very important. The development and conservation of global resources are being studied by various bodies (Trikoilidou et al. 2017; Tiwari et al. 2014; Bouaicha et al. 2017; Mishra et al. 2017; 2018; Sharma and Kansal 2011; Guettaf et al. 2017). The contribution of economists, environmentalists, and geographers in this study is significant. Today many scientists are studying resources like geography, environmental geography, biological geography, ecology (Zhu et al. 2013; Chen et al. 2017; Gu et al. 2013; Kumar et al. 2020; Bisen and Kudnar 2013a, b, 2019; Gadekar and Sonkar 2020).

Environmental water, air, sunlight, land, forests, minerals, fauna, these vital elements are called resource wealth. Human involvement is very important in this. Humans are extremely influential in the development of resources. The face of the earth has changed over the last several centuries because of human activity. Civilized human beings built culture by developing the resources available to them, which is why human labor, knowledge, intellect, skills, health, consciousness, and freedom. Social cohesion is a matter of material things (Karande et al. 2020; Lakshmi et al. 2020; Rajasekhar et al. 2019; Mogaji and Lim 2020; Mishra et al. 2014). Human intervention has led to many environmental disasters, including the extinction of plant and animal species. With the increasing use of resources, their qualitative and quantitative reserves are declining.

Water is a great source of economic development. If equitable, community-based, and practical development of water resources has to be ensured, then traditional systems shall have to be revitalized and developed. It studies the comparative position of the spatial distribution of all water resources in nature except groundwater, oceans, ground surface, and subsurface. However, in many cases, it is difficult for the local administration to implement the resource management plans at a time in the entire tehsil. Hence, multi-criteria (Dongare et al. 2013; Archibald and Marshall 2018; Pathare and Pathare 2020; Khan et al. 2001; Javed et al. 2009; Todmal 2020) analysis plays a key role in the identification of critical areas based on the assessment of human carrying capacity in the management of water resources on a sustainable basis. Apart from for own, man uses domestic use, water available in nature for economic use in various forms mainly for industrial and agricultural uses (Siddi Raju et al. 2018; Xu et al. 2020).

Geospatial modeling in GIS enables the integration of environmental resources, crop production, and consumption scenarios (Cannata et al. 2018; Chattaraj et al. 2017; Mahadevan et al. 2020; Kadam et al. 2020; Kudnar 2020a, b; Mishra ad Rai 2016) is that environmental flows must be maintained in case of all perennial rivers, and if there are impounding structures like dams, then sufficient water must be reserved for releasing it into the main river course in order to maintain the ecological functions and integrity of the river, dam, and pond system intact.

In the study, an attempt has been made to delineate distinct physiography, soil types, geology, slope, relief, climate and rainfall, water quality analysis, demographics, water resource management, resource units, inventory of existing medium/minor/lift irrigation schemes constructed the areas through geospatial modeling for sustainable development in Godia District, central India.

4.2 Study Area

Gondia District is situated in the Valley of Wainganga River (Kudnar, 2015a, b, 2017, 2019), and its axial extension is 20° 35′ to 21° 45′ North latitudes and 79° 45′ to 80° 45′ East longitudes (Fig. 4.1). To the East of this district is the state of Chhattisgarh (Rajnandgaon District) of the state is the land fair, Madhya Pradesh State to the North. To the South of this district is Gadchiroli District and West side Bhandara district in the state of Maharashtra. Gondia District is located on the northeastern side of the state of Maharashtra and is rich in ancient history, literature, agriculture, and natural beauty. The total area of Gondia District is 5234 km² which is 1.7% of the total area of Maharashtra. Due to its large rice production, Gondia District is also known as the city of rice, also known as the gateway to Maharashtra. The city of Gondia is named after a large number of tribal Gond community. The district has eight talukas namely Gondia Tirora Goregaon Deori Amgaon Saleksa Arjuni Morgaon Sadak-Arjuni. In this district, villages like Amgaon, Salekasa, Gangazari, Dareksa, Deori, Chichgad, Pimpar-Khari, Dongargaon, and Arjuni Morgaon are surrounded by Satpuda mountain ranges and Darekasa ranges. Pratapgad hill is the highest hill in this district. Gondiya District forms a part of the Wardha–Penganga–Wainganga plain at micro-level of the Deccan Plateau. The city of Gondia is located on the Mumbai–Howrah main railway line. It is 1006 km from Mumbai and 169 km from Nagpur. As the city is on Mumbai–Calcutta National Highway No. 6, transportation facilities are widely available. A population of 1,322,507 persons as per 2011 census, 1.18% of the total population of the state. The density of population is 253 persons per km².

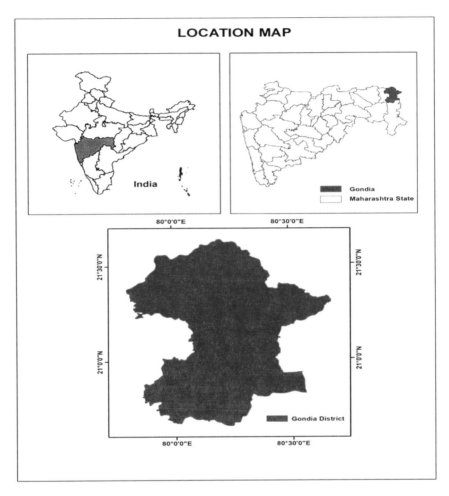

Fig. 4.1 Location map of the Gondia District

4.3 Data Collection and Methodology

The present study is grounded on the field based on primary as well as secondary data. Apart from the general observation in the field, various topographical sheets wear consulted for detailed environmental analysis. Environmental analysis of the Gondia District was carried out, using the SOI topographical maps on a 1:50,000 scale. The analysis includes relief, forest, soil, drainage, rainfall, and water analysis, Collection of rainfall and long-term rainfall data for the entire Wainganga basin as well as small tributaries, particularly for district area have been collected from Indian Meteorological Department (IMD). Water sample has been collected from the tehsil area and also at river Wainganga, and physio-chemical water has been analyzed, Satellite data products Multispectral imageries have been acquired for time series analysis of various hydrological as well geomorphological features of in study area. The following steps have opted for the study environmental network of the tehsil was analyzed. Using SOI topographic maps and Universal Transverse Mercator (UTM) zone 44 N projection was georeferenced using WGS 84 datum, in ArcGIS desktop 9.3. In this study, the Tirora tehsil was delineated and the geomorphic unit was extracted using Cartosat (1 arcsec) in conjunction with SOI toposheets, GPS location, and river hydrology, rainfall data, water availability, and discharge data. Water sample has been collected from the downstream as well as upstream of water at river Wainganga (Kudnar and Rajashekhar 2020) and environmental as well as geomorphology, soil types, geology, slope, relief, landforms, climate and rainfall, hydrology, water availability analysis water inflow water balance, measurement of water discharge, demographics, prioritization of land resource management has been analyzed (Fig. 4.2). To attempt this Water Quality analysis (WQA), water samples were analyzed for various parameters such as TDS, pH, EC.

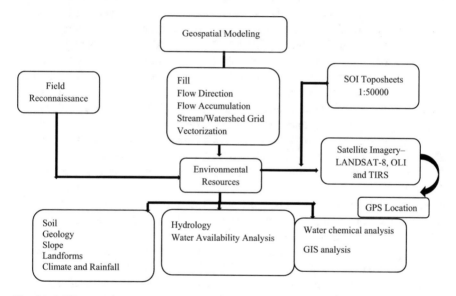

Fig. 4.2 Methodology flowchart

4.4 Results and Discussion

4.4.1 Physical Features

4.4.1.1 Physiography

Gondia District forms a part of the Wardha–Penganga–Wainganga plain at micro-level of the Deccan Plateau. Erosion and weathering together have locally shaped the surface in different landforms which may be grouped as (i) The Wainganga–Bagh River plain, (ii) Chulband Valley, (iii) Gaikhuri Hills, (iv) Chichgarh Plateau, (v) Chichgarh–Palasgarh Hills, and (vi) Mahadeo Hills. The Wainganga–Bagh River plain includes Amgaon and parts of Tirora, Gondiya, Goregaon, and Salekasa Tahsils. This plain is situated at a height varying between 250 and 300 m (Fig. 4.3). It has an even surface dotted with hillocks. The plain has very fertile soils and is well known for rice cultivation in the state. The soils are deep near the river bed but get shallow toward the South. The whole area slopes toward Wainganga River, which flows in a South-West direction within the district. Chulband Valley occupies major portion of Sadak-Arjuni and Arjuni Morgaon Tahsil in the South-West part of the district. This valley is almost flat with residual hillocks dotted here and there. It has altitude varying between 150 and 300 m and slopes gently toward South. Some residual hillocks in the valley rise more than 300 m. The valley is fairly covered with forests. Gaikhuri Hills spread over parts of Goregaon, Tirora, and Gondiya Tahsils in a cluster of elongated knolls and act as the water divide between Wainganga, Chulband, and Bagh Rivers. These hills attain a height of between 300 and 400 m and are covered with dry mixed deciduous forests. Chinchgarh Plateau covers only a small part of Deori Tahsil. It has an altitude varying between 300 and 400 m and slopes toward the North. It is drained by Satbahini Nadi. This area is partly covered by forests. Chinchgarh, Palasgaon Hills spread over parts of Sadak-Arjuni, Deori, and Arjuni Morgaon Tahsils. The height of these hills rises up to about 600 m. However, the highest point (710 m) is near Jhanda Dhongav. These hills have a dense cover of forest, and they are mainly reserved forests. From transport and communications point of view, the region is not well-developed. Mahadeo Hills cover a very small part in the extreme East of Gondia District in Salekasa Tahsil. The height of these hills rises up to 600 m and descends toward North-East. Small streams rise from these hills and are the source of Kandhas Nadi. The hills mainly consist of Rhyolites and esites and porphyrites (Kudnar et al. 2021; Salunke et al. 2020, 2021).

4.4.1.2 Drainage

The Wainganga River is the largest and most important river in Gondia District. Many tributaries of the river Wainganga are found in these districts. The major tributaries are Wainganga-Bagh, Pangoli, Gadhvi, Chor, Chandan, and Bawanthadi (Fig. 4.4). The total length of the river Bagh in these districts is 130 km, and it

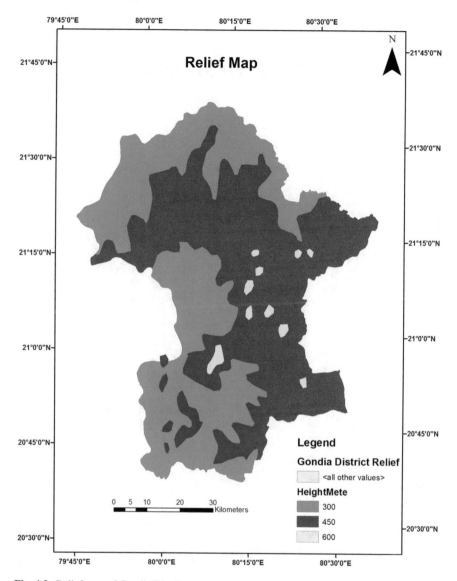

Fig. 4.3 Relief map of Gondia District

flows northwards from the Chichgad Hills. A medium water supply project has been set up on the River Bagh near Shirpur Village in Deori Taluka. Dhapewada Upsa Irrigation Project has been set up near Kavalewada Village on Wainganga River in Tirora Taluka. The southern portion of the district is mainly drained by the Chulband River and its tributaries Kapuri, Umarjhari, and Sasakuran. The Chulband originates in the Chinchgarh–Palasgaon Hills and flows southwards till it meets the Wainganga near Soni Village in Lakhandur Tahsil of Bhandara District. Another small stream

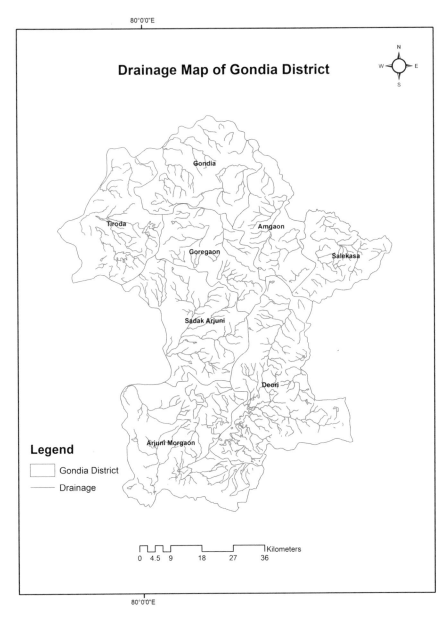

Fig. 4.4 Drainage map of Gondia District

known as Garhvi originates in the Chinchgarh Plateau and flows toward South-West through Arjuni Morgaon Tahsil and meets with Kobragarh Nadi a tributary of the Wainganga River in Gadchiroli District.

4.4.1.3 Climate

In this Gondia District, May is the hottest month of the year with the mean daily maximum temperature at 44.5 °C, December and January are the coldest months of the year with the mean daily minimum temperature at 13 °C.

4.4.1.4 Forests

Gondia District has about 52% forest cover of the total geographical area. Most of the forest cover in the district is found in Tirora, Deori, Arjuni Morgaon as well as a large area of Pratapgad Hills (Fig. 4.5). The forest reserve is divided into three protected and unclassified areas. Out of the total area received in the district, reserved forest area is 56.4%, protected forest area is 26.3%, and unclassified forest area is 17.3%. The forest area in the district is divided into dense, medium, and sparse areas. According to the total forest area, the percentage of dense medium and sparse forest area is 56.4%, 26.3%, and 17.3%, respectively. Navegaon Dam National Park has 129.55 km^2 of forest area and 426.66 km^2 of forest area below the sanctuary. The best quality teak forests are found in Gondia District. Bija, Halda, Tivas Sisav, Moh, Khair, Garadi, etc., are found in major trees. Tendu leaves are very useful for smoking in these forests. Along with this, Moha's flowers are useful for lakhs of trees like Palas Dink, Kadai Dhavda, and Katara, Khair. The forest development program is implemented in the district with the objective of protecting the soil from inundation and flooding and maintaining the fertility of the soil.

4.4.1.5 Soil

Soil is a basic resource. Soil is formed by the earthy topsoil, the most valuable natural resource. The issue is also the formation of environmental and biological reactions. Soil elements are found in the minerals and chemicals of Gondia District. It contains silica and aluminum in the topsoil. In addition, oxygen and aluminum are found in large quantities. In addition, nitrogen, potassium, calcium, phosphorus, and odor as well as copper are found in the soil. These soils supply the soil with organic matter from animal and plant matter. The soils of Gondia District (Fig. 4.6) are mainly black shore, sihar, morand khardi, and birdie type. Of these, Kali Kanar land is less and more valuable. Bardi is different from Kanar and is mixed with limestone. The red-yellow sea necklace is formed from the friction of the ground crystalline stone. This soil is eroded in summer. Light type of soil is called Kharadi. Kali Kanar soil is found in large numbers in the Wainganga river basin. It retains the hardness and

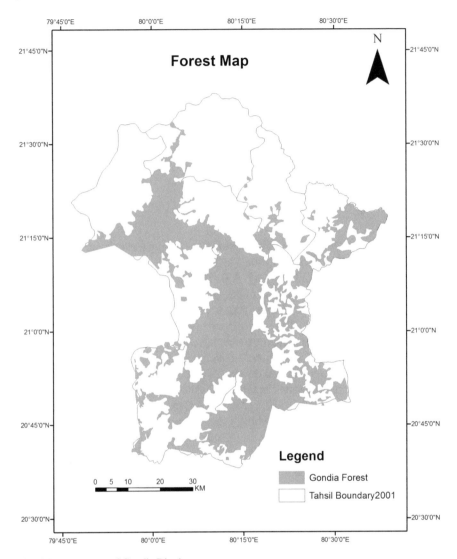

Fig. 4.5 Forest map of Gondia District

moisture of the deep layer and is grown twice a year. At the same time, some sandy soils are loosened. The district covers 71% of the total cultivable area. Wheat and linseed are the main crops grown in the Morad type of land, due to the mountain ranges of Satpuda mountain range as well as the mountainous terrain of Darekasha which forms a number of hilly areas in the district.

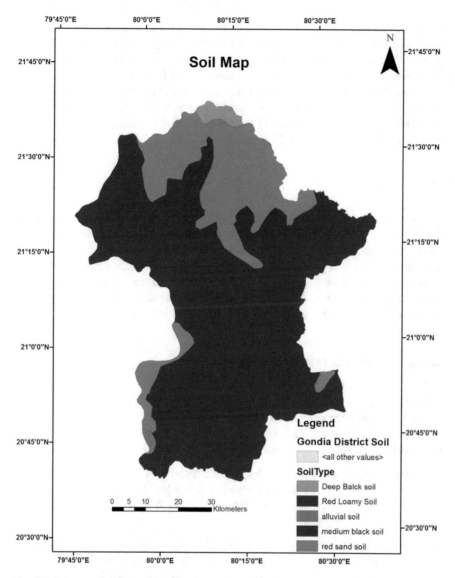

Fig. 4.6 Soil map of Gondia District

4.4.1.6 Agriculture

A total of 1748 former cattle ponds are recorded in Gondia District and in fact 1421 former cattle ponds exist. Their projected irrigation capacity is 28,730 ha. There are 192 small irrigated ponds below 100 ha, and its projected irrigation capacity is 10,897 ha. There are also 294 Kolhapuri type dams with a projected irrigation capacity of 10,075 ha and 1559 storage dams with a projected irrigation capacity of

14,817 ha. In addition, there are two large projects, nine medium projects, and 20 small projects in the district. To increase the irrigation capacity under this project, to provide irrigation facilities to the farmers, to raise their economic status by benefiting them and at the same time to raise the groundwater level and to develop fisheries business. Most of the people of this district are engaged in agricultural activities. As per 2011 census, 28.03% of the total workers are reported as cultivators and 42.29% of the total workers as agricultural laborers in the district. Together they constitute 70.32% of the total workers of the district.

4.4.1.7 Fisheries

Wainganga is a major river in Gondia District, and its tributaries are found in the district. The total length of rivers in the district is 250 km, and there are many lakes. These districts are known as lake districts. Rivers, lakes, and large reservoirs are conducive to fish production. In the year 2013–14, 9125 metric tons of fish were produced, out of which 54.75 crore was earned. How to change this district Dam Chorkhamara, Dam Chulband Katangi, Dam Khair Bandya Navegaon, Dam Pujaritola, large-scale fishing is done in dams like Shirpur Dam.

4.4.1.8 Tourist Places in Gondia District

Hazara Waterfall in Saleksa Taluka is a major tourist attraction during the monsoon season. At a distance of one kilometer from the railway station, this waterfall is overflowing with tourists during the rainy season. August to December is the best time to visit this place as it is home to natural vegetation and large mountains.

Chinchgad is a tourist destination located at a distance of 55 km from Gondia. The site is about 25,000 years old. The cave attracts tourists. This is a favorite place for trekkers and is a place of worship for the local tribals. It is worth visiting during the two months of January to February.

Nagzira Wildlife Sanctuary situated at a distance of about 60 km from Gondia, Nagzira Wildlife Sanctuary is home to many wild animals and birds and is a great destination for tourism. Today, as many species are endangered in India, many of them are found here. More than 166 species of birds are found here. At the same time, you can see many animals like tiger, bibel, rangve, sambar, belkar, ran, pig, monkey, chital, nilgai, cat, taras, fox, wolf. The Nagzira Sanctuary has about 36 species of snakes, six of which are on the verge of extinction, including Rockpouthan, Dhaman, Indian Cobra, Russell, Wifer Check, Kar, Killback, and Common Monitor.

Navegaon Dam National Park is located in the South of Gondia District and in the East of Maharashtra. It is of great importance in terms of conservation of biodiversity and also in terms of natural conservation. It is home to a wide variety of plants, 209 species of birds, 9 species of snakes, and 26 species of animals.

Suryadev and Mandav Devi temples situated on a hill, the temple of the Sun God, and the Mandav Goddess is considered to be the village deity of Gondia. It is an awakened temple and the reputation of this temple as a goddess who fulfills desires is spread all over the world. There is a cave in the vicinity of this temple which has a temple of Hanumana and Annapurna mother.

4.4.1.9   Rural Water Supply Scheme

The district has a total of four regional water supply schemes with 9675 hand pumps, 151 power pumps, 483 solar pumps, and 4446 simple wells. Various schemes are implemented from the central and state as well as district funds to provide clean drinking water to the general public as well as school children and Anganwadi boys and girls in the rural areas of the district.

Spatial–temporal variation of rainfall

The rainy season in Gondia begins with the arrival of southwest monsoon winds in June. This period lasts till September. Monsoon winds come from southwest in Gondia. These monsoon winds have several periods of moderate to heavy rainfall, including strong winds on the cloud cover surface in high heat high humidity extended areas. The arrival of southwest monsoon winds in Gondia District is usually on June 10. But in many cases, the arrival is sometimes early and sometimes late. About 60% of the monsoon arrives between June 10 and June 25. Looking at the date of arrival of monsoon in the last 100 years, it is noticed that the earliest arrival of monsoon is found on 29th May, 1918. The earliest arrival is on June 18, 1972. A branch of the Arabian Sea winds blows in eastern Maharashtra. Southwest monsoon winds arrive at Gondia around June 10 and continue to blow North. At the same time, another branch falls in the Bay of Bengal and comes to Gondia District. The Western Ghats, which encircle the West coast of India, are like a vertical wall and are more or less parallel to the coast.

1. **Average rainfall**

The average rainfall in these districts is 1323.27 mm (Fig. 4.7). The highest rainfall is recorded in Arjuni Morgaon Taluka at 1384.21 mm followed by Khamgaon at 1379.66 mm. The lowest average rainfall in Tirora is 1282 mm. At the same time, Deori 1328.16 mm Gondia millimeter Goregaon millimeter Saleksa 1291.56 mm road Arjuni 1301.20 mm of rain from 1971 to 2013 (Table 4.1).

2. **Monsoon rainfall**

The average monsoon rainfall in Gondia District from 1971 to 2013 was 1201.81 mm (Fig. 4.8). The highest rainfall is recorded at Amgaon at 1263.64 mm followed by Tirora at 1159.99 mm. Monsoon rainfall is recorded at Arjuni Morgaon with an average rainfall of 1262.17 mm, Deori 1205.91 mm, Gondia 1190.80 mm, Goregaon 1183.43 mm, Saleksa 1170.14 mm, and road Arjuni 1178.38 mm.

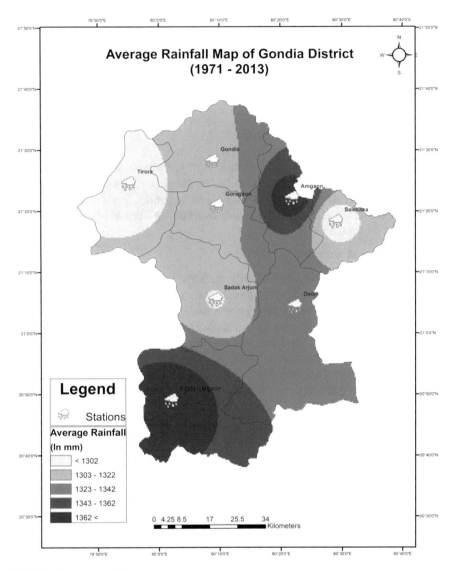

Fig. 4.7 Average rainfall

3. Post-Monsoon Rainfall

The average post-monsoon rainfall in Gondia District is 63.98 mm from 1971 to
2013 (Fig. 4.9). The highest rainfall is in Deoria taluka at 65.73 mm and the lowest
in Amgaon at 60.35 mm. Similarly, Arjuni Morgaon receives 65.94 mm of Deori
65.73 mm, Gondia 64 points 20 Goregaon 62.96 Saleksa 63.99 Tiroda 63.88 mm,
and Sadak-Arjuni 64.76 mm of rainfall.

Table 4.1 Rainfall in Gondia District (1971 to 2013)

S. No.	Station name	Average rainfall	Monsoon rainfall	Post-monsoon rainfall	Winter rainfall	Pre-monsoon rainfall
1	Amgaon	1379.66	1263.64	60.35	32.99	22.68
2	Arjuni Morgaon	1384.21	1262.17	65.94	31.12	24.99
3	Deori	1328.16	1205.91	65.73	30.18	26.33
4	Gondia	1314.77	1190.86	64.20	33.87	25.83
5	Goregaon	1305.38	1183.43	62.96	34.38	24.61
6	Salekasa	1291.56	1170.14	63.99	33.28	24.16
7	Tirora	1282.00	1159.99	63.88	34.10	24.03
8	Sadak-Arjuni	1301.20	1178.38	64.76	33.14	24.92

4. **Winter Rainfall**

The average winter rainfall in Gondia District is 32.88 mm (Fig. 4.10). The highest rainfall is 34.38 mm in Goregaon Taluka, and the lowest is 31.12 mm in Deori. Simultaneously, Amgaon receives 32.99 mm, Deori 30.18 mm, Gondia 33.87 mm, Saleksa 30.28 mm, Tirora 34.10 mm, and Sadak-Arjuni 30.14 mm.

5. **Pre-Monsoon Rainfall**

In Gondia District, the average pre-monsoon rainfall is 24.69 mm with maximum rainfall of 26.33 mm at Deori and lowest rainfall at Amgaon at 22.68 mm. 24.03 mm and Sadak-Arjuni received 24.92 mm of rainfall from 1971 to 2013 (Fig. 4.11).

4.4.2 Water Quality Analysis (WQA)

The investigation has been carried out to understand the hydrochemistry of the groundwater of semiarid region of Gondia District and its suitability for irrigation uses (Gadekar and Sonkar 2021; Rajasekhar et al. 2020; Kudnar 2018). A total of 41 groundwater samples were collected during pre-monsoon season. The water is neutral to alkaline in nature with pH ranging from 6.6 to 8.92 (Fig. 4.12) with high TDS range (Fig. 4.13) from 140 to 2184 (Table 4.2).

4.5 Conclusions

To evaluate the state of this resource, I used the physiology, soil types, slope, relief, climate and rainfall, hydrology, water quality analysis, demographics, can be sustained by judiciously using both the surface and subsurface waters. The forest area in the district is divided into dense, medium, and sparse areas. According to the

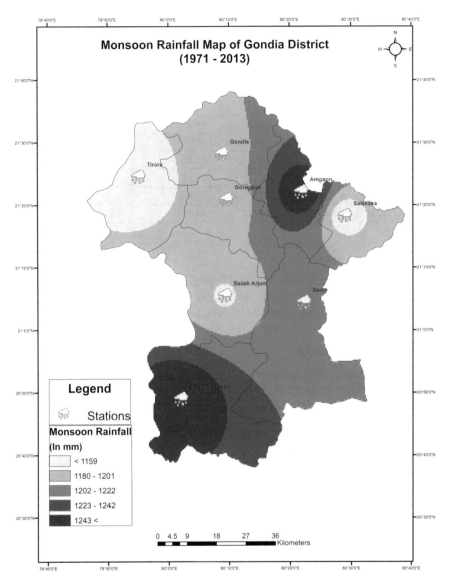

Fig. 4.8 Monsoon rainfall

total forest area, the percentage of dense medium and sparse forest area is 56.4%, 26.3%, and 17.3%, respectively. Most of the people of this district are engaged in agricultural activities. As per 2011 census, 28.03% of the total workers are reported as cultivators and 42.29% of the total workers as agricultural laborers in the district. The average rainfall from 1971 to 2013 in these districts is 1323.27 mm; average

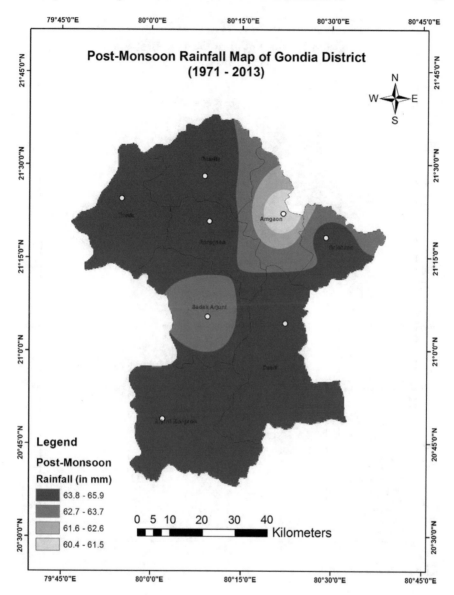

Fig. 4.9 Post-monsoon rainfall

monsoon rainfall was 1201.81 mm. The water is neutral to alkaline in nature with pH ranging from 6.6 to 8.92 with high TDS range from 140 to 2184 ppm.

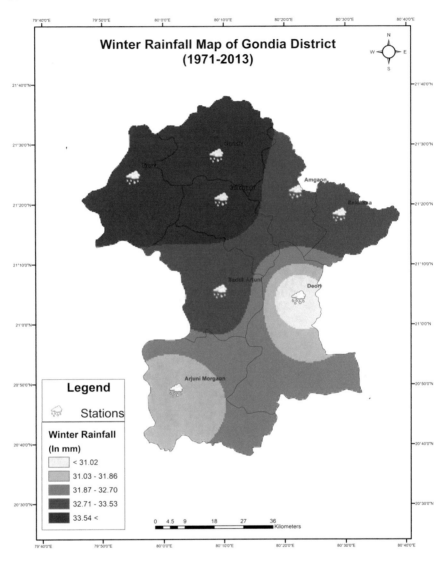

Fig. 4.10 Winter rainfall

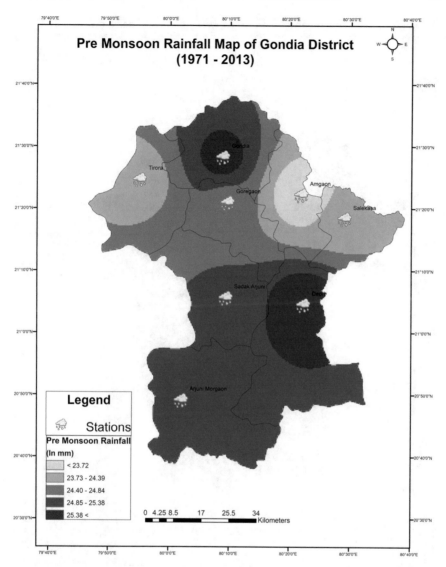

Fig. 4.11 Pre-monsoon rainfall

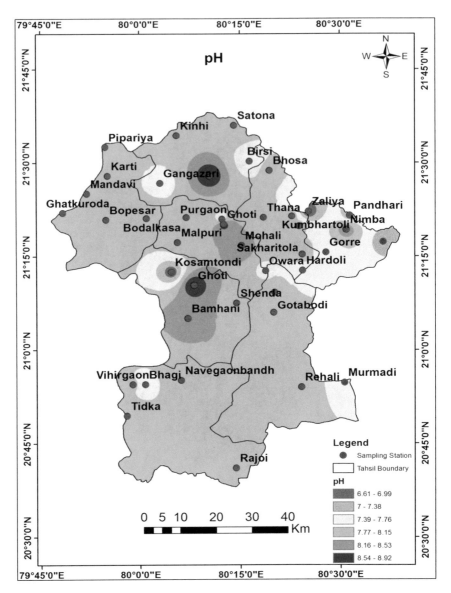

Fig. 4.12 pH

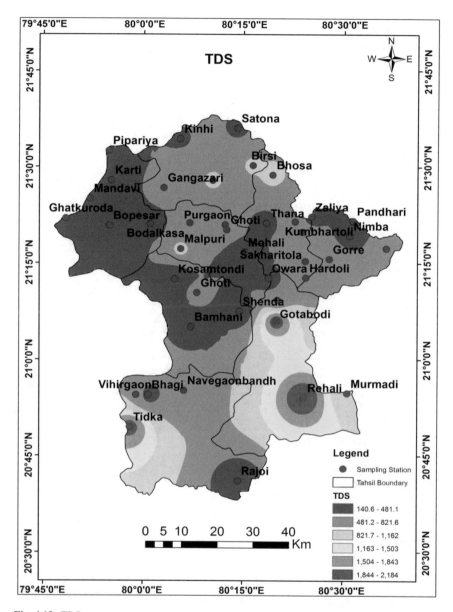

Fig. 4.13 TDS

Table 4.2 Water quality

S. No	Station name	pH	EC	TDS	S. No.	Station name	pH	EC	TDS
1	Mandavi	7.9	890	330	22	Gorre	7.5	1050	790
2	Pipariya	7.9	920	340	23	Pandhari	7.1	508	311
3	Bodalkasa	8.11	540	310	24	Kachargarh	7.2	509	711
4	Ghatkuroda	8.1	931	350	25	Zaliya	6.6	490	241
5	Karti	8.1	940	320	26	Nimba	8.51	5408	320
6	Suryatola	8.92	1387.2	880	27	Ghoti	6.7	495	255
7	Gangazari	7.43	989	620	28	Owara	7.6	509	256
8	Satona	8	1210.2	420	29	Shenda	7.92	285	140
9	Birsi	7.37	1588	860	30	Kosamtondi	6.8	496	256
10	Kinhi	7.9	981	455	31	Bamhani	8.51	3418	310
11	Mohali	8.51	540.8	320	32	Rehali	8.1	3152	2184
12	Purgaon	7.25	980	520	33	Salegaon	8.01	601.4	480
13	Sukhapur	7.8	1250	520	34	Gotabodi	8	3140	2100
14	Ghoti	8.92	775.2	560	35	Hardoli	7.34	1445	740
15	Malpuri	7.9	1332.8	910	36	Murmadi	7.47	1688	860
16	Thana	7.8	780	310	37	Navegaonbandh	7.98	785.7	480
17	Bhosa	7.9	811	925	38	Tidka	7.9	3139	2011
18	Borkanhar	7.9	1479	750	39	Bhagi	7.6	495	284
19	Kumbhartoli	7.9	790	830	40	Rajoi	7.9	511	249
20	Sakharitola	8.1	1005	730	41	Vihirgaon	7.8	1240	821
21	Bopesar	8	781	310					

Source Environmental Laboratory, APML

Conflict of Interest The authors declare no conflict of interest.

References

Aslam B, Ismail S, Ali I (2020) A GIS-based DRASTIC model for assessing aquifer susceptibility of Safdarabad Tehsil, Sheikhupura District, Punjab Province. Pak Model Earth Syst Environ 6:995–1005. https://doi.org/10.1007/s40808-020-00735-8

Akhtar MP, Roy LB, Vishwakarma KM (2020) Assessment of agricultural potential of a river command using geo-spatial techniques: a case study of Himalayan river project in Northern India. Appl Water Sci 10:8. https://doi.org/10.1007/s13201-020-1165-8

Archibald TW, Marshall SE (2018) Review of mathematical programming applications in water resource management under uncertainty. Environ Model Assess 23:753–777. https://doi.org/10.1007/s10666-018-9628-0

Azhar S, Ahmad ZA, Mohd KY, Mohammad FR, Hafizan J (2015) Classification of river water quality using multivariate analysis. Procedia Environ Sci 30(1):79–84. https://doi.org/10.1016/j.proenv.2015.10.014

Bisen DK, Kudnar NS (2013) A sustainable use and management of water resource of the Wainganga river basin: a traditional management systems. J Contrib. https://doi.org/10.6084/m9.figshare.663 573.v1

Bisen DK, Kudnar NS (2013b) Watershed development: a case study of drought prone village darewadi source, review of research [2249–894x] d-1-6

Bisen DK, Kudnar NS (2019) Climatology, Sai Jyoti Publication, Nagpur, pp 1–211

Bouaicha F, Dib H, Belkhiri L, Bouaicha F, Manchar N, Chabour N (2017) Hydrogeochemistry and geothermometry of thermal springs from the Guelma region Algeria. J Geol Soc India 90:226. https://doi.org/10.1007/s12594-017-0703-y

Cannata M, Neumann J, Rossetto R (2018) Open source GIS platform for water resource modelling: Freewat approach in the Lugano Lake. Spat Inf Res 26:241–251. https://doi.org/10.1007/s41324-017-0140-4

Chamine HI, Gómez GM (2019) Sustainable resource management: water practice issues. Sustain Water Resour Manage 5:3–9. https://doi.org/10.1007/s40899-019-00304-7

Chattaraj S, Srivastava R et al (2017) Semi-automated object-based landform classification modelling in a part of the deccan plateau of Central India. Int J Remote Sens 38(17):4855–4867. https://doi.org/10.1080/01431161.2017.1333652

Chen C, Ahmad S, Kalra A et al (2017) A dynamic model for exploring water-resource management scenarios in an inland arid area: Shanshan County, Northwestern China. J Mt Sci 14:1039–1057. https://doi.org/10.1007/s11629-016-4210-1

Dongare VT, Reddy GPO, Maji AK et al (2013) Characterization of landforms and soils in complex geological formations-a remote sensing and GIS approach. J Indian Soc Remote Sens 41:91–104. https://doi.org/10.1007/s12524-011-0195-y

Gadekar DJ, Sonkar S (2020) Statistical analysis of seasonal rainfall variability and characteristics in Ahmednagar District of Maharashtra, India. Int J Sci Res Sci Technol 2395–6011. https://doi.org/10.32628/IJSRST207525

Gadekar DJ, Sonkar S (2021) The Study of physico-chemical characteristics of drinking water: a case study of Nimgaon Jali Village. Int Adv Res J Sci Eng Technol 8:61–65

Gu JJ, Guo P, Huang GH (2013) Inexact stochastic dynamic programming method and application to water resources management in Shandong China under uncertainty. Stoch Env Res Risk Assess 27(5):1207–1219. https://doi.org/10.1007/s00477-012-0657-y

Guettaf M, Maoui A, Ihdene Z (2017) Assessment of water quality: a case study of the Seybouse River (North East of Algeria). Appl Water Sci 7:295–307. https://doi.org/10.1007/s13201-014-0245-z

Javed A, Khanday MY, Ahmed R (2009) Prioritization of sub-watersheds based on morphometric and land use analysis using remote sensing and GIS techniques. J Indian Soc Remote Sens 37(2):261. https://doi.org/10.1007/s12524-009-0016-8

Kadam AK, Umrikar BN, Sankhua RN (2020) Assessment of recharge potential zones for ground-water development and management using geospatial and MCDA technologies in semiarid region of Western India. SN Appl Sci 2:312. https://doi.org/10.1007/s42452-020-2079-7

Karande UB, Kadam A, Umrikar BN et al (2020) Environmental modelling of soil quality, heavy-metal enrichment and human health risk in sub-urbanized semiarid watershed of western India. Model Earth Syst Environ 6:545–556 (2020). https://doi.org/10.1007/s40808-019-00701-z

Khan MA, Gupta VP, Moharana PC (2001) Watershed prioritization using remote sensing and geographical information system: a case study from Guhiya India. J Arid Environ 49(3):465–475. https://doi.org/10.1006/jare.2001.0797

Kudnar NS (2015) Linear aspects of the Wainganga river basin morphometry using geographical information system. Mon Multidiscip Online Res J Rev Res 5(2):1–9

Kudnar NS (2015b) Morphometric analysis and planning for water resource development of the Wainganga river basin using traditional & GIS techniques. University Grants Commission (Delhi), pp 11–110

Kudnar NS (2017) Morphometric analysis of the Wainganga river basin using traditional & GIS techniques. Ph.D. thesis, Rashtrasant Tukadoji Maharaj Nagpur University, Nagpur, pp 40–90

Kudnar NS (2018) Water pollution a major issue in urban areas: a case study of the Wainganga river basin. Vidyawarta Int Multidiscip Res J 2:78–84

Kudnar NS (2019) Impacts of GPS-based mobile application for tourism: a case study of Gondia district. Vidyawarta Int Multidiscip Res J 1:19–22

Kudnar NS (2020a) GIS-based assessment of morphological and hydrological parameters of Wainganga river basin, Central India. Model Earth Syst Environ. https://doi.org/10.1007/s40808-020-00804-y

Kudnar NS (2020b) GIS-Based investigation of topography, watershed, and hydrological parameters of Wainganga river basin, Central India, Sustainable Development Practices Using Geoinformatics, Scrivener Publishing LLC, pp 301–318. https://doi.org/10.1002/9781119687160.ch19

Kudnar NS, Rajasekhar M (2020) A study of the morphometric analysis and cycle of erosion in Waingangā Basin India. Model Earth Syst Environ 6:311–327. https://doi.org/10.1007/s40808-019-00680-1

Kudnar NS, Padole MS et al (2021) Traditional crop diversity and its conservation on-farm for sustainable agricultural production in Bhandara District, India. Int J Sci Res Sci Eng Technol (IJSRSET) 2394–4099 8(1):35–43. https://doi.org/10.32628/IJSRSET207650

Kumar BP, Babu KR, Rajasekhar M et al (2020) Identification of land degradation hotspots in semiarid region of Anantapur district, Southern India, using geospatial modeling approaches. Model Earth Syst Environ https://doi.org/10.1007/s40808-020-00794-x

Lakshmi Keshava Kiran Kumar P, Veeraswamy G, Raghubabu K et al (2020) Geochemical characteristics of iron ore deposits and processing of Landsat-8 data (geology, geomorphology and lineaments) in semi-arid region and using geospatial techniques. Model Earth Syst Environ 6:1245–1252.https://doi.org/10.1007/s40808-020-00754-5

Mahadevan H, Krishnan KA, Pillai RR et al (2020) Assessment of urban river water quality and developing strategies for phosphate removal from water and wastewaters: integrated monitoring and mitigation studies. SN Appl Sci 2:772. https://doi.org/10.1007/s42452-020-2571-0

Mishra VN, Prasad R, Kumar P et al (2017) Dual-polarimetric C-band SAR data for land use/land cover classification by incorporating textural information. Environ Earth Sci 76:26. https://doi.org/10.1007/s12665-016-6341-7

Mishra VN, Rai PK (2016) A remote sensing aided multi-layer perceptron-Markov chain analysis for land use and land cover change prediction in Patna district (Bihar), India. Arab J Geosci 9:249https://doi.org/10.1007/s12517-015-2138-3

Mishra VN, Rai PK, Prasad R, Punia M, Nistor MM (2018) Prediction of spatiotemporal land use/land cover dynamics in rapidly developing Varanasi district of Uttar Pradesh, India using Geospatial approach: a comparison of hybrid models. Appl Geomat 10(3):257–276

Mishra VN, Rai PK, Mohan K (2014) Prediction of land use changes based on land change modeler (LCM) using remote sensing: a case study of Muzaffarpur (Bihar) India. J Geogr Inst Jovan Cvijic 64(1):111–127

Mogaji KA, Lim HS (2020) A GIS-based linear regression modeling approach to assess the impact of geologic rock types on groundwater recharge and its hydrological implication. Model Earth Syst Environ 6:183–199. https://doi.org/10.1007/s40808-019-00670-3

Pathare JA, Pathare AR (2020) Prioritization of micro-watershed based on morphometric analysis and runoff studies in upper Darna basin, Maharashtra, India. Model Earth Syst Environ. https://doi.org/10.1007/s40808-020-00745-6

Rajasekhar M, Gadhiraju SR, Kadam A et al (2020) Identification of groundwater recharge-based potential rainwater harvesting sites for sustainable development of a semiarid region of southern India using geospatial, AHP, and SCS-CN approach. Arab J Geosci 3–24. https://doi.org/10.1007/s12517-019-4996-6

Rajasekhar M, Sudarsana Raju G, Siddi RR (2019) Assessment of groundwater potential zones in parts of the semi-arid region of Anantapur District, Andhra Pradesh, India using GIS and AHP approach. Model Earth Syst Environ 5:1303–1317. https://doi.org/10.1007/s40808-019-00657-0

Salunke VS, Lagad SJ et al (2021) A geospatial approach to enhance point of the interest and tourism potential centers in Parner Tehsil in Maharashtra, India. Int J Sci Res Sci Eng Technol (IJSRSET) 8(1):186–196. Online ISSN : 2394–4099, Print ISSN : 2395–1990. https://doi.org/10.32628/IJSRSET218136

Salunke VS, Bhagat RS et al (2020) Geography of Maharashtra, Prashant Publication, Jalgaon, pp 1–229

Sharma D, Kansal A (2011) Water quality analysis of River Yamuna using water quality index in the national capital territory, India (2000–2009). Appl Water Sci 1:147–157. https://doi.org/10.1007/s13201-011-0011-4

Siddi Raju R, Sudarsana Raju G, Rajsekhar M (2018) Estimation of rainfall-runoff using SCS-CN method with RS and GIS techniques for Mandavi Basin in YSR Kadapa District of Andhra Pradesh, India, Hydrospatial Analy 2(1):1–15. https://doi.org/10.21523/gcj3.18020101

Soutter M, Alexandrescu M, Schenk C et al (2009) Adapting a geographical information system-based water resource management to the needs of the Romanian water authorities. Environ Sci Pollut Res 16:33–41. https://doi.org/10.1007/s11356-008-0065-5

Tiwari AK, Singh PK, Mahato MK (2014) GIS-based evaluation of water quality index of groundwater resources in West Bokaro Coal field India. Curr World Environ 9(3):843–850. https://doi.org/10.12944/CWE.9.3.35

Todmal RS (2020) Understanding the hydrometeorological characteristics and relationships in the semiarid region of Maharashtra (western India): implications for water management. Acta Geophys 68:189–206. https://doi.org/10.1007/s11600-019-00386-z

Trikoilidou E, Samiotis G, Tsikritzis L et al (2017) Evaluation of water quality indices adequacy in characterizing the physico-chemical water quality of lakes. Environ Process 4:35–46. https://doi.org/10.1007/s40710-017-0218-y

Tsihrintzis VA (2017) Integrated water resources management, efficient and sustainable water systems, protection and restoration of the environment. Environ Process 4:1–7. https://doi.org/10.1007/s40710-017-0271-6

Xu H, Cai C, Du H et al (2020) Responses of water quality to land use in riparian buffers: a case study of Huangpu River, China. GeoJournal. https://doi.org/10.1007/s10708-020-10150-2

Zhu Y, Li YP, Huang GH, Guo L (2013) Risk assessment of agricultural irrigation water under interval functions. Stoch Env Res Risk Assess 27(3):693–704. https://doi.org/10.1007/s00477-012-0632-7

Chapter 5
Hydrochemical Characteristics of Groundwater—Assessment of Saltwater Intrusion Along Krishna and Godavari Delta Region, Andhra Pradesh, India

R. Kannan, K. Appala Naidu, Abhrankash Kanungo, M. V. Ramana Murty, Kirti Avishek, and K. V. Ramana

Abstract The intrusion of saltwater into a freshwater aquifer is of particular concern to the coastal community. Removal of excess groundwater from the shallow aquifers is known to be the primary cause of contamination by saltwater. In this study, groundwater samples were obtained in the 2016 and 2017 pre-monsoon and post-monsoon seasons at 105 locations from dug wells and bore wells along the coast of Andhra Pradesh in the Krishna and Godavari deltas. Groundwater samples are tested for large ions to determine the infiltration of saltwater and to classify the salinity sources in the delta zone. The various hydrogeochemical parameters such as pH, electrical conductivity (EC), total dissolved solids (TDSs), Ca^{2+}, Mg^{2+}, K^+, Na^+, CO_3, HCO_3^-, Cl^-, and SO_4^{2-} are evaluated for the delineation of the intrusion of saltwater in terms of Ca^{2+}/Mg^{2+}, $Cl^-/(CO_3 + HCO_3^-)$, Na^+/Cl^- ratios. It is reported that the availability of fresh groundwater is 14% and 62%, respectively, during the pre-monsoon and post-monsoon seasons. The percentage levels of contamination in groundwater for slight, moderate, injuriously, highly, and severely categories are 43%, 22%, 12%, 8%, and 1%, respectively, for pre-monsoon season. However, during the post-monsoon season, the levels of contaminations in the above-mentioned categories are 22%, 9%, 4%, 1%, and 3%. The extent of contamination during the post-monsoon season is observed to be lower than during the pre-monsoon. The groundwater ratio of $Na^+ - Cl^-$ during pre-monsoon and post-monsoon seasons is 71% and 60%, respectively. The saltwater mixing index (SMI) is also measured, and extremely high is found.

R. Kannan (✉) · K. A. Naidu · A. Kanungo · M. V. R. Murty
Andhra Pradesh Space Applications Centre, Vijayawada, Andhra Pradesh 520008, India

K. Avishek
Birla Institute of Technology, Mesra, Ranchi, Jharkhand 835215, India

K. V. Ramana
National Remote Sensing Centre, ISRO, Hyderabad, Telangana 500037, India

Keywords Groundwater · Hydro-chemical characteristics · Saltwater intrusion · Krishna Godavari delta · Piper diagram · Seawater mixing index

5.1 Introduction

Freshwater is life's main tool. The growth of the population in water-scarce regions would only increase the demand for available resources (Rai et al. 2018; Nistor et al. 2020). While water is the most commonly used material available in the natural world, it is not spread evenly across the globe (Tulipano et al. 2008; Sudhakar and Narsimha 2013). Just 0.69% of aggregate water can be used human needs (Vasanthavigar et al. 2010; Anil Kumar et al. 2015). Coastal regions are generally vulnerable to infiltration by seawater, particularly in areas where over-pumped groundwater. Owing to the over-extraction of groundwater, saltwater intrusion is the major concern in coastal areas and it leads to the destruction of freshwater aquifers' excellence (Annapoorani et al. 2014; Chen and Jiao 2007; Mohan et al. 2011). The Ghyben–Herzberg ratio states that, for every meter of fresh water in an unconfined aquifer above sea level, there will be forty meters of fresh water in the aquifer below sea level. Recognizing the extent of the intrusion of seawater is necessary to prevent the depletion of groundwater quality. The main causes of the decline in groundwater levels in the coastal aquifer are high water demand for various domestic, agricultural, industrial purposes, and rainfall recharge lower than groundwater depletion (Nair et al. 2013). Groundwater is the water found under saturated conditions at the interstices of rock formations, and it is one of the earth's most significant natural resources. This is necessary for the earth's sustenance and growth and it forms an integral part of the natural environment (Venkateswaran 2001; Sreedevi 2004; Nistor et al. 2019). The consistency of groundwater is dependent on physical properties such as color, odor, turbidity, and chemical elements such as large cations and anions. Water's physical and chemical characteristics alter spatially and momentarily and are caused by natural causes, human action, and infiltration of seawater in coastal areas (Narayan and Natarajan 1993). The flow of seawater into freshwater raises the concentrations of electrical conductivity, sodium and chloride, and groundwater ratio Cl/HCO_3.

The intrusion of seawater is a significant source of increased salinity, and groundwater typically demonstrates high concentrations not only of total dissolved solids (TDS) but also in cations and anions (Richter and Krietler 1993) as well as selective trace element rises (Saxena et al. 2003; Mondal et al. 2010b). The intrusion of seawater is characterized as the movement of saline water from the sea into a freshwater aquifer which is hydraulically attached to the shore. In densely populated coastal areas with greater groundwater dependency, the depletion typically exceeds the rate of regeneration resulting in intrusion by seawater. Seawater densities are slightly greater than freshwater concentrations (Magesh et al. 2016). When intrusion of seawater is a significant cause of high salinity, groundwater typically exhibits high concentrations not only of total dissolved solids (TDS) but also of other different chemical constituents such as Cl^-, Na^+, Mg^{2+}, and SO_4^{2-} (Richter and Kreitler 1993) as well as aggregation of identified trace elements (Saxena et al. 2003; Mondal et al. 2010b).

The normal balance between freshwater and saltwater in waterfront aquifers is aggravated by groundwater withdrawals and other human exercises. Henceforth, individuals living in the lake zone are getting drinking water from far spots which demonstrates the accessible groundwater in the lake region is not consumable (Pujari and Soni 2009, Subba Rao et al. 2005). Comprehension of saline interruption into the beachfront aquifers is fundamental for productive arranging and the executives of seaside aquifers (Saha and Choudhury 2005; Chaurasia et al. 2013). It is likewise fundamental to portray and foresee the degree of saline water interruption into the aquifers in light of varieties in the segments of the freshwater mass-balance (Rajib Paul et al. 2019). Over-exploitation of groundwater results in the decline of water levels, leading to the intrusion of saltwater along the coastal region, which is a natural phenomenon (Brindha et al. 2016). Saltwater intrusion can pose serious problems to freshwater aquifers along the coastal areas having marine aquifer hydraulic interaction (Giuseppe Sappa and Maria 2012). Coastal aquifers are contaminated with saltwater intrusion particularly in and around the region, mainly because of the growing demand for seafood, has led many farmers to take up aquaculture as a more profitable source of income, where saltwater is used from the nearby creaks (Bear 1999; Ashraf Ali et al. 2019). The water quality in these aquaculture tanks is usually saline, which slowly infiltrates and reaches the water table. The aquifers in this region were contaminated with saltwater. Hence, the aim and objective of this study are to identify delineate the seawater intrusion places in the Krishna Godavari delta regions based on the groundwater sample collection.

5.2 Study Area

The Krishna–Godavari delta region is covered in four districts, i.e., Guntur, Krishna, West Godavari, and East Godavari district, Andhra Pradesh, India. The study area lies between 80°27′58.44″E, 15°54′37.88″N and 82°32′8.06″E, 17°15′16.85″N (Fig. 5.1). The total area of study stretches about 360 km long with a width of 12 km. The area is bounded by the Bay of Bengal in the east, the Guntur in the south, and the East Godavari on the north. The temperature ranges from 20 to 37 °C (winter seasons) and from 35 to 42 °C (summer seasons). The highest temperature occurs from May to June (Central Groundwater Commission, 2015). The average monthly humidity in June was 65%, and the average monthly humidity in November was 91%. The monthly evapotranspiration ranges from 4.2 to 8.2 mm/day. Geologically, the area consists of alluvial layers. Alluvial sediments are the youngest strata composed of sand and clay, deposited by river areas. Groundwater exists in unrestricted and confined aquifers in alluvial and weathered crystalline square-pore rock layers.

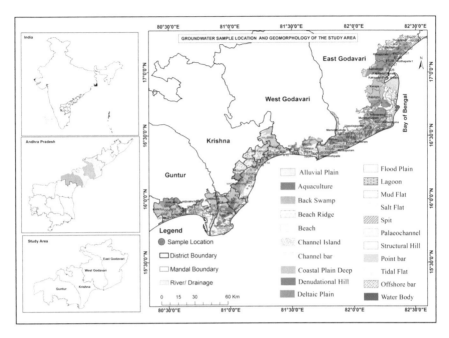

Fig. 5.1 Study area of Krishna–Godavari delta region

5.3 Sampling and Analytical Procedure

Groundwater samples representing the pre-monsoon and post-monsoon seasons were collected from 105 locations in June 2016 and January 2017 (Fig. 5.1). The collected samples were analyzed in the laboratory using the standard method of the American Public Health Association (APHA 1992) to measure the concentration of quality parameters. The samples were collected in a one-liter capacity high-density polyethylene (HDPE) bottle, and the pH, conductivity (EC), major cations, and anions of each groundwater sample were analyzed. The Ca^{2+}, Mg^{2+}, Na^+, K^+, HCO_3^-, CO_3, SO_4^{2-}, and Cl^- are the main ions in groundwater in the study area. The pH and EC were measured by using a Systronics micro pH meter. The concentrations of calcium and magnesium are determined by EDTA titration, using a chrome black T indicator. Determine the sodium and potassium concentrations by using a flame photometer. The chloride concentration was determined by silver nitrate titration. The carbonate and bicarbonate concentrations are measured by acid–base titration. The measuring concentration of sulfate and nitrate with a colorimetric spectrophotometer. The main water chemical phase was determined by Piper trilinear diagram (Piper 1944) using Aquachem Scientific v4.0 software. The analysis accuracy of the main ions was cross-analyzed from the ion balance of all samples within $\pm7\%$. The formulas proposed were used to calculate the total dissolved solids (TDSs) and total hardness (TH) of water, respectively.

$$TDS(mg/l) = EC(\mu S/cm) \times 0.64 \tag{5.1}$$

$$TH(mg/l) = 2.497\,Ca^{2+}(mg/l) + 4.115\,Mg^{2+}(mg/l) \tag{5.2}$$

5.4   Spatial Interpolation Methods

Spatial interpolation is the process of using points with known values to estimate values at other points. In GIS applications, spatial interpolation is typically applied to a raster with estimates made for all cells. Spatial interpolation is therefore a means of creating data from sample points. The ArcMap version 10.3 was utilized to foresee values in un-examined areas. Estimated values encompassing an unmeasured area are utilized for expectation. The inverse distance weighted (IDW) strategy was utilized for this investigation. After linking the spatial and non-spatial data together, the groundwater auxiliary data and spatial data (coordinates) collected by GPS are combined in GPS 10.3 software and the groundwater quality point layer is generated for further analysis. IDW is a local deterministic spatial interpolation method, which estimates continuous values by weighted averaging the values related to the values of known positions. In this technique, the weight of sampling points located within the average distance is higher than the weight of points farther from the average distance. The advantage of IDW is its instinct mastery, which works best when distracting the focus, and is extremely sensitive. IDW has two assumptions. When choosing this technique, one assumption must be kept in mind. It is assumed that the data sets are automatically linked not clustered. These studies will use IDW interpolation methods and IDW and barrier interpolation methods.

5.5   Results and Discussion

The analysis results are listed in Table 5.1 is the study area before and after the 2016 and 2017 monsoons. Table 5.1 summarizes the chemical analysis of groundwater and the percentage of Indian standards. The table lists the water quality parameters of 105 groundwater samples (BIS 2005; WHO 1997).

pH is a measure of the balance between the concentration of hydrogen ions and hydroxide ions in water. Before the monsoon season, the pH varied between 6.9 and 8.8 (average 8.0) (Fig. 5.2). Most samples are alkaline. After the monsoon, the pH (Fig. 5.3) varied between 6.8 and 8.0 with an average value of 7.3. It was observed that all samples were within the allowable range. When inverse distance weighting (IDW) interpolation technology was used to generate a pH map, it was found that the village of Pandurangapuram in Bapatla Mandal had the highest pH of 8.8. According to the BIS standard, the water exceeded the allowable limit. Total

Table 5.1 Water quality variables represented by BIS (2005) and WHO (1997) for drinking uses and the number of samples from the Krishna delta and Godavari delta surpassing the maximum permissible values

Chemical parameters	WHO standard	BIS standard	Number of sample exceed permissible limit	Percentage of sample exceeding permissible limit	Number of sample exceed permissible limit	Percentage of sample exceeding permissible limit
			Pre-monsoon (2016)		Post-monsoon (2017)	
pH	6.5–8.5	6.5–8.5	2	1.9	0	0.0
EC (μS/cm)	1500	1500	73	70.2	70	67.3
HCO_3^- (mg/l)	600	600	25	24.0	76	73.1
Cl^- (mg/l)	600	1000	32	30.8	27	26.0
SO_4^{2-} (mg/l)	400	400	0	0.0	0	0.0
Ca^{2+} (mg/l)	200	200	8	7.7	9	8.7
Mg^{2+} (mg/l)	150	100	26	25.0	24	23.1
Na^+ (mg/l)	200	200	73	70.2	69	66.3
K^+ (mg/l)	30	12	79	76.0	78	75.0
TDS (mg/l)	1500	2000	40	38.5	45	43.3

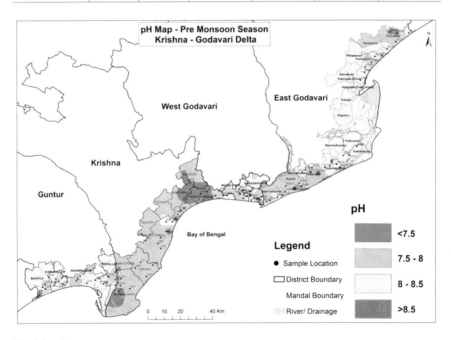

Fig. 5.2 pH map pre-monsoon season

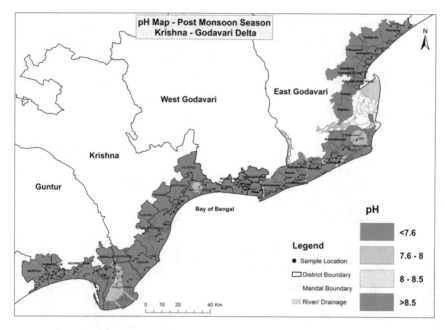

Fig. 5.3 pH map post-monsoon season

dissolved solid (TDS) is an important parameter in drinking water and other water quality standards. TDS stands for various minerals in dissolved form in water. In natural water, the dissolved solids are mainly composed of carbonate, bicarbonate, chloride, sulfate, phosphate, silica, calcium, magnesium, sodium, and potassium. The TDS is calculated by Eq. (5.1).

According to WHO (1997) specifications, the maximum allowable maximum TDS is 500 mg/l and the maximum allowable is 2000 mg/l. In the study area, the pre-monsoon TDS value varied between a minimum of 256 mg/l and a maximum of 18,560 mg/l which indicated that most groundwater samples exceeded the maximum allowable limit. According to Davis and De Wiest (1966), TDS-based groundwater classification (Table 5.2) was 29% before the monsoon and 24% was allowed to drink after the monsoon. In the post-monsoon season, 36% is suitable for irrigation before the monsoon and 39% is suitable for irrigation after the monsoon. As shown in Figs. 5.4 and 5.5, in both seasons, about 30 and 28% of the samples are not suitable for drinking and irrigation, before the monsoon, 5% of the samples are suitable for drinking and not 9% after the monsoon.

According to Freeze and Cherry (Table 5.3), TDS-based groundwater classification analysis of freshwater before and after the monsoon is 33–36% freshwater and brackish water before and after the monsoon is 63–59% of the season. The high TDS is due to the presence of large amounts of sodium and chloride ions. The high concentration of TDS in groundwater samples is due to the leaching of salt from the soil and domestic sewage may also penetrate the groundwater.

Table 5.2 David and Dewiest classification of groundwater based on TDS (mg/l)

TDS (mg/l)	Classification	Pre-monsoon (2016)		Post-monsoon (2017)	
		Number of samples	Percentage (%)	Number of samples	Percentage (%)
<500	Desirable for drinking	5	5	9	9
500–1000	Permissible for drinking	30	29	25	24
1000–3000	Useful for irrigation	38	36	41	39
>3000	Unfit for drinking and irrigation	32	30	29	28

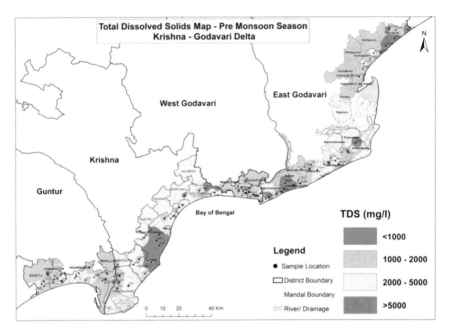

Fig. 5.4 Total dissolved solid map for pre-monsoon season

Total hardness (TH) depends on the calcium and magnesium content of water are calculated by the flowing Eq. (5.2). The maximum and minimum value of TH in the water samples are 140–15,935 mg/l with an average of 1258 mg/l in the pre-monsoon and 99–19,969 mg/ in the post-monsoon season. As per the WHO standards, the desirable limit of TH is 100 mg/l whereas the maximum permissible level is 300 mg/l.

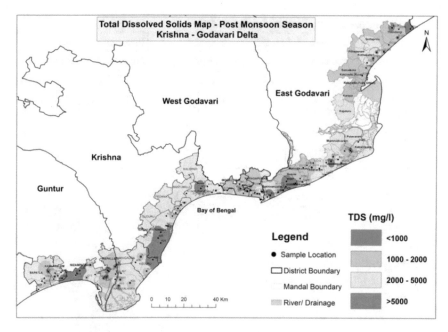

Fig. 5.5 Total dissolved solid map for post-monsoon season

Table 5.3 Freeze and Cherry classification of groundwater based on TDS (mg/l)

TDS (mg/l)	Pre-monsoon (2016)			Post-monsoon (2017)	
	Water type	Number of samples	Percentage	Number of samples	Percentage
<1000	Fresh	35	33	38	36
1000–10,000	Brackish	66	63	62	59
10,000–100,000	Saline	4	4	5	5
>100,000	Brine	0	0	0	0

TH was above the maximum permissible limit of 300 mg/l in 87 groundwater samples are considered as very hard type water (Table 5.4), 16% of the groundwater samples fell under hard type water in pre-monsoon season, and 83% of groundwater has very hard type water in post-monsoon, as well as 1% and 20% are moderately high, hard type in post-monsoon season shown in Figs. 5.6 and 5.7

The electrical conductivity (EC) value ranges from 400 to 29,000 μS/cm (pre-monsoon) and 400–45,000 μS/cm (post-monsoon) with an average of 4018 and 4451 μS/cm during pre-monsoon and post-monsoon season, respectively, are shown in Figs. 5.8 and 5.9. The high EC values in the Krishna area appear pre and post the monsoon. It is the measurement of all soluble salts in the sample, which is the most important water quality standard for crop productivity, that is, water and salt damage. The main effect of high EC water on crop productivity is that plants cannot

Table 5.4 Classification of groundwater based on TH (mg/l)

	Pre-monsoon (2016)			Post-monsoon (2017)	
Total hardness (mg/l)	Type of water	Number of samples	Percentage (%)	Number of samples	Percentage (%)
<75	Soft	0	0.0	0	0.0
75–150	Moderately high	1	1.0	1	1.0
150–300	Hard	17	16.2	20	19.2
>300	Very hard	87	82.9	83	79.8

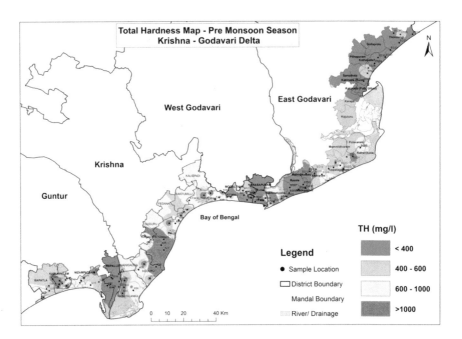

Fig. 5.6 Total hardness map for pre-monsoon season

compete with ions in soil solutions for water. Even though the soil may show wet, because plants can only transpire pure water, useable plant water in the soil solution decreases significantly as EC increases. The amount of water transpired through a crop was directly related to yield. Therefore, irrigation water with high EC reduces yield potential. The present study indicated that overall the water quality was medium to high in the EC category.

The carbonate (CO_3) value is a maximum of 48 mg/l in pre-monsoon with an average of 6.5 mg/l. The bicarbonate (HCO_3) value ranges from 73 to 1159 mg/l in pre-monsoon and 97–2293 mg/l during the post-monsoon period with an average of 458 and 947 mg/l. The high HCO_3 values toward the northeast direction are

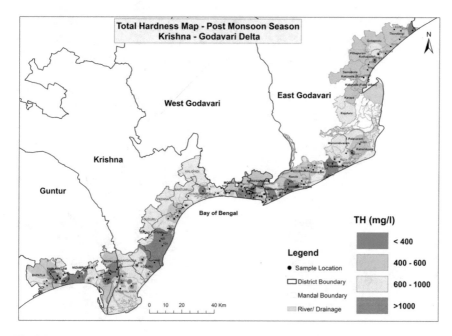

Fig. 5.7 Total hardness map for post-monsoon season

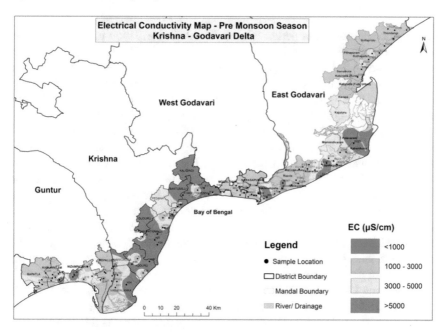

Fig. 5.8 Electrical conductivity map for pre-monsoon season

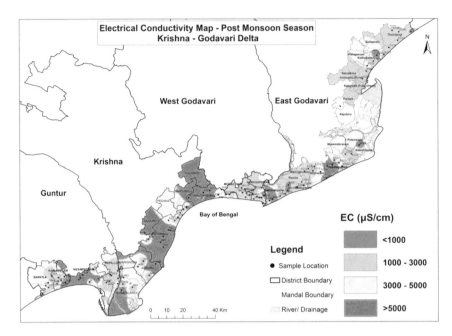

Fig. 5.9 Electrical conductivity map for post-monsoon season

seen in the post-monsoon period whereas isolated distribution is seen in the pre-monsoon season. The chloride (Cl^-) value ranges from 56.7 to 8820 mg/l in pre-monsoon and 71.0–11,928.0 mg/l during the post-monsoon period, with an average of 947.5 and 957.6 mg/l during the pre-monsoon and post-monsoon, respectively. The high chloride values are shown in Nizampatnam Mandal in a post-monsoon period, whereas a northward trend is seen in pre-monsoon (Figs. 5.10 and 5.11).

The calcium (Ca^{2+}) value ranges from 20 to 528 mg/l in pre-monsoon and 32–464 mg/l during post-monsoon period, with an average of 99.4 and 122 mg/l during pre-monsoon and post-monsoon, respectively. The high calcium values are seen near coastal areas toward the stream mouth area in both pre-monsoon and post-monsoon periods. The magnesium (Mg^{2+}) value ranges from 4.8 to 651 mg/l in pre-monsoon and 4.8–849.6 mg/l during the post-monsoon period, with an average of 80.6 and 98.2 mg/l during pre-monsoon and post-monsoon, respectively. The post-monsoon period high magnesium values are seen in Machilipatnam Mandal, Sakthinethipalle river mouth area, Uppalaguptam, and Nizampatnam near the coastal region and the north parts of the Godavari area where low magnesium occur.

The sulfate (SO_4) value ranges from 2.4 to 224.8 mg/l (pre-monsoon) and 2–43.7 mg/l (post-monsoon) with an average of 62.3 and 10.3 mg/l. The low SO_4 values are seen entire study area in the post-monsoon period, whereas this trend is seen isolated in pre-monsoon. The sodium (Na^+) value ranges from 40.8 to 5006.8 mg/l (pre-monsoon) and 9.3–8285 mg/l (post-monsoon) with an average of 634 and 670.

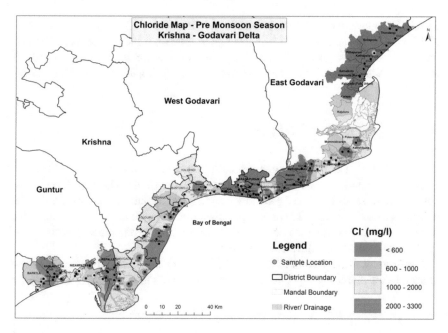

Fig. 5.10 Chloride map for pre-monsoon season

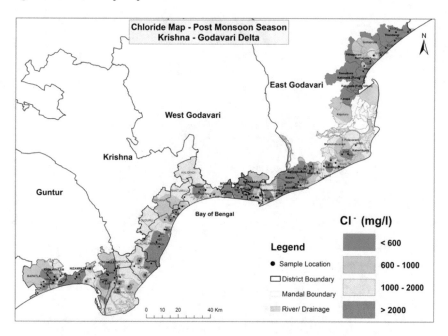

Fig. 5.11 Chloride map for post-monsoon season

8 mg/l. The potassium (K^+) value ranges from 1.6 to 443 mg/l (pre-monsoon) and 1.9–712.2 mg/l (post-monsoon), with an average of 60.4 mg/l and 83.1 mg/l during pre-monsoon and post-monsoon, respectively. In pre-monsoon high potassium, values are seen in Machilipatnam mandal and it is slightly varying in post-monsoon season.

5.5.1 Piper Trilinear Diagram

The concentration of major anions and cations can be plotted on the Piper trilinear graph to understand the geochemical evolution of groundwater. Use AquaChem 2014 software to transpose different groundwater chemical evolution paths and freshwater composition fields (Fig. 5.12) onto Piper diagrams. Piper diagram is a combination of anion and cation triangles on a common baseline. The diamonds between them are used to characterize different types of water.

The graphical illustration of the ion signature helps to reveal the main ions that control water chemistry. Piper divides water into four types by placing water near the

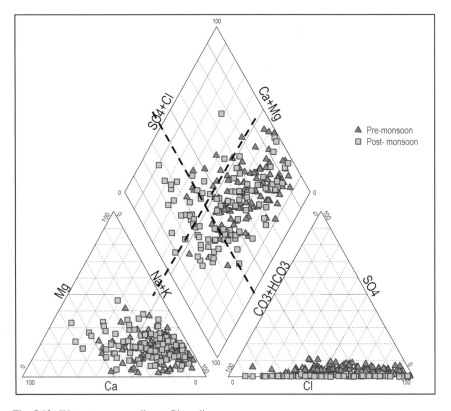

Fig. 5.12 Water-types according to Piper diagram

four corners of the diamond. The water plotted at the top of the diamond is considered as high with $Ca^{2+} + Mg^{2+}$ and $Cl^- + SO_4^{-2}$, which is the area of permanent hardness. The water plot near the right side corner is rich in $Ca^{2+} + Mg^{2+}$, this water region is temporary hardness. The water plot at the lower corner is composed of alkali carbonates ($Na^+ + K^+$ and $HCO_3^- + CO_3^{-2}$). The water near the left-hand side may be saline water ($Na^+ + K^+$ and $Cl^- + SO_4^{-2}$). It has been observed in Piper's diagram that the nature of groundwater exists in the $Na^+ - Cl^-$ type study area. Therefore, the Piper diagram can not only identify the nature of water samples but also reveal the relationship between water samples. It is possible to predict and classify geological units and chemically similar water and then to analyze the trend and flow path of water chemical analysis. In pre-monsoon piper diagram, it can not only identify the nature of water samples but also reveal the relationship between water samples. It is possible to predict and classify geological units and chemically similar water and then to analyze the trend and flow path of water chemical analysis. The diagram shows whether the salinity is derived solely from mixing with seawater.

5.5.2 Common Indicators of Sea Water Intrusion (SWI)

Calcium/Magnesium Ratio

A high Ca/Mg ratio may indicate that the brine is contaminated because the magnesium content in seawater is much higher than calcium. Therefore, the Ca/Mg ratio (>1) is regarded as a parameter to determine seawater contamination. According to the Ca/Mg ratio, many samples in the study area showed a higher Ca/Mg ratio. Figure 5.13 shows that the calcium and magnesium contents in the pre-monsoon season are higher than in the post-monsoon season.

Cl/($CO_3 + HCO_3$) Ratio

In the seawater, Cl^- is the dominant ion and it is only available in small quantities in groundwater, while HCO_3 that is available in large quantities in groundwater occurs only in very small quantities in seawater. The Cl/($CO_3 + HCO_3$) known as Simpson's ratio that is important as evidence for seawater intrusion into the freshwater aquifer. The Cl^-/($CO_3 + HCO_3$) ratio was used as a criterion to evaluate the saltwater intrusion. The chloride is the dominant ion of ocean water and normally occurs in only a small amount in groundwater, while HCO_3 is usually the most abundant negative ion in groundwater, but it occurs in only minor amounts in seawater.

There is 14% fresh groundwater in the pre-monsoon season with <0.5 value. In pre-monsoon, majority of the area shows slightly contaminated groundwater with a value of 0.5–1.30 and 22% be moderately contaminated groundwater, 12% water with injuriously contaminated, 8% highly contaminated, and 1% of water severely contaminated groundwater. In the Post-monsoon season, there is 62% fresh groundwater. The majority of the area shows 43% slightly contaminated groundwater, moderately contaminated groundwater is 9%, injuriously contaminated groundwater is 4 and 3%

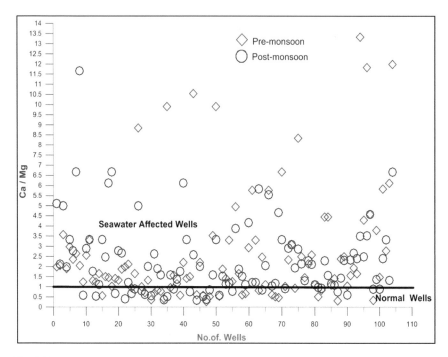

Fig. 5.13 Ca/Mg ratio values in the study area

is highly contaminated groundwater in the study area according to this ratio details are shown in Fig. 5.14

Sodium/Chloride Ratio

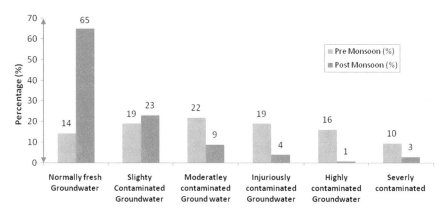

Fig. 5.14 Contamination in the study area based on $Cl/(CO_3 + HCO_3)$ in pre-monsoon season (2016) and post-monsoon season (2017)

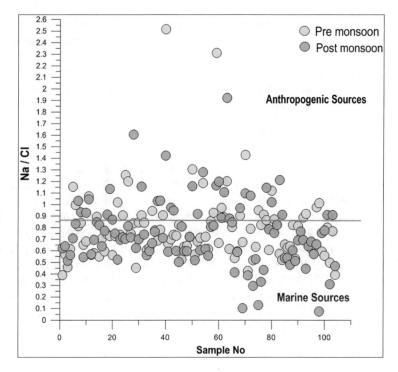

Fig. 5.15 Source of contamination based on the Na/Cl ratio

The Na/Cl ratio of saltwater intrusion is usually lower than the ocean values (i.e., <0.86, molar ratio). On the other hand, high (more than 1) Na/Cl ratios typically characterize anthropogenic sources like domestic water. Thus low Na/Cl ratios combined with other geological parameters can foretell the arrival of saltwater intrusion. According to the Na/Cl ratio, in the pre-monsoon and post-monsoon seasons, approximately 71% and 60% of the study area was contaminated by the ocean (Fig. 5.15) and the remaining areas were 29% and 40% contaminated due to human pollution. Domestic sewage may be controlled by water–rock interaction. The combination of high Na/Cl molar ratio and high salinity value can prove that seawater intrusion may be the main reason for salinization near the coastline.

5.5.3 Seawater Mixing Index (SMI)

The seawater mixing index was developed by Park et al. (2005) to identify seawater intrusion based on the concentration of magnesium, sodium, chlorine, and sulfate. Mondal et al. (2011) revised this to understand the salinization process and SMI stated that it would quantitatively evaluate the mixed grade of brine in residential water. The seawater mixing index of groundwater aquifers is a quantitative assessment used to identify seawater mixing based on four major ions such as Na^+, Mg^{2+}, Cl^-, and SO_4^{2-}. These ions are highly used in the study of saltwater and groundwater mixing index. SMI is calculated by the following equation, where a, b, c, and d are Na^+ (a = 0.31), Mg^{2+} (b = 0.04), Cl^- (c = 0.57), and SO_4^{2-} (d = 0.08) (Park et al. 2005) and the concentration of ions. A calculated SMI value greater than 1 will indicate the effect of mixing seawater and freshwater.

C = Concentration of groundwater (mg/l)
T = Regional threshold values.

Cumulative probability (CP) curves can be used to estimate seawater mixing and thresholds for selected parameters (such as Na^+, Mg^{2+}, Cl^-, and SO_4^{2-}), which exhibit a log-normal density distribution. For the influence of the mixing of seawater and freshwater aquifers, the cumulative probability of the above parameters at the intersection of regional thresholds can be considered. The CP considers the four main ions in groundwater chemistry with regional thresholds such as Mg^{2+} = 17 mg/l, Na^+ = 112 mg/l, Cl^- = 142 mg/l, and SO_4^{2-} = 41 mg/l in post-monsoon season, pre-monsoon Mg^{2+} = 36 mg/l, Na^+ = 81 mg/l, Cl^- = 152 mg/l, SO_4^{2-} = 90 mg/l area. The SMI in the post-monsoon season is between 0.4 and 72.9, and in pre-monsoon season is between 0.4 and 53.1; about, 78% of the groundwater is post-monsoon period, and about, 87% of the SMI is greater than 1 pre-monsoon season.

5.5.4 Integrated Groundwater Quality

Based on the comparison of the conductivity and the concentration of Na^+ and Cl^- with the allowable limits of BIS (2005) and WHO (1997), the quality of groundwater was evaluated. According to the recommended drinking water quality of Na^+, Cl^-, Na^+/Cl^-, and Cl^-/HCO_3^-, the groundwater quality is roughly divided into two categories, suitable and unsuitable (Figs. 5.16 and 5.17). In the pre-monsoon and post-monsoon seasons, groundwater in rivers (Godawari, Krishna) and other areas was affected by backwater, salt pans, and canals and was found to be unsuitable for drinking, while groundwater in other areas was suitable for drinking. There are very few locations with good portable drinking water during the pre-monsoon season. They include Mamidikuduru, Sakhinetipalle, in East Godavari district, near Narsapuram in West Godavari district, Avinigadda, in Krishna district, and Karlapalem Mandal in Guntur district. The significant parts of the Krishna and Godavari delta region are portable during the post-monsoon season, except Allavaram and Sakhinetipalle in East Godavari district, Koduru in Krishna district, and Nizampatnam in Guntur district. The spatial distribution of sodium in the groundwater indicating the suitability for drinking purposes is shown in Figs. 5.16 and 5.17.

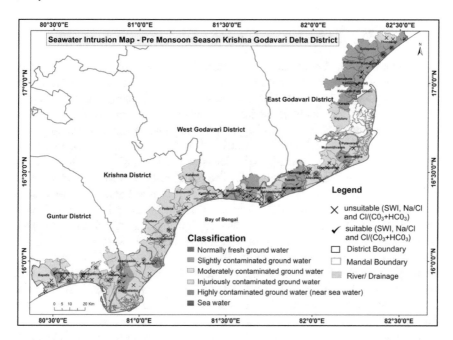

Fig. 5.16 Classification based on suitability of groundwater for drinking purposes pre-monsoon (2016)

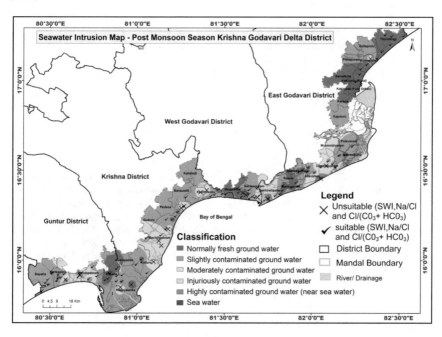

Fig. 5.17 Classification based on suitability of groundwater for drinking purposes post-monsoon (2017)

5.6   Conclusion

The groundwater hydro-chemical characteristics of the Krishna and Godavari deltas indicate that seawater plays a very important role in groundwater quality/composition. In the pre-monsoon season, close to Makiripatnam, Alavaram, Sahnetipal, and many parts of the northern region, low resistivity was affected. About 71% of groundwater is $Na^+ - Cl^-$ type, the SMI concentration post-monsoon is between 0.4 and 72.9, and the SMI concentration pre-monsoon is between 0.4 and 53.1. In the present area, about 82% of freshwater chemicals with high SMI concentrations are contaminated with saltwater. The pre-monsoon season is more suitable for drinking than the post-monsoon season. The few locations of this area were identified with seawater intrusion, due to the influence of geomorphology of the study area and salt flat. During the monsoon season, the quality of groundwater is improved due to rainfall replenishing the aquifer which makes the freshwater level higher and reduces seawater intrusion in the delta. To overcome this situation, it is necessary to change the pumping method of coastal areas to inland areas by constructing new wells field and effectively control the intrusion of seawater and the restoration of groundwater quality. The conclusion is that the process of controlling groundwater salinity revealed by chemical, and statistical analysis is mainly related to seawater intrusion and ion exchange processes related to seawater intrusion. This research may be useful to help planners and decision-makers protect our limited groundwater resources for future generations.

Acknowledgements   The authors express their sincere gratitude to the Chairman, APSAC for his continuous support for the research activity and saline water testing lab at Agricultural University, Bapatla; also the authors are grateful to the editor and anonymous reviewers for suggestions and comments which improved the manuscript significantly.

References

Ashraf S, Harue M, Ryo A, Bhattacharya P, Yoriko Y, Shimizu Y (2019) Hydrogeochemical and isotopic signatures for the identification of seawater intrusion in the paleo beach aquifer of Cox's Bazar city and its surrounding area, south-east Bangladesh. J Groundwater Sustain Dev

Annapoorani A, Murugesan A, Ramu A, Renganathan NG (2014) Hydrochemistry of groundwater in and around Chennai, India—a case study. Res J Chem Sci 4(4):99–106. ISSN 2231-606X

Anil Kumar KS, Priju CP, Narasimha Prasad NB (2015) Study on saline water intrusion into the shallow coastal aquifers of Periyar River Basin, Kerala using hydrochemical and electrical resistivity methods. Aquatic Procedia 4

APHA (1992) Standard methods for the examination of water and wastewater. American Public Health Association, Washington, DC, USA

BIS Recommendations (2005) Guidelines for drinking water quality. 1

Bear J (1999) Seawater intrusion in coastal aquifers: concepts, methods, and practices. In: Seawater Intrusion in Coastal Aquifers. Kluwer Academic, Boston, Mass

Brindha K, Paul P, Touleelor S, Somphasith D, Elango L (2016) Geochemical characteristics and groundwater quality in the Vientiane Plain, Laos. J Water Qual Exposure Health

Chen K, Jiao JJ (2007) Seawater intrusion and aquifer freshening near the reclaimed coastal area of Shenzhen. Water Sci Technol Water Supply 137–145

Childs C (2004). Interpolating surfaces in ArcGIS spatial analyst. ESRI Education Services

Chaurasia J, Rai PK, Singh AK (2013) Physico-chemical status of groundwater Near Varuna River in Varanasi City, India. Int J Environ Sci (integrated Publication Association) 3(6):2114–2121

Davis NS, DeWiest RJM (1966) Hydrogeology. Wiley, New York, p 463

Freeze RA, Cherry JA (1979) Groundwater. Prentice-Hall, New Jersey

Giuseppe S, Maria T (2012) Seawater intrusion, and salinization processes assessment in a multistrata coastal aquifer in Italy. J Water Resour Protec 4:954–967

ISI (1983) Indian standard specification for drinking water. ISI 10500

Mondal NC, Singh VS, Puranik SC, Singh VP (2010b) Trace element concentration in groundwater of Pesarlanka Island, Krishna Delta, India. Environ Monit Assess 215–227

Mondal CN, Singh PV, Singh S, Singh SV (2011) Hydrochemical characteristic of the coastal aquifer from Tuticorin, Tamil Nadu, India. Environ Monit Assess 531–550

Magesh NS, Chandrasekar N, Elango L (2016) Occurrence and distribution of fluoride in the groundwater of the Tamiraparani River basin, South India: a geo-statistical modelling approach. Environ Earth Sci 1483

Mohan K, Shrivastava A, Rai PK (2011) Ground water in the city of Varanasi, India: present status and prospects. Quaestiones Geographicae 30(3):47–60. https://doi.org/10.2478/v10117-011-0026-9

Narayan SM, Natarajan K (1993) Seawater-freshwater interface studies in coastal areas with special reference to Tamilnadu. In: International proceedings volume on workshop on artificial recharge of groundwater in coastal aquifers

Nair IS, Parimala Renganayaki S, Elango L (2013) Identification of seawater intrusion by Cl/Br ratio and mitigation through managed aquifer recharge in aquifers North of Chennai, India. J Ground Water Res

Nistor MM, Rai PK, Dugesar V, Mishra VN, Singh P, Arora A, Kumra VK, Carebia IA (2019) Climate change effect on water resources in Varanasi District, India. Meteorol Appl 1–16. https://doi.org/10.1002/met.1863

Nistor MM, Rai PK, Carebia IA, Singh P, Shahi AP, Mishra VN (2020) Comparison of the effectiveness of two budyko-based methods for actual evapo-transpiration in Uttar Pradesh, India. Geographia Technica 15(1):1–15 (2020)

Rajib P, Brindha K, Gowrisankar G, Tan M, Mahesh K (2019) Identification of hydro-geochemical processes controlling groundwater quality in Tripura, Northeast India using evaluation indices, GIS, and multivariate statistical methods. Environ Earth Sci 78:470

Rai PK, Mishra VN, Singh P (2018) Hydrological inferences through morphometric analysis of Lower Kosi River Basin of India for water resource management based on remote sensing data. Appl Water Sci (Springer) 8(15):1–16. https://doi.org/10.1007/s13201-018-0660-7

Pujari PR, Soni AK (2009) Seawater intrusion studies near Kovaya limestone mine, Saurashtra coast. Indian J Environ Monitoring Assess 109

Piper AM (1944) A graphical procedure in the geochemical interpretation of water analysis. Trans Am Geophy Union 25

Park SC, Yun ST, Chae GT, Yoo IS, Shin KS, Heo CH, Lee SK (2005) Regional hydro-chemical study on salinization of coastal aquifers, a western coastal area of South Korea. J Hydrol 313

Richter BC, Kreitler CW (1993) Geochemical techniques for identifying sources of groundwater salinization. CRC Press, Boca Raton

Saxena VK, Singh VS, Mondal NC, Jain SC (2003) Use of chemical parameters to delineation fresh groundwater resources in Potharlanka Island, India. Environ Geol 516–521

Saha DK, Choudhury K (2005) Saline water contamination of the aquifer zones of Eastern Kolkata. J Ind Geophys Union 9(4):241–247

Sreedevi PD (2004) Groundwater quality of Pageru River Basin, Cuddapah District, Andhra Pradesh. J Geol Soc India 619–636

Sudhakar A, Narsimha A (2013) Suitability and assessment of groundwater for irrigation purpose: a case study of Kushaiguda Area, Ranga Reddy District, Andhra Pradesh, India. Adv Appl Sci Res 75

Subba Rao N, Saroja Nirmala I, Suryanarayana K (2005) Groundwater quality in a coastal area: a case study from Andhra Pradesh, India

Tulipano L, Fidelibus MD, Sappa G, Coviello MT (2008) Evolution of seawater intrusion in coastal aquifers of Pontina Plain. In: 20th salt water intrusion meeting, pp 278–281

Vasanthavigar M, Srinivasamoorthy K, Vijayaragavan K, Ganthi R, Chidambaram S, Anandhan P, Manivannan R, Vasudevan S (2010) Application of water quality index for groundwater quality assessment, Thirumanimuttar sub-basin, Tamilnadu, India. Environ Monitoring Assessment 595–609

Venkateswaran S (2001) Hydrogeology and geochemical characterization of groundwater with special emphasis on agricultural development in Vaniyar sub-basin, Ponnaiyar River, Tamilnadu. Int J Rec Sci Res 1(12):213–221

WHO (1997) Guidelines for drinking water quality, vol 1, Recommendations, 2nd edn. WHO, Geneva

Chapter 6
Microlevel Planning for Integrated Natural Resources Management and Sustainable Development: An Approach Through a Micro Watershed Using Geospatial Technology

L. Prasanna Kumar

Abstract "Village" is the microlevel or the baselevel where the development starts. It may be social, economic, political, or environmental. Proper planning, monitoring, and managing natural resources are always a challenging task for decision-makers and experts. In the present scenario, the use of remote sensing (RS), geographical information system (GIS), global positioning system (GPS), and other information technology has the potential to manage the natural resources for sustainable livelihood. Implementation of these technologies in microlevel planning through the micro watershed approach is always encouraging and for these, we need a storehouse of the different geospatial database. In the present study, an attempt has been made to use the geospatial technology in the land and water resources management of a village enclosed within a micro watershed. The geospatial database has been generated at the cadastral level or plot level, i.e., at 1:4000 scale. The resource inventory includes land use/land cover, digital elevation model (DEM), slope, geomorphology, ground water prospect, soil, and land capability maps. The site suitability analysis technique is used to develop the land and water resources management action plan. In the land resource management plan, seven types of alternative land use practices are suggested for the study area. On the other hand, the water resource management action plan suggests suitable sites for the construction of bore well and dug well. Further, the action plan suggests the site for the construction of different farm ponds, check dam, nala bund, water harvesting structure, etc. The integration of the action plan map on the DEM gives a 3D perspective which will be an add-on in terms of visualization to the administrators and decision-makers.

Keywords Geospatial technology · Resource inventory · Site suitability analysis · Land and water resource action plan · Decision making

L. Prasanna Kumar (✉)
MIS-GIS Expert, Water Resources Department, Government of Odisha, Bhubaneswar, Odisha 751001, India

© The Author(s), under exclusive license to Springer Nature Singapore Pte Ltd. 2022
P. K. Rai et al. (eds.), *Geospatial Technology for Landscape and Environmental Management*, Advances in Geographical and Environmental Sciences,
https://doi.org/10.1007/978-981-16-7373-3_6

6.1 Introduction

The development of a village depends upon the availability of resources in the area, and it has to be self-sustained and capable of managing its resources. Therefore, an integrated approach to microlevel planning or the grassroot planning of natural resources is very crucial for decision making. The natural resource management and socio-economic condition of the region can be improved through these approaches. Comprehensively, microlevel planning acts as one of the key factors in sustainable rural development and optimal management and utilization of natural resources for sustainable livelihood (Rai and Kumra 2011; Ghosh et al. 2011; Page 2016; Wang et al. 2016; Mishra et al. 2016).

In microlevel planning, the watershed management perspective has been developing enormously. Rajesh Rajora (1998), Ramamohana Rao and Suneetha (2015) have mentioned that watershed development and management are a holistic and bottom-up planning approach toward integrated resource management and sustainable development. Sangameswaran (2008) has discussed that an ideal village depends upon the construction of community feeling within a watershed. As a part of integrated development, the land and water resources are taken based on micro watershed. This is mainly due to the shape and pattern of micro watershed that helps to study and control natural resources like land, water, vegetation, etc.

To make sustainable planning at the microlevel, we need to have diverse geographical information. Geospatial technologies have the potential to address the solution. The advancement of geospatial technologies like remote sensing (RS), global positioning system (GPS), geographical information system (GIS), and information technology has provided the best tool for mapping natural resources (Melesse et al. 2007). Gebre et al. (2015) has analyzed the watershed attributes for water resource management through an approach of micro watershed using GIS. Mishra and Panda (2015) discussed the role of geo-informatics in developing a watershed management plan.

In the present study, a micro watershed is taken as the unit for grassroot level or microlevel planning of a village. The village is the unit of microlevel planning, and it is to be planned based on the watershed for the overall development of the village. The major objectives of the present study are to delineate the micro watershed boundary that encloses the village to be implemented and to generate geospatial and non-spatial digital cadastral level datasets at 1:4000 scale on GIS environment. Further, this spatial and non-spatial information are used to develop a natural resource management action plan at the cadastral level to evaluate the land and water resources management of that village.

6.2 Study Area

The Kumakhal micro watershed is located between 19°48′34.19″ and 19°50′33.17″ North latitude and 83°20′18.605″ and 83°22′45.208″ East longitude. The areal extent of the micro watershed is 730.79 ha. In this study, Kumakhal village of Bandhapari Grampanchayat, Lanjigarh block of Kalahandi district of Odisha state is taken as the study area for the microlevel planning. The aerial extent of Kumakhal village is from 19°48′52.39″ to 19°49′48.01″ N latitude and 83°20′53.52″ to 83°22′20.21″ E longitude. It covers an area of 197.07 ha. Kumakhal village is surrounded by Hatisal reserved forest in the northern and eastern part. In the western part, the major River Ret is flowing in the southwestern direction. The location map of the Kumakhal village insets within Kumakhal micro watershed overlaid on the high-resolution satellite is shown in (Fig. 6.1).

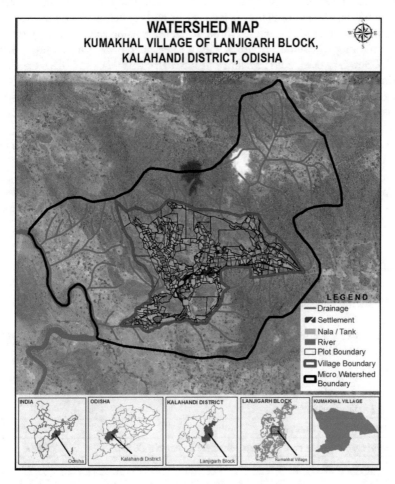

Fig. 6.1 Location map of the study area insets within the watershed boundary and overlayed on the high-resolution satellite image

6.3 Materials and Methods

6.3.1 Geospatial Technology in Natural Resource Management

Broadly, geospatial technology is explained in three major domains: remote sensing (RS), geographical information system (GIS), and global positioning system (GPS). Geospatial technology plays a vital role in natural resource monitoring and management. Satellite remote sensing provides reliable, near real-time, temporal, as well as very minute baseline information on land, water, and vegetation of the earth's surface. Temporal data receiving facilities through remote sensing satellite makes it useful for monitoring of natural resources within a reasonable short time frame (Sabins 1997; Chowdary et al. 2001; Mishra and Rai 2016; Mishra et al. 2018). Geographical information system (GIS) on the other hand, with its capability of integration and analysis of spatial, multi-layered information obtained from remote sensing and other sources have proved to be an effective tool in planning and management of natural resources. The technology has been an instrumental breakthrough that permits examining natural resources and environmental issues in a geographic context (Foresman 1998). Further, the field information collected through GPS technology strengthens the results and helps us to provide more accurate and authentic information. Thus, geospatial technologies demonstrate an effective management tool that is used by various domain specialists for different land and water resources development plans.

6.3.2 Database and Methodology Used

The following datasets (Table 6.1) are used to achieve the objective of the research paper.

The datasets are further processed in the GIS environment by using the ArcGIS 10.0 software. The detailed methodological approach (Fig. 6.2) shows the outline of the study.

Table 6.1 Data used in this study

Data used	Source	Scale/resolution
Cadastral map	Survey and map publication, Cuttack, Odisha	1:4000
Google earth image	Map data: Google, 2020 maxar technologies	60 cm–15 m
Toposheets	Survey of India, Bhubaneswar (Toposheet No.65M5SW)	1:25,000

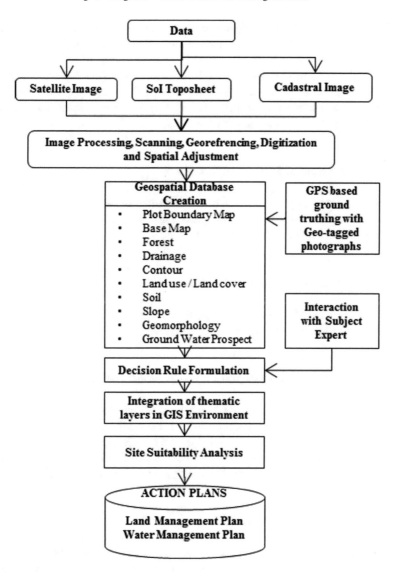

Fig. 6.2 Methodology outline

6.3.3 Development of Geospatial Database at Cadastral Level (1:4000)

There is a need to develop a geospatial database at the microlevel or grassroot level. Microlevel planning through the micro watershed approach is being considered as the most appropriate approach, and the augmentation of modern geospatial technologies such as remote sensing (RS) and geographical information system (GIS) solves the problem of creation of the geospatial database. High-resolution satellite images, SOI topo map, and cadastral database coupled with the ground or field observation has been utilized to create layer-wise natural resources spatial information at 1: 4000 scale matching with the individual plot of Kumakhal village. An intensive field survey was done to update the thematic maps. The geo-tagged photographs of different land features are shown in Fig. 6.3.The following paragraphs depict in detail the geospatial environment of the natural resources of the village (see Fig. 6.4).

6.3.3.1   Delineation of Watershed Boundary

The watershed boundary is delineated based on the contour and drainage/waterbodies generated from the SOI topo map and updated in high-resolution satellite data. The DEM is generated using ArcGIS software.

6.3.3.2   Drainage Map and Waterbodies

The major drainage of the nearby village is Ret River. Three water bodies exist in Kumakhal village along with the drainage/streams.

6.3.3.3   Contour Map and Digital Elevation Model (DEM)

The contour map has been generated from the SOI topo map at ten meters contour interval. A further 5 m interval contour map has been extracted through the interpolation method. The highest contour value in this area is 595 m, whereas the lowest value is 500 m. The valleys remain within the contours of 530–535 m above MSL. Using the contour values, the DEM has been prepared for Kumakhal village.

6.3.3.4   Land Use/Land Cover Map

Land use and land cover are the most valuable natural resource information that is required for the overall planning of the village. The land use and land cover of Kumakhal village have been generated through the interpretation of high-resolution satellite image coupled with ground truth data. The map was prepared at 1:4000 scale, and the plot boundaries have been matched or spatially adjusted with the high-resolution satellite image.

In total, nine land use and land cover units have been delineated (NRSC 2007). Most of the village is covered by scrubs which covers 47.24% (93.11 ha) area. These lands are treated as wastelands inside the village. The total surface water source in the shape of a tank, pond inside the village is nearly 0.79 ha. Village settlements occupy nearly 2.78 ha of lands along with village roads of 2.92 ha. The cropland covers an area of 90.71 ha which constitutes 46.03% of the total area of the village.

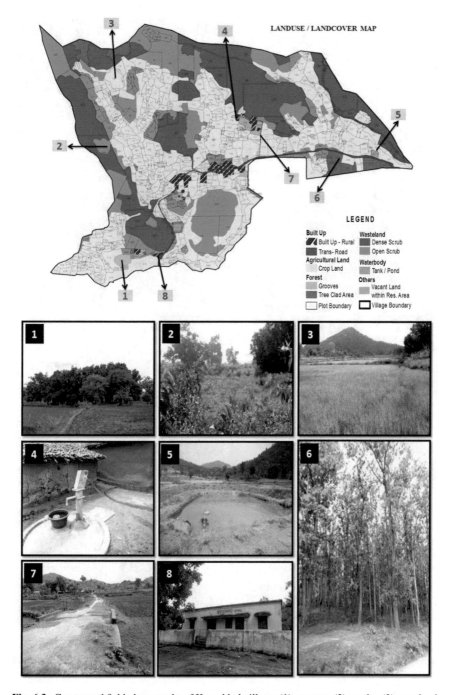

Fig. 6.3 Geo tagged field photographs of Kumakhal village: (1) grooves, (2) scrubs, (3) crop land, (4) tube well, (5) pond, (6) tree clad area, (7) earth road, (8) health center

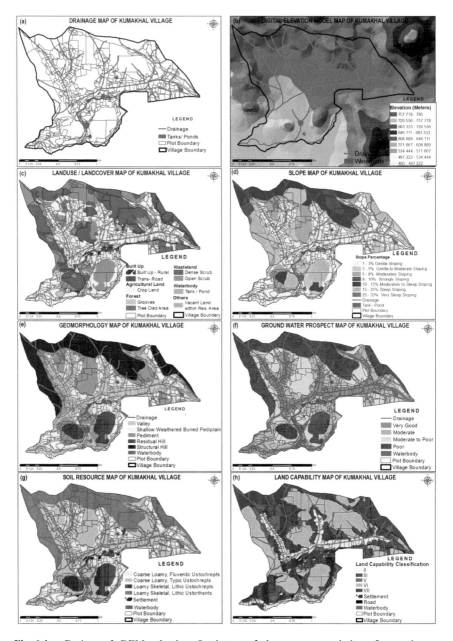

Fig. 6.4 **a** Drainage, **b** DEM, **c** land use/land cover, **d** slope, **e** geomorphology, **f** ground water prospect, **g** soil, **h** land capability map of Kumakhal village of Lanjigarh block in Kalahandi district

Table 6.2 Land use/land cover of Kumakhal Village

Class I	Class II	Area (ha)	% to total
Built-up	Built-up—rural	2.78	1.41
	Transportation—road	2.92	1.48
Agricultural land	Crop land	90.71	46.03
Forest	Grooves	0.97	0.49
	Tree clad area	4.61	2.34
Wasteland	Dense scrub	50.91	25.83
	open scrub	42.20	21.41
Others	Vacant land within residential area	1.20	0.61
Waterbody	Tank/pond	0.79	0.40
Total		197.07	100.00

Table 6.3 Slope statistics

Slope type	Slope in percentage	Area (Ha.)
Gently sloping	1–3%	92.07
Gently to moderate sloping	3–5%	20.32
Moderate sloping	5–8%	22.27
Strongly sloping	8–10%	8.89
Moderately steep sloping	10–15%	25.78
Steep sloping	15–25%	27.36
Very steep sloping	>25	0.35
Total		197.07

Table 6.2 depicts the spatial extent and percentage of the total village area of each identified land use and land cover unit.

6.3.3.5   Slope Map

The spatial information on the slope of the village has been derived from the contour lines and the area. It has been divided into seven slope categories, i.e., 1–3, 3–5, 5–8, 8–10, 10–15, 15–25, and >25%. Table 6.3 illustrates the spatial distribution of slope categories of Kumakhal village.

6.3.3.6   Geology, Geomorphology Map

Geologically, Kumakhal village is exhibited to the Khondalite group of rock. Using the satellite image, interpretation coupled with field check establishes the spatial distribution of Khondalite rocks in the village is shown. The geomorphic development

Table 6.4 Spatial distribution of geomorphology of Kumakhal village

Geomorphology	Area (ha)	% To total area
Structural hill	50.83	25.79
Residual hill	11.82	6.00
Pediment	42.22	21.42
Shallow weathered buried pediplain	46.83	23.77
Valley	45.37	23.02
Total	197.07	100.00

Table 6.5 Ground water prospect map of Kumakhal village

Ground water prospect	Area (ha.)	% To total area
Very good	45.37	23.02
Moderate	46.83	23.77
Moderate to poor	42.22	21.42
Poor	62.65	31.79
Total	197.07	100.00

of an area depends mainly on lithology, tectonics, and climate. Broadly, Kumakhal village is divided into five geomorphic classes, i.e., structural hill, residual hill, pediments, valley, and shallow weathered buried pediplain. Table 6.4 shows the spatial distribution of geomorphological units in the Kumakhal village.

6.3.3.7 Ground Water Prospect Map

Based on the integration of the thematic data, a comprehensive groundwater prospects map has been prepared. Broadly, Kumakhal village is divided into four zones of groundwater prospect, i.e., very good, moderate, moderate to poor, and poor. Table 6.5 shows the statistical information on the groundwater prospect zone of the village. It is observed that 45.37 ha is under a very good zone of groundwater, whereas 62.65 ha of the area is under the zone of poor groundwater prospect.

6.3.3.8 Soil Map

The spatial distribution of the variety of soils occurring in Kumakhal village is shown in Table 6.6. The map has been generated concerning the Odisha Sampad Web-GIS portal of Odisha Space Applications Center (ORSAC). The soil map was georeferenced and modified concerning the field survey and high-resolution satellite image. The village is occupied with coarse loamy, typic ustochrepts which cover 86.35 ha which constitutes 43.82% of the total area. 53.62 ha of land which constitutes

Table 6.6 Soil taxonomy map of Kumakhal village

Soil taxonomy	Area (ha)	% To total area
Coarse loamy, fluventic ustochrepts	44.5	22.58
Coarse loamy, typic ustochrepts	86.35	43.82
Loamy skeletal, lithic ustochrepts	11.82	6
Loamy skeletal, lithic ustorthents	53.62	27.21
Waterbody	0.79	0.4
Total	197.07	100

27.21% of the total area are comprised of loamy skeletal, lithic ustorthents. Mostly, such type of soil is seen in the hilly region of the village.

6.3.3.9 Land Capability Map

Land capability classification is the basis of watershed management. The basic theme of soil and water conservation in watershed management is to use the land according to its capability and treatment of the land according to its needs. Thus, knowledge of land capability classification is a prerequisite and important for planning, implementation, and evaluation of soil and water conservation measures in an integrated watershed management program. There are two broad groups of land capability classification, namely:

(i) Lands suitable for cultivation (Class I–IV)
(ii) Lands not suitable for cultivation, but suitable for forestry, grassland, and wildlife (Class V–VIII lands).

Land capability analysis of Kumakhal village is carried out during this study. The integrated GIS analysis has resulted that the lands of the village are divided into five land capability classes, i.e., II, III, V, VI, and VII. The details of land capability classes are discussed at Table 6.7. From the land capability analysis of the Kumakhal village, the following facts have been established.

6.4 Results and Discussion

A natural resource management plan for sustainable rural development is achieved through the integration of spatial databases on various natural resource themes, through analysis to develop alternate natural resource management plans. The generation of action plans at the cadastral level action plan constitutes the generation of

Table 6.7 Land capability classification of Kumakhal village

Land capability class	Area (ha)	% To total area	Land capability description
II	26.50	13.45	Slight limitation for field crops
III	60.98	30.94	Moderate limitation for field crops
V	15.37	7.80	Adverse for field crops
VI	67.90	34.46	Adverse for field crops
VII	19.84	10.07	Adverse for field crops requires protection
Road	2.92	1.48	Road
Settlement	2.78	1.41	Settlement
Waterbody	0.79	0.40	Waterbody
Total	197.07	100.00	

thematic layers at the cadastral level, development of decision rules for site suitability analysis, and development of land and water resource management plans.

6.4.1 Land Resource Management Plan

The action plan for the land resources management of the Kumakhal village is generated by integrating the thematic layers like land use/land cover, soil, slope, hydrogeomorphology, drainage, groundwater potential, etc. The overlay technique is used to identify the suitable sites for different land management action plan. While generating the action plan, several observations are made like pertinent site information concerning natural vegetation, present land use, local farming practices, etc., and are taken into consideration and further alternate land use practice is suggested. With this detail, it is possible to assess whether a particular parcel of land is over/underutilized or optimally used, considering the need for sustainable production, and the quality of the ecosystem. While suggesting alternate land use practice, future consideration should be taken care of like exploitation of groundwater, soil–water conservation, etc., was also considered.

6.4.2 Decision Rules for Land Management Plan

The logic developed under this study for the generation of a land management action plan is described in Table 6.8. Seven types of alternative land use practices are suggested for the Kumakhal village, and the characteristic of each action plan item is discussed in the following section.

Table 6.8 Decision rules for land management solution

S. No.	Suggested action plan	Existing land use	Geomorphology	Ground water potential	Slope	Soil taxonomy	Land capability	Area (ha)	% To total area
1	Optimum used land	Tree clad area/grooves	**	**	**	**	**	5.58	2.83
2	Extensive agriculture	Crop land	Valley/shallow weathered buried pediplain/pediment	Very good/moderate	1–3%/3–5%	Coarse loamy, typic ustochrepts/coarse loamy, fluventic ustochrepts/coarse loamy, typic ustochrepts	III	55.16	27.99
3	Extensive agriculture with ground water exploitation	Crop land	Valley/shallow weathered buried pediplain	Very good/moderate	1–3%	Coarse loamy, typic ustochrepts/coarse loamy, fluventic ustochrepts	II	26.50	13.45
4	Fodder and fuel wood plantation	Vacant land within residential area	Shallow weathered buried pediplain/pediment	Moderate/poor to moderate	1–3%/3–5%	Coarse loamy, fluventic ustochrepts	III	1.02	0.52
5	Agro horticulture with land labeling	Crop land/vacant land within residential area	Pediment	Poor to moderate	3–5%/5–8%	Coarse loamy, typic ustochrepts	III	5.46	2.77

(continued)

Table 6.8 (continued)

S. No.	Suggested action plan	Existing land use	Geomorphology	Ground water potential	Slope	Soil taxonomy	Land capability	Area (ha)	% To total area
6	Forestplantation	Dense scrub/open scrub/crop land	Structural hill/residual hill/pediment/shallow weathered buried pediplain	Poor/poor to moderate/moderate	3–5%/5–8%/8–10%/10–15%/15–25%/25–33%	Loamy skeletal, lithic ustorthents/loamy skeletal, lithic ustochrepts/coarse loamy, fluventic ustochrepts/coarse loamy, typic ustochrepts	V/VI/VII	96.98	49.21
7	Pisiculture/duckery with Rennovation of waterbody	Tank/pond	**	**	**	**	**	0.79	0.40
8	Settlement	Settlement	**	**	**	**	**	2.79	1.42
9	Road	Road	**	**	**	**	**	2.79	1.42

6.4.3 Land Management Solutions

Based on the decision rules, seven types of alternative land use practices are suggested for land resource management. The characteristic of each action plan item is discussed below. Figure 6.5a, b show the land management plan and 3D perspective map of the land management plan for Kumakhal village.

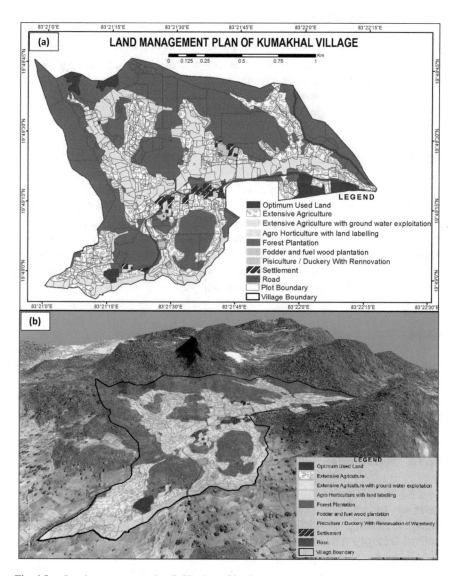

Fig. 6.5 a Land management plan, **b** 3D view of land management plan

6.4.3.1   Optimum Used Land

The tree-clad area and the grooves within the Kumakhal village are categorized as optimum used land. This land needs no action, only thing is to preserve it. This is 2.83% (5.58 ha) of the village.

6.4.3.2   Extensive Agriculture

Agriculture is the major occupation of Kumakhal villagers. Agriculture is the major action item in the land management plan. It is highly suitable for extensive agriculture. Extensive agriculture can be done mostly on the valley with very less slope, i.e., 0–1% to 1–3% and the groundwater prospect should be very good to moderate. 55.16 ha, i.e., 27.99% of the village is suggested for extensive agriculture.

6.4.3.3   Extensive Agriculture with Ground Water Exploitation

In Kumakhal village, a single crop is grown and due to low rainfall, the annual crop production reduces. So, extensive agriculture with groundwater exploitation is planned in the valleys with good groundwater availability. The groundwater of the intermontane valley is to be exploited through the deep or shallow tube/bore wells and dug wells. It is observed that field bunding is highly suggested for groundwater exploitation. This bunding will also solve the problem of soil erosion. An area of 26.50 ha is suggested for this item in the action plan.

6.4.3.4   Fodder and Fuel Wood Plantation

In Kumakhal village, most families possess some kind of livestock (cows, goats, etc.) and they are an important source of livelihood. In this action item, the vacant land within the residential area is used for fodder and fuelwood plantation. It covers an area of 1.02 ha which constitutes 0.52% of the total area.

6.4.3.5   Agro Horticulture with Land Labeling

As the land is having moderate sloping and the groundwater condition is poor to moderate, so agro-horticulture is suggested with proper land labeling. This land management is suggested to an extent of an area of 5.46 ha which constitutes 2.77%. Contour bunds/field bunds/earthen bunds are suggested to check soil erosion along with the moderate sloping lands.

6.4.3.6   Forest Plantation

Forest plantation is highly recommended in Kumakhal village. This action plan covers 96.98 ha which constitutes 49.21% of the land area. The existing land use pattern in this region is mainly scrubland. The groundwater condition in this zone is moderate to poor. Different species, like sal, acacia, can be planted in this area.

6.4.3.7   Pisiculture/Duckery with Renovation of Waterbody

The existing tanks and ponds located inside the village can be used for pisiculture and duckery. Fish farming is called pisiculture, and duck farming is called duckery. The suggested water harvesting structures may also be used for pisiculture. The earthen bunds surrounding the tanks/ponds can be used with the plantation of horticultural species. This will add to the net return from the same piece of land. An area of 0.79 ha is suggested for pisiculture and duckery in the village.

6.4.4   Water Resource Management Plan

Water resource development and management influence land resource management to a great deal. Therefore, it is necessary to develop, conserve, and efficiently utilize the available water on the surface, soil profile, and groundwater. Here, both surface and groundwater resources are considered to generate action plans to utilize the existing resources. Different thematic layers like land use/land cover, drainage, slope, geology, geomorphology, etc., are used for the action plan preparation. Small structures like farm pond, check dams, nala bund, and water harvesting structures are suggested to address local needs for providing supplementary water for irrigation, drinking, etc. The water management plan generated for Kumakhal village (Fig. 6.6a, b) envisages optimal utilization scenario both for ground and surface water.

6.4.5   Decision Rules for Water Management Plan

The thematic layers like land use/land cover, geomorphology, slope and soil resources, etc., are integrated with the GIS environment to prepare the decision rules for the water management action plan. Different thematic layers are superimposed, and decision rules are prepared for suggesting suitable sites for surface water and groundwater exploitation.

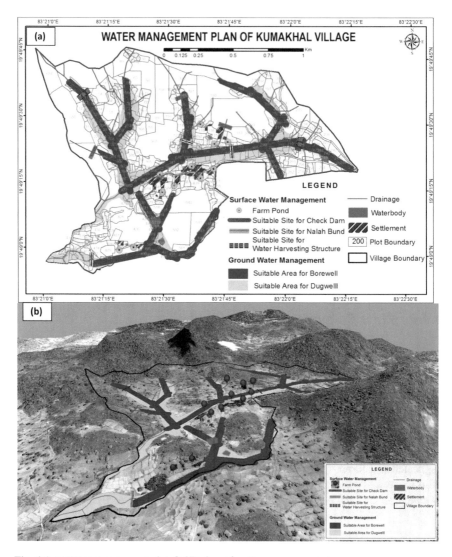

Fig. 6.6 **a** Water management plan, **b** 3D view of water management plan

6.4.6 Water Management Solution

The water management action plan (Fig. 6.6a) generated for Kumakhal village envisages optimal utilization scenario both for ground and surface water. Broadly, the water management solutions are categorized into two parts, i.e., groundwater and surface water. Details are explained in the following section.

Table 6.9 Proposed water management plan

Action plan	Area (ha)
Suitable for bore well	34.41
Suitable for dug well	18.31

6.4.6.1 Ground Water Solution Suitable for Bore Wells and Dug Wells

The groundwater solution in Kumakhal village is suggested based on the availability of water resources. Broadly, there are two types of groundwater solution based on the suitable sites: Suitable site for bore wells and suitable sites for dug well. The fracture zones or the zones of lineaments and lineament intersections are suited for exploitation through deep bore well. Borewells are suitable in valleys and buried pediments regions having good lineaments. Suitable areas for bore wells have been suggested in 25-m buffer zones of the lineaments in the valley area. Based on the field survey, it was identified that bore wells can be driven up to 60–80 m depth to encounter ground water-saturated zone. In Kumakhal village, 34.41 ha area is suggested for bore wells (Table 6.9). Dug wells are suggested in the valley area. The site for dug wells is demarcated considering the weathered thickness with joints and fractures and the nature of alluvium deposits. The formation with a mixture of fine, medium, and coarse sand as well as gravels in some locations in different proportions holds and easily yields sufficient water to support the installation of dug wells. Dug wells are advisable for irrigation purpose. In Kumakhal village, 18.31-ha area is suggested for dug wells in the valley region (Table 6.9).

6.4.6.2 Surface Water Solution

The main source of surface water is rain, but it is known that rainwater is running water. So steps should be taken to obstruct the flowing water so that it can be used for a different purpose. Therefore, suitable structures are constructed to control the running water, e.g., check dam, percolation tank, etc. In this area, suitable sites are identified for the construction of check dam, nala bund, water harvesting structure, farm pond, etc., at Kumakhal village.

6.4.6.3 Farm Pond

Farm ponds are constructed as an embankment across a watercourse. Normally, such structures are provided within an individual farm. Farm ponds are generally constructed, where there is a settlement with nearby cultivation. In Kumakhal village, 11 numbers of farm ponds are suggested for suitable water resource management.

6.4.6.4   Check Dam

In Kumakhal village, the check dams are constructed across a drainage line to reduce the velocity of runoff and to help in the percolation of retained water. Such structures not only check the erosive velocity of runoff but also prevent the gullies and stores water for optimum utilization. These structures are suggested on the first/second-order streams in moderately sloping pediment areas and they should be constructed nearby the settlement and agriculture area. In Kumakhal village, three numbers of check dams are proposed for sustainable water resources management.

6.4.6.5   Nala Bund

Nala bund is another important structure for controlling or checking the water flowing in a nala. Nala bunds are suggested on first/second-order streams or at the confluence of third-order streams. Three numbers of nala bunds are proposed in Kumakhal village for sustainable water resources management.

6.4.6.6   Water Harvesting Structures (WHSs)

WHSs are the major structure in water management solutions. WHS is constructed across the stream which harvests surface runoff during the monsoon rains. These structures are suggested on higher-order drainage and in moderately sloping areas. In Kumakhal village, two numbers of WHS are proposed in the water resources management plan. The details of suggested water management action items are listed in Table 6.10.

Table 6.10 Water management sites in Kumakhal village

S. No	Action item	Plot number
1	Check dam	105
2	Check dam	313
3	Check dam	689
4	Nala bund	319
5	Nala bund	471
6	Nala bund	135
7	WHS	3
8	WHS	225

6.5 Conclusion

Management of natural resources at the microlevel through a micro watershed is not a new approach, but conventional methods of planning were adopted in these approaches. Under the present research study, both high-resolution satellite image and digital cadastral image have been utilized for developing the resource inventory in the GIS environment. The inventory includes different thematic layers, land and water resource management action plans, and the geospatial database. Nowadays, the geospatial technology has been used widely for microlevel or village level planning. Therefore, generating plot level resource inventory will be a big challenge for the field experts. In the present study, Kumakhal village is taken as the case study or model village where the micro watershed approach has been used for the development of a geospatial database and land and water resource management action plan. Such action plans will be very helpful for the experts and the decision-makers in monitoring and managing the natural resources of a village. Apart from that, there should be awareness creation, appraisal, orientation, and training program among local people (villagers) and the official people about the usage of thematic maps. Thus, a continuous effort is required to educate the local people and make people's participation essential for the success of microlevel planning through the micro watershed approach.

Acknowledgements The author gratefully acknowledge Survey and Map Publication, Cuttack, Odisha for the cadastral dataset. The author also acknowledge the "Google Earth" platform for high-resolution satellite images. I also acknowledge the Odisha Sampad Web-GIS portal developed by Odisha Space Applications Center (http://www.odishasampad.orsac.gov.in) for providing other thematic information.

References

Chowdary VM, Saikat P et al (2001) Remote sensing and GIS approach for watershed monitoring and evaluation: a case study in Orissa State, India. Paper presented at the 22nd Asian conference on remote sensing, 5–9 Nov, Singapore

Foresman TW (1998) GIS early years and the threads of evolution. In: The history of geographic information systems: perspectives from the pioneers. Prentice Hall, pp 3–18

Gebre T, Kibru T, Tesfaye S, Taye G (2015) Analysis of watershed attributes for eater resources management using GIS: the case of Chelekot micro-watershed, Tigray, Ethiopia. J Geogr Inf Syst 7:177–190. https://doi.org/10.4236/jgis.2015.72015

Ghosh A, Ghosh PK, Datta D (2011) Development of spatial database for sustainable micro-level planning of Chandanpur Mouza, Purilia, West Bengal. Ind J Spat Sci II(2)

Melesse AM, Weng QS, Thenkabail P, Senay GB (2007) Remote sensing sensors and applications in environmental resources mapping and modelling. Sensors (basel) 7(12):3209–3241

Mishra P, Panda GK (2015) Geo informatics based sustainable cadastral level watershed planning—a case study of Kundeimal micro-watershed, Bolangir District, Odisha, India. Int J Sci Res Publ 5(7):1–14

Mishra VN, Rai PK, Kumar P, Prashad R (2016) Evaluation of land use/land covers classification accuracy using multi-temporal remote sensing images. Forum Geogr J 15(1):45–53

Mishra V, Rai PK (2016) A remote sensing aided multi-layer perceptron-Marcove chain analysis for land use and land cover change prediction in Patna district (Bihar), India. Arab J Geosci 9(1):1–18. https://doi.org/10.1007/s12517-015-2138-3

Mishra VN, Prashad R, Rai PK, Vishwakarma AK, Arora A (2018) Evaluation of textural features in improving land use/land cover classification accuracy of heterogeneous landscape using multi-sensor remote sensing data. Earth Sci Inform 2019 12(1):71–86. https://doi.org/10.1007/s12145-018-0369-z

NRSC (2007) National land use land cover mapping using multi-temporal satellite data manual. National Remote Sensing Centre, Department of Space, Government of India, Hyderabad

Page (2016) Integrated planning and sustainable development: challenges and opportunities. Partnership for Action on Green Economy (PAGE), UNDP

Prajakta A, Adinarayana J, Sunil G, Suryakant S (2014) Information system for integrated watershed management using remote sensing and GIS. 17–34

Rai PK, Kumra VK (2011) Role of geoinformatics in urban planning. J Sci Res 55:11–24

Rajora R (1998) Integrated watershed management. Rawat Publication, Jaipur

RamamohanaRao P, Suneetha P (2015) Micro level planning and rural development through remote sensing & GIS. LAP Lambert Academic Publishing, Page, p 200

Sabins FF (1997) Remote sensing-principles and interpretation, 3rd edn. W.H. Freeman Publisher, New York

Sangameswaran P (2008) Community formation, 'ideal' villages and watershed development in western India. J Dev Stud 44(3):384–408

Wang G, Mang S, Cai H, Liu S, Zhang Z, Wang L, Innes JL (2016) Integrated watershed management: evolution, development and emerging trends. J For Res 27(5):967–994

Chapter 7
Ecohydrological Perspective for Environmental Degradation of Lakes and Wetlands in Delhi

Anindita Sarkar Chaudhuri, Nischal Gaur, Pragya Rana, Pallavi, and Pradipika Verma

Abstract The lakes are important ecological units in urban ecosystem which preserve local climate, groundwater, and biodiversity. The unplanned continuous population growth in urban area causing severe destruction to the urban ecosystem across the world. The lake in urban areas provides many functional advantages, they play a vital role in flood management, and are related to the underground water quantity and quality, they preserve the biodiversity and habitat of surrounding area, and they are huge areas of urban heat sinks within incessant built-up area. Delhi the capital of India is around 97% urbanised and home for around 26 million people, which is situated over the Yamuna watershed. To make this city sustainably resilient it is important to study about the hydrological sustainability of the city, which is deteriorating with every passing year. Encroachment, pollution, rapid urban growth, dispersal of solid waste, pumping for drinking water is some of the anthropogenic activities which intensifies the reduction of these hydrological units. Between 1970–2009 around 108 lakes were lost, in between 2009–2013 around 230 lakes have been lost and cannot be revived. Those which still exists are in terrible condition, all the lakes are dying silently which results in serious depletion of groundwater and impacting the ecology. Taking into the consideration of all above facts the present study is an attempt to create an inventory for current status of the lakes and wetlands using remote sensing and GIS techniques and their ecohydrological consequences on the surrounding environment. For the study Six lakes have been identified from different part of Delhi, which are on verge to devastating if not provided with immediate attention. Time series satellite images from LISS-III and Survey of India maps of 1:50,000 were used for the study.

Keywords Urban Lakes · Delhi · Groundwater · Remote sensing · GIS

A. S. Chaudhuri (✉) · N. Gaur · P. Rana · Pallavi
Indraprastha College for Women Delhi University, New Delhi, India

P. Verma
Amity Institute of Geoinformatics and Remote Sensing, Amity University, Sector-125, Noida, India

7.1 Introduction

Over the surface of globe, the lake ecosystems cover only approximately 5 million km^2 (Verpoorter et al. 2014). Though, due to widespread human exploitation of environment, their quality has been drastically deteriorated. According to 2018 Living Planet Index since 1970 about 83% of freshwater genus have disappeared. Particularly, the freshwater fish trawl rates were extremely high which led to extinction of many fish species. so there is urgent need to look for these surface water bodies security and biodiversity. They play a critical role in global biogeochemical cycles and to maintain sustainability of human society. Based on a lake inventory survey by Verpooter et al. (2014) in his study imperilled lakes he said approximately 117 million lakes are on Earth, which area larger than 0.002 km^2, these lakes are main source of water resource for industrial applications, irrigation, hydroelectricity generation and fisheries. Because of increasing anthropogenic uses from these freshwater bodies, impacts on lake ecosystems are increasingly reported. It is projected estimate that, by 2030, more than 60% of global human race, and the resources it consumes, will be clustered in a metropolitan setting (United Nation Report 1996). The impact that these large cosmopolitan areas will ascertain on the environment has given rise to a new branch of ecological study is the urban ecology. Urban ecosystems are characterised by highly disintegrated, heterogeneous surroundings dominated by built up environment and man-made structures like buildings, roads, and pavement, and often deficient in dense vegetation cover. The remaining vegetation in mostly part of plantation for beautification of landscape. The underground basement infrastructure and other constructions deliberately reduces the ecosystem services in urban areas (Jokimäki 1999). The fiscal aspect of human derived changes in land use and land cover has contributed to serious degradation of the ecosystem. Climate changes observed in recent years has adverse effect on this infrastructural development, built-up areas shows UHI effects as a result the city surroundings experienced higher evapotranspiration and less wind speed, additionally deepen, and accelerate this process. Increased rate of urbanisation and unplanned development leads to the destruction of water bodies like lakes and ponds which are an essential part of the city landscape and ecosystem. They absorb flood water, store water for dry months, replenish groundwater, enhance the water quality, and support biodiverse habitats. Based on land use data of India water bodies across India are shrinking as they are filled up and put to other use by individuals and governments alike. There are many examples in, Assam, Bihar, Telangana, Uttar Pradesh, Kerala, Gujarat, and Karnataka. In Lucknow capital of Uttar Pradesh's losing its ponds and lakes due to encroachment for urban land use (Verma et al. 2020; Chaudhuri et al. 2018). Such encroachments are taking place despite of strict laws and court orders to preserve water bodies to avoid scarcity of water in the future.

From the hydrological point of view to quench the thirst of every increasing population is the matter of concern. In the last 100 years world population has tripled whereas demand for water consumption went up by 6 times and UN statistics show that every 20 years demand for water is doubling, on the other hand we are losing water bodies due to this expansion of human society. Water become one of the reasons for international tensions over trans-boundary rivers and aquifers, inter-provincial disputes, urban rural conflict, and fierce inter-sectoral competition.

According to IPCC the factors climate change and global warming are just the new entrants among other factors may enhance the temporal and spatial variation in water resource availability. The major reasons remain the population growth and change in land uses, all forecasts' studies by various organisation point towards increasing water stress with exponential demand, especially urban areas, putting pressure on unevenly distributed, limited variable resources.

Urbanisation in India is witnessing an exponential growth with more than 300 million people already residing in cities and towns. In the coming two decades another 300 million people will get add up to the present urban population. If the trend in urbanisation is not managed properly, Indian cities will be under severe constrain for providing basic services for sustainable living. Water is among one of the basic resources to sustain life on the earth. Lakes and ponds are one of the landscape features that significantly contribute to increase the quality of life in urban centres, providing recreational and economic activities, and even contributing to mitigate the urban climate are major source of drinking water (Martinez-Arroyo and Jáuregui 2000; Kumar et al. 2012; McFeeter 2013).

The lake ecosystems are presently endangered by the anthropogenic disturbances and have been heavily degrading due to pollution, encroachment, and eutrophication. Since last two decades due to intense migration from rural to urban, density of urban areas has increased triple folded, due to the lack of waste disposal urban water bodies are suffering because of pollution and are mostly used for disposing untreated local sewage and solid waste, and in many cases the water bodies have been ultimately turned into landfills. Encroachment is the next major threat to water bodies particularly in urban areas. There is huge land scarcity in urban areas for the reason these urban water bodies are no more acknowledged for their ecosystem services but as real estate. These dyeing lakes serves as economically potential zones for residential development in urban areas, there are many cases in recent past where the lakes have been taken over to turn into residential; complexes Ousteri lake in Puducherry (Dhanam et al. 2016; Irfan et al. 2018), Charkoplake in Maharashtra (Bhateria and Jain 2016), Deeporbeel in Guwahati (Envis report 2016, Sur et al. 2019; Bera et al. 2008) Pallikaranai marshland in Bangalore (Bhaskar et al. 2017; Visnu and Purosotanam 2018) are some well-known examples of lake encroachment and pollution. Lakes are closed ecosystems and are very fragile in terms of chemical composition, little change in water quality effects its biodiversity. Hence, whatever enters in the lake as surface runoff in the lakes become a permanent part of the system as only a part of that can be removed depending on the water exchange system.

In 2008, according to Delhi Revenue Department there were 640 water bodies were there in Delhi under the jurisdiction of various government agencies. According to the same report in 1970, the total number of water bodies was 807, which reduced to 640 by 2008, i.e., an absolute decline of 167 water bodies. It is stated there that around 21% of the total water bodies are lost and the area under water bodies dropped from 14.41 to 8.51 km^2, a total loss of around 5.9 km^2. This amount contributes to a loss of 41% of the area under water bodies. According to the survey done by revenue department in 20 villages around Delhi during 1960–1970 there were 87 water bodies in total. In 2009, they had reduced to 54. Out of these 54 water bodies, 27 are wet and 27 are dry. During the time period between 1970 and 2009, 33 water bodies ceased to exist due to various reasons. This scenario states how every passing year we are losing Delhi water bodies due to extreme urbanisation. In 2010–2011 survey by MoEF in 2010–2011 identifies 0.2556 km^2 of wetland in Delhi which comprises of 0.86% of the total geographical area there is total loss of 67 water bodies in between this time period. MoEF identified around 11 natural lakes and around 352 manmade ponds and tanks in total of 573 lakes or wetlands in the periphery of Delhi the major portion of the wetland of Delhi consist of river, ponds, and waterlogged areas. INTACH's identified total 44 lakes and 355 village ponds in Delhi as recharge water points and as most of the water bodies constructed by historic Delhi rulers are defunct. Based on change detection of these lakes it has been recorded that 21 out of 44 lakes have been either been encroached or were permanently dry. The water bodies which are located as parts of protected areas have survived but their water quality is generally poor (Fig. 7.1).

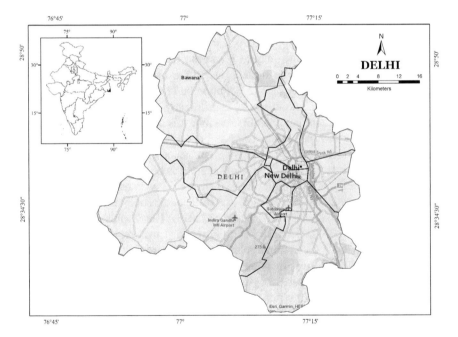

Fig. 7.1 location of study area

7.2   The City of Delhi-the Water Resource and its Challenges

The surface area of Delhi is approximately 1483 km^2, according to 2011 census, Delhi's population was 16.78 million and is projected to attain 26 million by 2030 (Bhagat 2011). Apart from the rivers Ganga and Yamuna, the other sources of surface water serving the needs of the city, are canals i.e., Agra, Western Yamuna, and Hindon—fed by Bhakra-Nangal Dam, which is the third largest reservoir in the country. Delhi is highly dependent on groundwater sources for meeting its water requirements (Rumi 2020). Figure 7.3 shows the distribution of raw water resource which is used by Delhi to meet its demand. The Himalayan rivers Yamuna and Ganga supply 389MGD and 253MGD respectively, Bhakra reservoir provides around 221 MGD and ground water supplies 90MGD yearly. The net annual water availability (the annual recharge) is 0.33 billion cubic meters (BCM), whereas the annual groundwater extraction is 0.39 BCM as reported by Central Ground Water Board (CGWB), and the depth to the water table has been increasing rapidly. The city water supply is mostly depended on external water resources. Rapid urbanisation and population increase led to plundering its aquifers, fire-fighting summer crises, showing indifference at the first sign of monsoon showers and generally engaged in service related and billing issues (Rumi Aijaz). Most of the Delhi water supplies from Yamuna River, ganga river Bhakra storage and ground water, where around 91% of the water is received from surface water bodies (Figs. 7.2 and 7.3).

As per existing data, the demand–supply gap for raw water in Delhi is 323 mgd. Delhi does not obtain enough raw surface water from its neighbouring states, and during the peak summers, these states often restrict supply, as these states have their own constraints to meet. The supply of raw water for Delhi is also impacted due to the depletion of its water table because of unrestricted and unregulated extraction (Akshit 2019). Due to the peripheral encroachment in rural areas and establishment of farmhouses, industries and residential apartments are generating stress on groundwater. The extraction of sub surface water in farmhouses and at other places by private

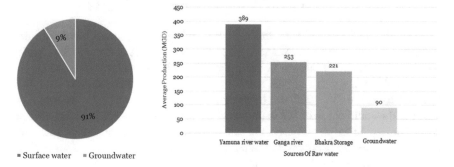

Fig. 7.2   Sources of raw water, 2020. *Source Planning Department, Economic Survey of Delhi 2020–20*

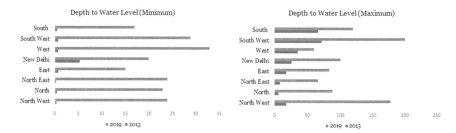

Fig. 7.3 Changing depth to water level in Delhi (Minimum and Maximum). *Source* CGWB, Water Yearbook, 2014, 2020

companies is causing groundwater to decline by about 20–30 m in several parts of Delhi (Delhi jal board). These encroachments at inappropriate places obstruct the natural flow of rainwater to the natural depressions on the land surface and thus prevent groundwater from being adequately recharged.

7.3 Data Used and Methodology

For the purpose of the study LISS III has been used for calculating NDVI, NDWI and NDBI around the lakes within buffer area of 5 km (Fig. 7.4 and Table 7.1).

Normalised Difference Vegetation Index (NDVI) is a dimensionless index which is calculated by subtracting red wavelength from NIR wavelength (Eq. 7.1) the resulted values showed in positive 1 to negative 1 (Weier and Herring 2000). Where the values above 0.66 to 1 show very dense vegetation and of healthy kind values between 0.33–0.66 shows moderate health of the vegetation, from 0 to 0.33 shows unhealthy vegetation and below zero and negative show no vegetation (NASA). Equation for NDVI has been given below.

$$NDVI = \frac{NIR(0.77 - 0.86) - RED(0.62 - 0.68)}{NIR(0.77 - 0.86) + RED(0.62 - 0.68)} \tag{7.1}$$

Normalised Difference Water Index (NDWI) this is used to demarcate the surface water bodies and to monitor changes related to water content in water bodies. Water bodies strongly absorb the wavelength of visible to infrared electromagnetic spectrum. NDWI uses green and NIR bands to highlight the surface water bodies. It is sensitive to built-up area which at times overestimates the water bodies so for urban area the threshold value of 0.3 is proposed by McFeeters (1996) to detect surface water bodies and for clean waters the values beyond 0.7 is the threshold. This indicator also shows the quality of the water bodies. As the polluted water shows lower values than clean water. Equation for NDWI has been given below.

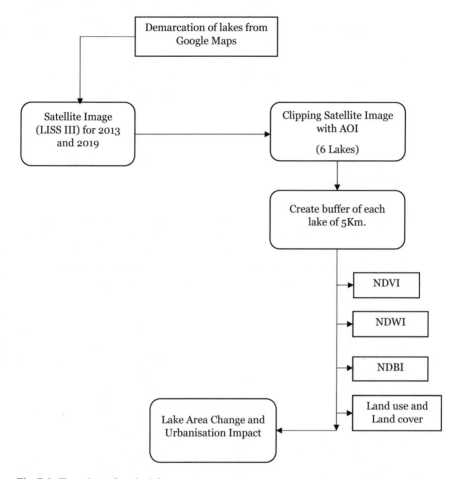

Fig. 7.4 Flow chart of methodology

Table 7.1 Details of satellite images used

Lakes	Bhuvan grid no
Bhalsawa	H43 × 01, H43 × 02
Hauz Khas	H43 × 03
Naini	H43 × 02
Najafgarh	H43W14, H43W15, H43 × 02, H43 × 03
Old fort	H43 × 02

$$NDWI = \frac{GREEN(0.77 - 0.86) - NIR(1.55 - 1.70)}{GREEN(0.77 - 0.86) + NIR(1.55 - 1.70)} \quad (7.2)$$

Normalised Difference built-up Index (NDBI) is an index which is used for monitoring the spatial distribution and growth of urban built-up. Values above zero shows built up area, though higher values shows dense built up area. Equation for NDBI has been given below.

$$NDBI = \frac{NIR(1.55 - 1.70) - GREEN(0.77 - 0.86)}{NIR(1.55 - 1.70) + GREEN(0.77 - 0.86)} \quad (7.3)$$

7.4 Urban Lake-Wetland Ecosystem and Reasons for Its Degradation

The lake-wetland ecosystem is hotspots region of urban areas which are constantly degrading with increasing urbanisation. They are the important hydrological units in urban hydrological cycle which maintains the water balance of the region. These surface water bodies serve major role in providing environmental, social, and economic services (Richard et al. 2011; Bassi et al. 2014). They are the source for drinking water, also, they are great receivers of storm water run-offs from urban land use. With persistent growth in cities infrastructure and increasing land use change is continuously modifying the urban ecosystem and impacting the freshwater ecosystem (Gavrilidis et al. 2019). These urban wetlands somehow represent the reflection of the underlying watershed, these watersheds are defined by topographical environment, in which the high areas drain into the low areas. The deteriorating ground water levels and contamination through urban waste disposal are the contemporary major issues related to the wetlands in urban areas. Although lakes deliver essential possessions and will have critical importance as hydrological reservoirs for human civilizations, the major side effects of urbanisation are the reduction in quantity of the groundwater to sustain these water bodies and the quality due to the discharge of affluents in water bodies. The major problems to these water bodies are tabulated in the Table 7.2.

Table 7.2 Factors effecting urban lake-wetland ecosystem

Factors	Details
Temperature	Effects of increasing impact of UHI on the surface water bodies, due to the rising temperature of the concreted surface the daytime water temperature also rises and effects the biotic and abiotic components
Nutrient enrichment	Eutrophication, arsenic contamination, fluoride contamination
Modification of hydrology	Building of hydrological structures, canal construction, installation of irrigation pumps
Climate change	Erratic rainfalls, local warming, cloud bursts
Land use change	Conversion of land to built-up areas, urbanisation, shrinking vegetation

Lakes and wetlands are the region which exists within a watershed, where its shape, size, soil types, land use, topography, and geology, all influence the quantity of water, its temporal distribution and the pollutant load associated with inflows to a lake. Majorly, these inland water bodies are deteriorating due to the disposal of solid waste, eutrophication, run off which includes industrial effluent load, soap detergents, domestic wastes, and sewage into lake emitting nutrients are also frequent issues of eutrophication of lakes in urban areas. For instance, Sankey Lake situated Bengaluru, has become a dormant water body with high-level eutrophication due to sewage influx from various points, storm water choked drains and leakage of sewage pipes. The dominant and unceasing conversion of natural vegetated areas to built-up, industrial, and developed land use brings about an expansion of impervious surface, which serves to modify the hydrological cycle through ever-increasing surface water runoff while reducing groundwater permeation the major effects of urbanisation are discussed in Table 7.3. The sources of pollution could be categorised into point sources and non-point sources. For urban the point source pollution for the lakes and

Table 7.3 Urbanization effects on lake-wetland ecosystem

Hydrology/Geomorphology	Ecology (Flora, Fauna)
Hrdrology	Vegetation
• Decreased surface storage of storm water results in increased surface runoff	• Large number of exotic species present; large and continuous sources of re-invasion
• Increased storm water discharge relative to base flow discharge results in increased	• Restricted pool of pollinators and fruit dispersers
• Erosive force within stream channels, which results in increased sediment input to recipient waters	• Chemical changes and physical impediments to growth associated with the presence of trash
• Changes occur in water quality (increased turbidity, increased nutrients, metals, organic pollutants, decreased O2 etc.)	• Small remnant patches of habitat not connected to other natural vegetation
• Decreased groundwater recharge results in decreased groundwater flow, which reduces base flow and may eliminate dry season flow	• Human enhanced dispersal of some species
	• Trampling along wetland edges and periodically unflooded areas
• Increased floodwater frequency and magnitude result in, or scour of wetland surface, physical disturbance of vegetation	Fauna
• Increase in range of flow rates (low flows are diminished high flows are augmented) may deprive wetlands of water during dry weather	• Species with small home ranges, high reproductive rates, high dispersal rates favoured 'edge' species favoured over forest-interior species
Geomorphology	• Absence of upland habitat adjacent to wetlands
• Decreased sinuosity of wetland / upland edge reduces amount of ecotones habitat	• Absence of wetland/upland ecotones
• Decreased sinuosity of stream and river channels results in increased velocity of stream water discharge to receiving wetlands	• Human presence disruptive of normal behaviour
• Alterations in shape of slopes (e.g., convexity) affects water gathering or water disseminating properties	

Source Ehrenfeld 2000; Indian National Trust for Art and Cultural Heritage 2010

wetlands is the disposal of solid waste. The non-point factors may be numerous like leakages of industrial pipelines, open sanitation, motor vehicle leakages. Associated with increased surface water runoff is increasing amounts of phosphorus, nitrogen, and sediments get delivered to these urban lakes. These pollutants stimulate excessive algal and weed growth within the receiving lake. For instance, Bhalswa lake located at northwest Delhi district has been strongly affected by the by the landfill site standing next to it, due to the underground seepage of toxic chemicals from solid waste disposal the water quality deteriorated and the release of toxic gases reduced the lake's ability to satisfy its ecological, recreational, and aesthetic functions.

7.5 Ecohydrology Approach for Urban Lake-Wetland Management

Ecohydrology has been developed by UNESCO under the international hydrological programme. This term has been derived from conjunction of two words ecology and hydrology. It is interdisciplinary field which studies the interaction and interdependence of ecological system with hydrological system of the region. These interaction takes place within the water bodies like lakes wetlands, rivers with surrounding terrestrial ecosystem, these two components seek feedback which influences the water dynamics and quality (Zalewski 2000; Breshears et al. 2005).It is more oriented to study anthropogenic impacts on hydrology with altered ecosystem and provide understanding to meet increasingly tight challenges facing sustainable water resources management (Nuttle 2002; Jackson et al. 2009; Singh and Bhatnagar 2012).

The concept of ecohydrology is based on three basic assumptions that first, the hydrology of a region is outcome of the natural flora and fauna, second, the flora and fauna can be used as tool to regulate the hydrological characteristics and third these two can be collaborated to develop hydrotechnical infrastructure for sustainable water management. Using this tool, the regional hydrological cycle can be quantified to study the bio-geo chemical imbalance and water balance of a region.

Based on above mentioned assumptions the following are the basic principle of ecohydrology approach to reduce ecosystem degradation using integrated terrestrial and aquatic processes across the globe. The framework defines the boundary of the hydrological cycle for basin or a watershed. It includes the physical characteristics of the basin and its relation to biotic life. For instance, the changes in land use in watershed, changes in surface temperature, vegetation coverage etc. all impacts the hydrological cycle. Target involves the ecosystem processes integrated with river basin characteristics to enhance the basins carrying capacity and ecosystem services. The target is to build more resilient and resistance ecosystem with environmental dynamics. The last principle deals with the methodology to regulate the ecosystem at basin level the new system approach, is Integrated Water Basin Management. This method integrates the hydrological framework and ecological targets to improve

Fig. 7.5 Principles of
ecohydrology

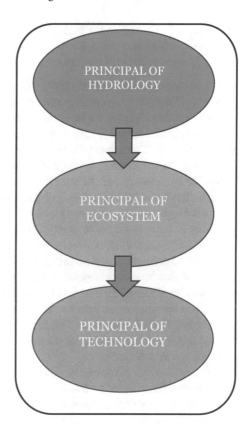

water quality and ecosystem services, using engineering methods such as levees, biomanipulation, reforestation, and other management strategies (Fig. 7.5).

The ecohydrological approach to water resources management uses the functions of various ecosystems available in a catchment to sustain natural flow patterns, year-round water accessibility, flood protection and quality (Hunt and Wilcox 2003). Ecosystems give resiliency to a watershed from extremes of flood and drought, thereby buffering against insecurity linked with climate change. Hence the understanding of the ecosystem with hydrology links in a basin area provide application of ecosystem services to gives sustainability to water resources management (McClain et al. 2012; Ahmed and Ismail 2017).

7.6   Result and Discussion: Environmental Degradation of Lakes and Wetlands

The loss of wetlands in metropolitan areas is a clear indication that wetlands have been occupied to meet the needs of greater land for urban development. With the majority of lakes and water bodies either drying up, or being contaminated, the issue is signalling towards a greater threat that humans shall not be able to handle (Chaudhuri et al. 2017). The major threats to wetlands in NCT Delhi are:-

- Unplanned Development of Urban Villages: The urban expansion in NCT Delhi has swallowed the surrounding rural areas. The land under agriculture is captured for urban development whereas the residential areas which were marked with red lines that remain rural, are overcrowded.
- Urbanization: The rapidly increasing urbanization has resulted in encroachment of rural areas including water bodies and has converted them into urban land. Therefore, a major part of the area, which was previously allocated as rural or agricultural has lost its original character, whereas the existing land use pattern of the city has also undergone sharp changes.
- Urban Expansion on Ecologically Sensitive Areas: Policy makers of Delhi-NCR have suggested that natural water bodies and ecologically sensitive areas of Delhi should be conserved, and no urban development should take place in these areas, but new developments have taken place significantly in these areas.
- Land Use Change: Urban development has resulted in significant changes in the land use pattern of the city. Transformation of the physical form of the city including the redesigning of building pattern, road pattern, drains and sewerage lines, open space, waste land and forest area, has led to the deteriorating condition of the water bodies.
- Climate Change: Changes in climatic conditions of an area largely affect the nature and functions of wetlands of that region. Delhi-NCR has an extreme climate with high temperatures, uncertain precipitation including flood and drought and frequent dust storms in summers.

Besides these, there are a number of other factors that contribute to the degradation of wetlands. The following are the few instances lake wetland environmental degradation.

7.7   Najafgarh Jheel

A mighty lake in the southwest of Delhi, in Najafgarh district from which it takes its name, it occupies around 300km^2 area. It was connected to the river Yamuna by a natural shallow 'nala' or drain called the Najafgarh nullah. However, after the 1960s the Flood Control Department of Delhi kept widening the Najafgarh drain in the pretext of saving Delhi from floods and eventually quickly drained the once huge and ecologically rich Najafgarh lake completely. Within the 5 km periphery of the lake the dense and sparse vegetation has reduced 43 and 2% respective also the area is showing increase in built up area and the NDWI values have also reduced in last 5 years from 0.41 to 0.12 which depict the deteriorating water condition. The main sources of pollution of Najafgarh jheel are domestic and municipal waste, agro-chemicals, industrial effluents and solid waste disposal. High levels of variation in the analysed parameters such as dissolved oxygen, biological oxygen demand, Phosphate, TDS, chloride and conductivity due to heavy discharge of industrial and domestic sewage were recorded during analysis which was attributed to human activity (Bhat 2013) and discharge of large amounts of waste into the water body. The water quality of Najafgarh jheel has deteriorated with respect to these water qualities determining parameters (Table 7.4).

7.8 Hauz Khas Lake

Hauz Khas is a fourteenth century water body that was dug up to serve as a tank during Alauddin Khilji's reign, located in Hauz Khas region in Southern Delhi ranges. The lake had lost 80% of its catchment area ever since1936. The hauzkhas lake buffer area shows increase in dense vegetation cover by 35% where on the other hand the moderate and sparse vegetation has reduced by 32 and 2% respectively. The lake area is always located around well-developed urban area, though within the 5 km buffer region of this lake between 2013–2019 there addition of urban area around 4%.

7.9 Bhalswa Lake

Bhalswa Horseshoe Lake, or BhalswaJheel, is a lake in northwest Delhi. It was originally shaped like a horseshoe. However, half of it was used as a landfill area. Now a low-income housing colony, an extension of the nearby town of Bhalswa Jahangir Puri has been built on it, destroying the once excellent wetland ecosystem and wildlife habitat of the region which once was a home to local and migratory wildlife species, especially water birds, including waterfowl, storks and cranes. Drainage water of its near colonies and dung of animals gets mixed in its water. Huge amount of spiritual waste is added to the lake water. Its water is getting very polluted (poisonous) and water creatures are dying due to poisonous water from the lake and a dung layer is spread under the water and comes out after some time and starts floating. Around the Bhalswa lake the dense and sparse vegetation is reduced round 4 and 3% respectively, and there is increase in built up area by 10% during last five years (Figs. 7.6, 7.7 and 7.8).

7.10 Naini Lake

One of the biggest lakes of the city, in Model Town, Naini Lake is located in the middle of dense neighbourhoods in north Delhi. The lake has been deteriorating over time due to dereliction of the authorities and lack of maintenance. Naini Lake has also been at the centre of a scuffle between different political parties as well

Table 7.4 Areal changes of NDVI, NDWI and NDBI in percentage

Lakes	Parameters		Area in % of 5 km buffer		
			2013.00	2019.00	Change %
Bhalsawa	NDVI	Dense	38.68	33.70	−4.98
		Moderate	55.14	63.29	8.15
		Sparse	2.93	0.00	−2.93
	NDWI		0.23	0.16	
	NDBI	0.3–0.5	13.89	23.68	9.79
Hauz Khas	NDVI	Dense	10.65	45.74	35.09
		Moderate	71.17	38.36	−32.81
		Sparse	2.28	0.00	−2.28
	NDWI		0.38	0.05	
	NDBI		18.56	22.57	4.01
Naini	NDVI	Dense	26.80	34.54	7.74
		Moderate	51.07	47.56	−3.51
		Sparse	4.23	0.00	−4.23
	NDWI		0.23	0.12	
	NDBI		6.90	10.44	3.54
Najafgarh	NDVI	Dense	61.22	17.38	−43.84
		Moderate	53.56	99.74	46.18
		Sparse	2.34	0.00	−2.34
	NDWI		0.41	0.12	
	NDBI	0.4–0.5	27.34	32.08	4.74
Old fort	NDVI	Dense	17.59	55.56	37.97
		Moderate	60.01	27.40	−32.61
		Sparse	5.36	0.00	−5.36
	NDWI		0.33	0.11	
	NDBI		43.16	40.44	−2.72
Sanjayvan	NDVI	Dense	39.65	56.47	16.82
		Moderate	53.80	41.54	−12.26
		Sparse	4.56	0.00	−4.56
	NDWI		0.33	0.11	
	NDBI		6.05	2.32	−3.73

Note NDWI values are DN numbers

as two groups of residents. Revival work has been at a standoff for a long time despite several efforts by a National Green Tribunal committee and residents giving recommendations. Naini lake shows (Fig. 7.4) that there is an increase in dense vegetation around 8% since 5 years, also there is an increment in built-up area by 4%.

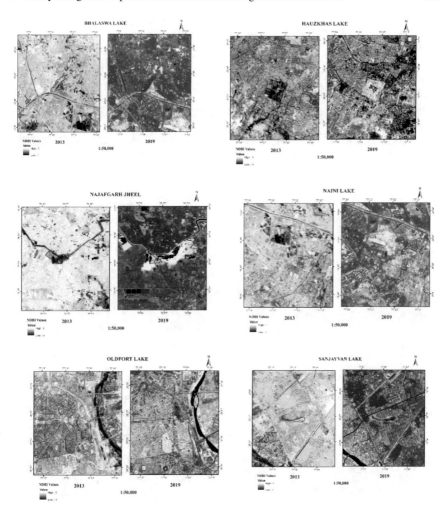

Fig. 7.6 NDBI around the lakes

7.11   Old Fort Lake

This lake dating back to the 1520 s dried up in 2017. Earlier the Yamuna used to help the lake sustain, however, since the Yamuna shifted to the east, which was the lake's primary source of water was the groundwater, after that the lake start declining. The National Buildings Construction Corporation (NBCC) had been assigned the task of restoring the lake. The buffer region comparison for the year 2013 and 2019 shows the NDWI values has reduced the region has experience increase in dense vegetation area 38% and the built up has reduced by 3% which may be because of cleaning of the roadside structures.

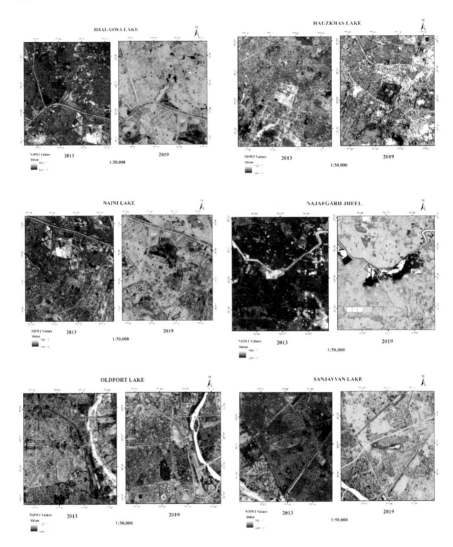

Fig. 7.7 NDWI around the lakes

7.12 Sanjay Van Lake

It is an artificial lake developed by DDA which occupies and area of 17 hectare. There
are many studies which states that there is a severe lake water pollution the lake is
dying with every passing year. The changes in the lakes buffer zone of 5 km shows
that there is increase in green tree cover by 17%. Also, due restoration work there is
reduction in built up area around 4%. The water quality conditions are worsening as
the NDWI values have reduced drastically from 0.3 to 0.1.

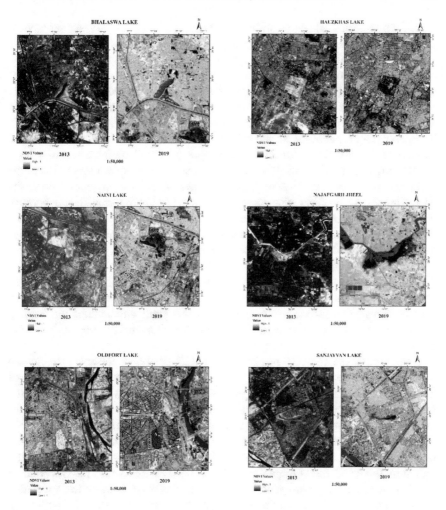

Fig. 7.8   NDVI around the lakes

7.13   Restoration of Lakes and Management

Efficient management and restoration of lakes is a major concern due to the alarming rate of deduction and degradation in number and water quality of lakes. It is quite obvious that ecologically and economically, degraded lakes are not as valuable as a healthy and efficiently working lake and represent loss. Effective measures can help in restoring these lakes and bringing back these small ecosystems back to life.

Eutrophication is one the most common and major problems observed in several lakes. High loading of nutrients, particularly phosphorus, is a major cause of eutrophication and lake restoration has been used to combat the effects of it.

In deep stratified lakes, restoration has been conducted by introducing oxygen into the hypolimnetic waters during summer with the aim to increase the redox potential and to reduce the accumulation of phosphorus released from reduced iron compounds. Oxygen is usually added through diffusors placed on the bottom of the lake at the deepest points. The ability of nitrate to oxidize the sediment surface in the absence of oxygen has been implemented in chemical lake restoration techniques to reduce the internal loading of phosphorus (Vuglinsky 1998; Yang et al. 2016). By injecting a concentrated nitrate solution into the surface sediment or the water immediately above the sediment the redox potential will be increased to a level maintaining iron in its oxidized form. This increases the phosphorus retention of the sediment and reduces the overall phosphorus availability for phytoplankton in the lake. Nitrate may, however, also contribute to increased phosphorus release because of the increased mobilization of organically bound phosphorus (Jørgensen 2009).

Manipulating the natural food webs using certain predominantly herbivorous fishes, will save the cost of other treatments for reclamation, and fishes capable of cleaning algal blooms can control eutrophication.

Water quality degrading components must be excluded from regular practices. DDT, an organochlorine chemical, must be abandoned and other natural options must be considered, such as neem, which increases yield without affecting fertility. Use of fossil fuels has to be controlled and alternate sources of energy like solar energy.

Apart from these, the USA has launched W.R.P. (Wetlands Reserve Programme) in order to restore wetlands, thereby indirectly restoring the lakes. Its objectives include Restoring hydrology and vegetation on wetland areas that have been used in the past for cropping and forage production, Protecting the functions and values of wetlands, Helping achieve the nal goal of no net loss of wetlands, Managing the restored wetlands with an active participation of the public. On similar lines, many organisations exist in India to prevent the degradation of lakes and to control water pollution. For eg. Central Ganga Authority has been established by the Government of India to prevent the pollution of the river Ganga that protects watershed vegetation and facilitates proper land use. The Water Act (Prevention and Control of pollution), promulgated and passed in parliament in 1974, could be an effective legal measure in the control of pollution provided concerned agencies implement it.

7.14   Conclusion

Wetlands are more important because of unparalleled value in maintaining the environmental stability; biodiversity, groundwater replenishment, water purification, flood control, wildlife protection and offering substantial benefits to people. The city expansion and other developments have destroyed the various types of wetlands of NCT Delhi. As a result, the area beneath the wetlands has declined dramatically over the past 40 years. A significant number of wetlands has been converted into high value urban land and some of which is threatened by rapid urbanisation. Besides urbanisation, uncontrolled development in environmentally sensitive areas, including human activities and various types of pollution, poses a major threat to natural resources.

Overall, with all the physical changes over the land surface of Delhi, all the lakes have show similar kind of development, with the proposal of redevelopment and respiration lakes the vegetation index in most of the lakes have been increased in dense category, where the moderate and sparse are reduced. All the water in the lakes is polluted and in last 5 years the condition have only worsened the NDWI values shows the reflectance values of water, the clearer the water values are high, but over Delhi the values are only as much as high of 0.4 in Huazkhas and Najafgarh after initiatives of resorting those lakes. The restoration of the Delhi water bodies are important for saving the city from severe drought, as the lakes also represents the ground water condition and are of great service provider during urban storms. The functions and services provided by natural wetlands to society far outweigh the perceived benefits of converting them into high value urban land for buildings and industries. Increased pollution, reduced groundwater levels, increased vulnerability to flooding, ecological imbalance and loss of natural habitats are just some of the consequences of the destruction of wetlands. If the loss of wetlands continues, the residents of NCT Delhi could face its catastrophic consequences, which could lead to annihilation of the urban system. As a result of the court's intervention, the government not only took a number of steps towards the protection and conservation, but also towards the revitalization of wetlands. More such efforts are needed, including both legal action and scientific planning to eliminate these major problems.

References

Ahmed S, Ismail S (2018) Water pollution and its sources, effects and management: a case study of Delhi. Int J Current Adv Res 7(2):10436–10442

Akshita N (2019) Delhi is running out of water, and it is everybody's problem. The Wire. https://thewire.in/urban/delhi-water-crisis-problem

Anindita SC, Singh P, Rai SC (2017) Assessment of impervious surface growth in urban environment through remote sensing estimates. In: Environmental earth science, Springer.https://doi.org/10.1007/s12665-017-6877-I

Anindita SC, Singh P, Rai SC (2018) Modelling LULC change dynamics and its impact on environment and water security-geospatial technology based assessment. Ecol Environ Conserv EM Int. ISSN 0971765X

Bera S, Tripathi S, Basumatary S, Gogoi R (2008) Evidence of biological degradation in sediments of Deepor Beel Ramsar Site, Assam as inferred by degraded palynomorphs and fungal remains. Current Sci 95:178–180

Bhagat RB (2011) Emerging pattern of urbanisation in India. Econ Political Wkly 34:10–12

Bhaskar A, Rao B, Vencatesan J (2017) Characterization and management concerns of water resources around Pallikaranai Marsh. South Chennai. https://doi.org/10.4018/978-1-5225-1046-8.ch007

Bhat P (2013) Water quality and pollution status of Najafgarh Jheel(Delhi) in contemporary urban scenarios. Asian Acad Res J Multidiscipl 1:326–336

Bhateria R, Jain D (2016) Water quality assessment of lake water: a review. sustain. Water Resour Manag 2:161–173. https://doi.org/10.1007/s40899-015-0014-7

Breshears D, Cobb N, Rich P, Price K, Allen C, Balice R, Romme W, Kastens J, Floyd M, Belnap J, Anderson J, Myers O, Meyer C (2005) Regional vegetation die-off in response to global-change-type drought. Proceedings of the National Academy of Sciences of the United States of America. 102:15144–15148. https://doi.org/10.1073/pnas.0505734102

Conservation of Ousteri lake in Puducherry–India environment portal India environmentportal.org.in

Delhi Jal Board, Budget 2019–20 (Delhi: Government of NCT of Delhi)

Dhanam, Sathya A, Elaya R (2016) Study of physico-chemical parameters and phytoplankton diversity of Ousteri lake in Puducherry. World Scientif News 54:153–164

Ehrenfeld JG (2000) Defining the limits of restoration: the need for realistic goals. Restor Ecol 8:2–9. https://doi.org/10.1046/j.1526-100x.2000.80002.x

Gavrilidis AA, Niță MR, Onose DA, Badiu DL, Năstase II (2019) Methodological framework for urban sprawl control through sustainable planning of urban green infrastructure. Ecol Indic 96:67–78. https://doi.org/10.1016/J.ECOLIND.2017.10.054

Irfan ZB, Ling V, Shan J (2020) (2018) Ecological health assessment of the Ousteri wetland in India through synthesizing remote sensing and inventory data. Lakes & Reserv 25:84–92. https://doi.org/10.1111/lre.12300

Jørgensen SE (2009) "Redox Potential." Science Direct

Jokimäki (1999) Occurrence of breeding bird species in urban parks: effects of park structure and broad-scale variables - Urban ecosystems, 1999

Kumar S, Hari MM, Kavita V (2012) Water pollution in India: its impact on human health: causes and remedies. Int J Appl Environ Sci 12(2):275–279

Martinez-Arroyo A, Jáuregui E (2000) On the environmental role of urban lakes in Mexico City. Urban Ecosyst 4:145–166. https://doi.org/10.1023/A:1011355110475

McFeeters SK (1996) The use of the Normalized Difference Water Index (NDWI) in the delineation of open water features. Int J Remote Sens 17(7):1425–1432. https://doi.org/10.1080/014311696 08948714

Naselli-Flores L (2008) Urban lakes: ecosystems at risk, worthy of the best care. In: Materials of the 12th World lake conference, Taal 2007. pp 1333–1337

Nitin B, Dinesh Kumar M, Sharma A, Pardha-Saradhi P (2014) Status of wetlands in India: a review of extent, ecosystem benefits, threats and management strategies. J Hydrol: Regional Stud 2:1–19. ISSN 2214–5818

Richardson E, Irvine E, Froend R, Book P, Barber S, Bonneville B (2011) Australian groundwater dependent ecosystems toolbox part 1: assessment framework. National Water Commission, Canberra

Rumi A (2020) Water supply in Delhi: five key issues. ORF Occasional Paper No. 252, June 2020, Observer Research Foundation

Singh R, Bhatnagar M (2012) Urban lakes and wetlands: opportunities and challenges in Indian cities—case study of Delhi. In: World wide workshop for young environmental scientists 12 (May)

Stuart KMF (2013) Using the normalized difference water index (NDWI) within a geographic information system to detect swimming pools for mosquito abatement: a practical approach. Remote Sens 5:3544–3561. https://doi.org/10.3390/rs5073544, ISSN 2072–429

Sur S, Mandal J (2019) Deepor Beel: a weeping and blemishing ramsar site. 9:48–58

United Nation Report (1996) United Nations Conference on Environment & Development (sustainabledevelopment/blog/2019/05/nature-decline-unprecedented-report)

Verma P, Singh P, Srivastava SK (2020) Impact of land use change dynamics on sustainability of groundwater resources using earth observation environment. Develop Sustain 22(6):5185–5198

Verpoorter C, Kutser T, Seekell DA, Tranvik LJ (2014) Imperiled lake ecosystems. Geophys Res Lett 41:6396–6402

Vishnu DPB, Purushothaman S (2018) Ecology of Pallikaranai wetland: an antidote for water scarcity in Chennai City 2018. IJRAR 5: E-ISSN 2348–1269, p-ISSN 2349–5138

Vuglinsky VS (1998) Hydrology: lakes and reservoirs. In: Hydrology and lakes. Springer Netherlands, pp 407–407. https://doi.org/10.1007/1-4020-4513-1_123

Weier J, Herring D (2000) Measuring vegetation (NDVI & EVI). NASA Earth Observatory, Washington DC

Yang G, Qi Z, Rongrong W, Xijun L, Xia J, Ling L, Xue D, Guangchun L, Jianchi C, Yongjun L (2016) Lake hydrology, water quality and ecology impacts of altered river-lake interactions: advances in research on the middle Yangtze river. Hydrol Res 47(7). https://doi.org/10.2166/nh.2016.003

Zalewski M (2000) Ecohydrology – The scientific background to use ecosystem properties as management tools toward sustainability of water resources. Ecol Eng 16:1–8. https://doi.org/10.1016/S0925-8574(00)00071-9

Chapter 8
Prioritization and Quantitative Assessment of Dhundsir Gad Using RS and GIS: Implications for Watershed Management, Planning and Conservation, Garhwal Himalaya, Uttarakhand

Ashish Rawat, M. P. S. Bisht, Y. P. Sundriyal, Pranaya Diwate, and Swapnil Bisht

Abstract The utilization of morphometric parameters in watershed management and conservation has reduced the work, time, manpower and expenditure immensely. Similarly, the quantitative assessment of Dhundsir Gad (*Gad* means Stream) watershed has been carried out to study the relationship between hydrological characteristics, lithological and structural antecedents for management, planning and conservation activity. The data for morphometric parameters is extracted from remotely sensed images and processed in geographical information system (GIS) platforms for quantitative analysis. Geological field investigation of the watershed is done to investigate inter-relationship between the hydrological characteristics, lithology and structural attributes. Morphometric characterization was measured from linear, relief and areal aspects for four subwatersheds of Dhundsir Gad. The geomorphic parameters quantified reveal the role of hydrology in association with lithology and structure in modifying the watershed and loss of natural resources. The prioritization of the subwatershed has been done after evaluating and ranking morphometric parameters. The subwatersheds have been given priority from low to high suggesting the state of urgency it is in for conservation. Regional patterns of hypsometric integral (HI) and hypsometric curves in the watershed have also been computed to understand the role of tectonics and soil erosion in shaping the relief of the watershed. The hypsometric HI and hypsometric curves also suggest the stage of development and total mass lost from the watershed.

A. Rawat (✉) · Y. P. Sundriyal · S. Bisht
Department of Geology, H N B Garhwal University, Srinagar Garhwal, Uttarakhand 246174, India

M. P. S. Bisht
Uttarakhand Space Application Center, Dehradun, Uttarakhand 248001, India

P. Diwate
Centre for Climate Change and Water Research, Suresh Gyan Vihar University, Jaipur, Rajasthan 302017, India

© The Author(s), under exclusive license to Springer Nature Singapore Pte Ltd. 2022   165
P. K. Rai et al. (eds.), *Geospatial Technology for Landscape and Environmental Management*, Advances in Geographical and Environmental Sciences,
https://doi.org/10.1007/978-981-16-7373-3_8

Keywords Watershed · Imagery · Conservation · Geographic information system · Morphometry · Prioritization · Hypsometry · Geology

8.1 Introduction

In the Himalayan terrain, the perennial and seasonally flowing rivers are in abundance and vary in size depending on source. The rivers act as an active agent in denudation and erosion process which is why it is vital to study the hydrological characteristics for planning and conservation beforehand. The hydrological characteristics of river basins in the Himalaya play a significant role in developing and modifying the morphology and topography of the basin (Rawat et al. 2021). The development of Himalayan mountain chain is a combined outcome of several processes which are either slow or rapid like glacier, river, ground water, etc. These hydrological processes primarily depend upon the behavior of the drainage, climate, lithology, structure and to a small extent on anthropogenic activity (Kumar and Chaudhary 2016). The major sources of fresh water in the Himalaya are rivers and groundwater (Rawat et al. 2021). The Dhundsir Gad (DG) a tributary of river Alaknanda along with several smaller streams fed by either rain or natural springs is the primary source of water in the watershed. The major part of population in the watershed is residing along the DG and its tributaries. The local population depends on DG for water and alluvium to sustain livelihood. The active erosion during heavy precipitation has contributed to decline of landmass and also produced a sense of worry for the population residing in the watershed.

The hydrological characteristics of watershed are the vital component for quantitative assessment of morphometric parameters. The influence of lithology, structure and geomorphic processes is expressed by hydrological characteristics of the watershed. For planning and management of a watershed using quantitative morphometric analysis, it is essential to study the behavior of lithology toward hydrological processes (Singh et al. 2014; Chaudhary and Kumar 2017, 2018). Several geomorphic aspects such as areal, linear and relief are studied in detail for DG watershed in order to derive features of the Dhanari watershed. The hypsometric assessment of the watershed is also has been done to derive the total loss of landmass. The remote sensing (RS) coupled with GIS tools have become an efficient and accurate means of understanding the influence of drainage, lithology and tectonic on a local or regional level basis for natural resource management, soil conservation, etc., especially if working on watershed (Chaudhary et al. 2021). Detailed geological and geomorphological study of the watershed has also been conducted to understand the influence of the structure on the drainage characteristics. The planning and conservation method can be applied to the DG watershed by keeping all the above things in mind for better results.

The watershed management and planning emphasize upon reducing or stopping the erosion from the watershed area. The quantitative analysis of the landscape shape is defined as morphometry (Keller and Pinter 1996; Biswas et al. 2014). The morpho-metric parameters present a simple approach to ascertain hydrological processes and characteristics of watershed (Rawat et al. 2021). The relationship of morphometric parameters with hydrology, climate, lithology, structure, vegetation and soil was first introduced by Horton (1932, 1945), to understand the origin of drainage networks. In later half of the century morphometric parameters were studied, modified and evolved by several workers like Strahler (1952, 1964), Schumm (1956), Melton (1957), Keller and Pinter (1996), etc. Though in recent times Sreedevi et al. (2005), Biswas et al. (2014), Rai et al. (2018) and Rawat et al. (2021), etc., have used RS and GIS with morphometric parameters for much faster, cost-effective and reliable results.

In recent times as the human population is increasing in explosive manner, it is accompanied by industrial demand for food and fodder; as the demand increases so does the need for more and more land for farming and irrigation, burdening our handful of land resources. The goal of the present study is to (a) provide accurate and relevant information for systematic planning and management of DG watershed; (b) prioritization of subwatersheds for planning and management; (c) role and influence of geological structure and lithology in the reduction of landmass from the watershed.

8.2   Salient Features of Study Area

The study area falls within the Tehri district of the Garhwal Lesser Himalaya, the area is covered in Survey of India (SOI) Topographical sheets No. 53 J/11, 12 and 15. The watershed is elongated in nature and is confined between 30°13′ and 30° 23′ N latitudes and 78° 44′ to 78° 49′ E longitudes (Fig. 8.1a). The DG watershed has the maximum flow length of 18.1 km and covers an area of approximately 50 km^2 (Fig. 8.1a). The altitudinal variation in watershed varies from 530 m near Kirtinagar to 2277 m above mean sea level (Fig. 8.1a). The DG watershed shows significant difference in rainfall, temperature, vegetation cover, topography, etc., when observed at the different altitudes. It is observed during our field visits that the temperature dips significantly when we start climbing toward higher elevation in the watershed. The higher elevation of the watershed receives maximum rainfall during monsoons and occasional snowfall during winters, while the lower elevations have a comparatively higher temperature and also receives lesser rainfall. The geomorphology of the DG watershed is varied and includes drainages, ridges, valleys, dissected hills, landslides, fluvial terraces and several other anthropogenic features (Fig. 8.1b).

8.3   Regional Geological Setup

The study area falls in between the Pratapnagar and Badiyargarh zone of Garhwal Himalaya. The area has significant variation geologically which has been verified

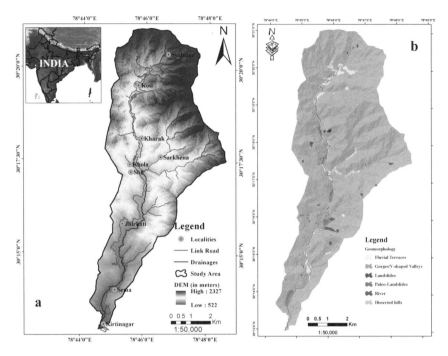

Fig. 8.1 Showing **a** location map and Digital elevation model (DEM), and **b** geomorphology of Dhundsir Gad watershed

during field visits. Geologically, the rock of the area has been placed in Garhwal Group, Chandpur Group and Dudhatoli Group (Kumar and Agrawal 1975). The Garhwal group of rocks consists mainly of arenaceous and calcareous formation having metamorphic rocks like quartzites, limestone, gneiss, metabasic, etc., and deposited alluvium as overburden. Whereas autochthonous Chandpur group in the watershed predominantly have phyllite and schist as major lithology; while the Dudhatoli Group is represented by the phyllites and quartzites forming the northern limb of the Dudhatoli syncline. The variability in the alteration of the rocks suggests that the area is tectonically modified and has evidenced different phases of deformation (Sandilya and Prasad 1982), which is confirmed by detailed field study. The rocks in the watershed have undergone three phases of deformation viz. D1 (NW–SE), D2 (NE-SW) and D3 (NNW-NNE) (Sandilya and Prasad 1982). The regional structure like North Almora Thrust (NAT) is encountered around 9–10 km using link road from Kirtinagar between 700 and 2300 m of altitudes. NAT is a south dipping low angle fault in which the quartzite is overlain by the phyllites. The thrust runs across the watershed in NW–SE direction and is evident by the presence of highly sheared, jointed and crushed rocks as fault gauge. Previous researchers have reported neotectonic activities either along the NAT or in the vicinity (Sati et al. 2007); though no neotectonic marker or feature has been evidenced during the present investigation. The lineament drawn for the DG watershed suggests control of tectonics on

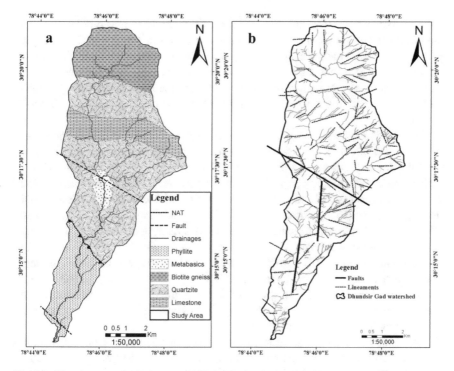

Fig. 8.2 Showing **a** geological setup (NAT is North Almora Thrust) with major lithology, and **b** lineament and drainage is drawn from satellite imagery suggesting tectonics in Dhundsir Gad watershed and their control on hydrological characteristics

drainages (Fig. 8.2b). Several other mesoscale structures like Kirtinagar fault and small lineaments have been evidenced on the basis of field study, morphological changes in drainage course and study of satellite imageries (Fig. 8.2b).

8.4 Materials and Methods

In present study, the primary data to achieve the objective has been collected via active and passive mediums. The Survey of India (SOI) topographical toposheet (53 J/11, J/12 and J/15) in 1:50,000 scales have been used to infer several geomorphic features and drainage extracted from the DEM. To generate drainages, we used freely available SRTM-DEM (Shuttle Radar Topographic Mission-Digital elevation model) https://earthexplorer.usgs.gov/ with spatial resolution of 30 m. The drainage extracted has also been verified by overlaying the extracted drainages in Google Earth and georeferenced SOI toposheet. The thematic layers for the study has been made by using the Bhuvan online platform of NRSC (National Remote Sensing Centre) http://bhuvan.nrsc.gov.in and by supervised classification of freely available

LISS-III (Linear Imaging Self Scanning sensors) dataset to generate land use and land cover (LULC) map. Base map has been prepared in ArcGIS-10.5 for detailed geological study of the watershed. The field survey of watershed is done all along the road section which runs almost parallel to the DG, the collection of structural data from the different lithological section is done with the help of requisite geological tools. To perform all the operations software like ArcGIS-10.5 and Corel Draw-12 were used.

Quantitative analysis of hydrological characters is done using morphometric parameters (Table 8.1). The geomorphic parameters used in the study are delineated from the drainages data extracted from DEM using GIS platform ArcGIS-10.5 with well-planned methodology (Fig. 8.3). The morphometric indices used are classified in (i) areal aspect, which has geomorphic parameters like drainage intensity (Di), drainage density (Dd), infiltration number (If), texture ratio (Rt), drainage texture (Dt), circularity ratio (Rc), elongation ratio (Re), stream frequency (Fs), form factor (Ff), compactness coefficient (Cc), basin shape (Bs), length of overland flow (Lg) and area (A) and perimeter (P) (Table 8.1); while (ii) linear aspect, have stream length (Lu), bifurcation ratio (Rb), stream order (Su), RHO co-efficient (RHO), stream length ratio (Lur) (Table 8.1) as geomorphic parameters; whereas, (iii) relief aspect, has parameters like ruggedness number (Rn), gradient ratio (Gr), dissection index (Di), and the relief ratio (Rr) (Table 8.1).

The removal of landmass from the watershed is calculated using hypsometric model which is based upon the area and relief of the watershed. Singh et al. (2008) suggested that the model also helps in understanding the stage of development and upliftment rate of a watershed. The hypsometric model is first introduced by Langbein (1947); where, the value of hypsometric integral (HI) is the area under the curve generated by plotting cumulative elevation against cumulative area produced from the DEM and by using elevation–relief ratio formula suggested by Pike and Wilson (1971). It is explained as a dimensionless model for comparing watersheds irrespective of their scale by Dowling et al. (1998). Hypsometric integral (HI) and hypsometric curve quantifies the erosion proneness, geological stage of watershed or watershed condition (Ritter et al. 2002). The variation in HC shape and HI value are associated with extent of uncertainty in tectonic and erosional processes (Weissel et al. 1994).

Insufficient resources have always been a concern in most of the conservation initiative so it becomes necessary to prioritize our area so that maximum results can be attained. Prioritization of the DG watershed is done to classify subwatersheds on the basis of severity of land degradation. The first step of the process includes quantitative analysis of geomorphic indices for subwatersheds. The next step involves distribution of rank for each geomorphic parameter on the basis of its influence on erosion; geomorphic parameters like Dt, Dd, Fs, Rb and Lg are associated with erodibility, suggesting low and high value are indicators of low and high erodibility, respectively (Singh et al. 2014; Nooka et al. 2005); while Er, Cc, Rc, Bs and Ff have inverse relationship, suggesting lower values are associated with higher rate of erosion and high values with lower rates, respectively (Nooka et al. 2005; Javeed et al. 2009). The final step of the method is to take the mean of all the ranks which

Table 8.1 Methodology followed to calculate morphometric parameters

S. No	Morphometric parameters	Formulas	References
Linear aspect			
1	Stream order (Su)	Based on hierarchical rank	Strahler (1952)
2	Stream number (Nu)	$Nu = N1 + N2 + N3 + ... + Nn$	Horton (1945)
3	Stream length (Lu)	$Lu = L1 + L2 + L4 + ... + Ln$	Strahler (1964)
4	Stream length ratio (Lur)	$Lur = Lu/Lu - 1$	Horton (1945)
5	Bifurcation ratio (Rb)	$Rb = Nu/Nu + 1$	Strahler (1964)
6	Mean bifurcation ratio (Rbm)	$Rbm = Rb/n$	Strahler (1964)
7	Rho coefficient (ρ)	$\rho = Lur/Rb$	Horton (1945)
Areal aspect			
8	Basin length (Lb) Kms	GIS analysis	Schumm (1956)
9	Basin area (A) Sq Kms	GIS analysis	Schumm (1956)
10	Basin perimeter (P)	GIS analysis	Schumm (1956)
11	Drainage density (Dd)	$Dd = Lu/A$	Horton (1932)
12	Stream frequency (Fs)	$Fs = Nu/A$	Horton (1932)
13	Drainage intensity (Di)	$Di = Fs/Dd$	Faniran (1968)
14	Infiltration number (If)	$If = Fs*Dd$	Faniran (1968)
15	Length of overland flow (Lg)	$Lg = A/2*Lu$	Horton (1945)
16	Constant of channel maintenance (C)	$C = 1/Dd$	Schumm (1956)
17	Form factor (Ff)	$Ff = A/Lb^2$	Horton (1932)
18	Elongation ratio (Re)	$Re = (2/Lb) *\sqrt{(A/\pi)}$	Schumm(1956)
19	Texture ratio (Rt)	$Rt = N1/P$	Schumm(1956)
20	Circularity ratio (Rc)	$Rc = 12.57*A/P^2$	Miller (1953)
21	Drainage texture (Dt)	$Dt = Nu/P$	Horton (1945)
22	Compactness coefficient (Cc)	$Cc = 0.2841*P/A^{0.5}$	Gravelius(1914)
Relief aspect			
23	Maximum height of the basin (Z)	GIS analysis	-
24	Basin relief (H)	$H = Z - z$	Strahler (1952)
25	Relief ratio (Rhl)	$Rhl = H/Lb$	Schumm(1956)
26	Gradient ratio (Rg)	$Rg = (Z - z)/Lb$	Sreedevi et al. (2005)
27	Dissection index (Dis)	$Dis = H/Ra$	Singh and Dubey (1994)

(continued)

Table 8.1 (continued)

S. No	Morphometric parameters	Formulas	References
28	Hypsometric integral (HI)	$E \approx \text{HI} = \frac{\text{Elev}_{\text{mean}} - \text{Elev}_{\text{min}}}{\text{Elev}_{\text{max}} - \text{Elev}_{\text{min}}}$	Pike and Wilson (1971)

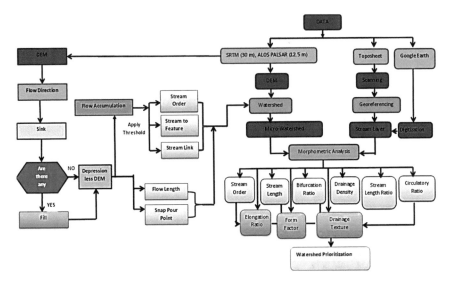

Fig. 8.3 Flowchart showing the steps involved for morphometric analysis

will give the compound priority (*Cp*) of subwatersheds; the subwatersheds with the lowest value of *Cp* are assigned the highest priority, medium priority is assigned to the next higher value, while the lowest priority is given to subwatersheds with highest *Cp* (Meshram et al. 2020).

8.5   Result and Discussion

The DG watershed is divided into four subwatersheds for geomorphic analysis. The geomorphic response in the watershed is dependent upon certain factors like geology, precipitation, external and endogenic processes going on in the watershed. The land use practices influences the hydrological characteristics as certain anthropogenic activity like plantation, construction of building, check dams, safety walls, etc., reduces erosion, while some activities like road cutting, grazing, mining, etc., escalates the erosion. Keeping it in mind a comprehensive map of land use and land cover (Fig. 8.6a) is prepared for DG watershed using satellite images. The analysis of drainage characteristics in DG watershed is calculated by using various geomorphic aspect viz. linear, areal and relief.

8.5.1 Linear Aspects

8.5.1.1 Stream Order (Su)

The hierarchical arrangement of streams according to their position in watershed for quantitative analysis is done by ordering stream. The method is important to index the size and scale of the basin. The stream arrangement done according to Strahler (1952) suggests that the DG watershed is a 5th order watershed with the main stream is fifth order stream. The arrangement of streams or stream pattern in the watershed is of dendritic type typical of the Himalayan river system (Fig. 8.2b). The dendritic arrangement of the streams and higher frequency of lower stream orders than the higher orders suggests a durable basement of the watershed with high channel flow.

8.5.1.2 Stream Number (Nu)

The Nu is defined as "sum of all stream segments in each order in a watershed." The Nu is an essential tool to assess the surface runoff (Rawat et al. 2021). The sum of Nu in each order forms an inverse geometric relationship with each order (Horton 1945) (Table 8.2). DG has 483 stream segments, out of which 354 segments are of 1st, 95 of 2nd, 26 of 3rd, 7 of 4th and 1 is of 5th order, respectively. The high percentage (73%) of first order streams suggests good infiltration and permeability in the area; it also confirms the perennial nature of stream as most of the water contributed is from springs.

8.5.1.3 Stream Length (Lu)

Stream length is total length of stream segments in an order; stream length of lower order have inverse relation with stream order where the total stream length decreases with increasing order (Table 8.2). Stream length helps in understanding surface runoff characteristics; stream segments with short lengths suggest a steep slope, while stream flowing for a longer period of time are associated with gentle topography. The Lu of the DG watershed is evaluated in GIS and 1st, 2nd, 3rd, 4th and 5th order stream have a total length of 106.5, 38.8, 22.05, 115.5 and 10.6 km, respectively, which computes to 193.45 km (Table 8.2). The Lu calculation for DG watershed supports the statement "geometrical similarity is preserved in watershed of increasing stream order" (Rai et al. 2018). In the present study after evaluation of Lu for DG watershed; it suggests that the highest order flows in a gentler topography while others are associated with steeply inclined slopes.

Table 8.2 Table showing results of linear aspects

S. No	Basin length (Km)	Stream length ratio (Lur)	RHO	Stream length (Lu) (in Km)	No. of stream of each order (Nu)					Total (Nu)	Mean bifurcation ratio (Rbm)
					1	2	3	4	5		
SW1	3.9	2.04	0.52	24.39	57	16	4	1	0	78	3.85
SW2	3.8	2.733	0.82	20.51	35	11	4	1	0	51	3.31
SW3	7.5	0.56	0.10	78.69	117	31	9	1	0	158	5.39
SW4	8.3	1.9	0.54	69.815	145	37	9	4	1	196	3.5
DG	15.25	0.57	0.13	193.45	354	95	26	7	1	483	4.23

Mean stream length (Lum) of DG watershed varies from 0.22, 0.40, 0.85, 2.21 and 10.6 for 1st, 2nd, 3rd, 4th and 5th order, respectively. The Lum of watershed suggests a direct relationship of stream length with stream order. The calculation of Lum reveals the characteristic size of the drainage elements and their contributing surfaces (Rawat et al. 2021). The alteration in Lum values suggests a variation in slope and topography of the watershed (Fig. 8.4c).

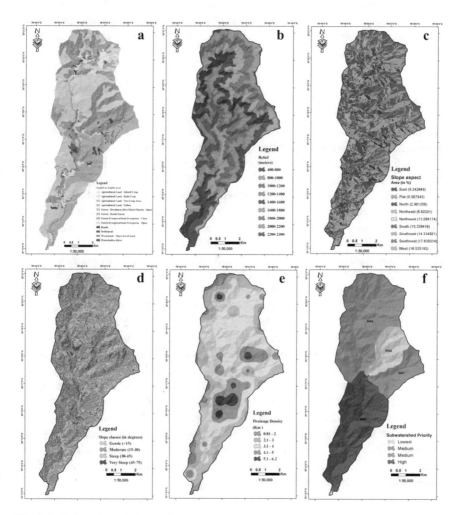

Fig. 8.4 Various thematic layers showing **a** Land use and land cover map, **b** Relief map, **c** slope aspect, **d** slope category, **e** drainage density, and **f** priority map of Dhundsir gad watershed

8.5.1.4   Stream Length Ratio (Lur)

The Lur is defined as "the ratio of the mean length of a higher stream order to the mean length of the lower stream order;" which tends to be successive throughout a basin (Horton 1945). The stream length ratio varies for four subwatersheds vary from 0.56 to 2.73 and for DG watershed is 1.80. The difference in the values of Lur in different stream order is the result of variation in the gradient and topography. This variation in Lur suggests that the watershed is moving toward mature phase of geomorphic development (Soni 2017). Lur is associated with surface flow discharge and the erosional phase of the watershed.

8.5.1.5   Bifurcation Ratio (Rb)

The ratio of stream number of higher order to the stream number of lower order is defined as bifurcation ratio (Horton 1932; Strahler 1964). The Rb helps in understanding the control of structure in drainage characteristics and also helps in justifying the influence of geological attributes in a watershed. The Rb value ranges between 1 and 10 where the low Rb suggests homogenous lithology in basement, while higher value are indicator of structural control on drainage characteristics (Strahler 1964). The mean Rb value of subwatersheds varies from 3.31 to 5.39; while the mean bifurcation ratio (Rbm) for DG watershed is 4.23 (Table 8.2). The Rb value for each order in subwatersheds ranges between 2.25 and 9; the values remain below 4 for most of the lower and higher stream orders in every subwatershed, though in case of SW3 the Rb value is 9 for highest order stream. The lower value of Rb in SW1, SW2 and SW4 suggests that lithology play a dominant role in these subwatersheds, except in case of SW3 where the higher value suggests the drainages are structurally controlled. The Rb values are also associated or related with shape of the watershed, more circular a watershed higher will be the Rb more will be structural control on the streams; while more elongated the watershed is lower will be the Rb and drainages are controlled by homogenous rocks (Soni 2017). In SW1, SW2 and SW4, the shape of the subwatersheds is elongated, whereas SW3 has more or less circular shape this variation in Rb values of different order is the result of substantial difference in frequency of successive order. The DG watershed has elongated shape though the high order show fairly large value of 7 but most of the lower order suggests a low value. The low Rb values, elongated nature and sheared rocks in the subwatersheds suggests high susceptibility to erosion.

8.5.1.6   RHO Coefficient (P/Rho)

Horton (1945) defined it "as the ratio between stream length ratio and bifurcation ratio." Rho is an indicator of storage capacity of drainages in a watershed. The Rho values of 4 subwatersheds vary from 0.10 to 0.82; whereas DG watershed has Rho value of 0.13 (Table 8.2). Subwatersheds SW1, SW2 and SW4 except SW3 have high

Rho values interpreting high-hydrological storage efficiency during floods making these SW prone to erosion.

8.5.2 Areal Aspect

8.5.2.1 Drainage Density (Dd)

Drainage density is defined as "the ratio of sum of stream length in all orders per unit watershed area." (Horton 1945; Strahler 1952) The parameter provides better quantitative expression to the change in landform and also the runoff potential (Chorley 1969). The factor influencing Dd are vegetation, climate, lithology and soil. The Dd values of subwatersheds influences permeability, surface runoff and the sediment yield indicates a direct relationship with groundwater potential and soil erosion. The Dd value for 4 subwatersheds ranges from 3.41 to 4.106 km^{-1} and 3.9 km^{-1} for DG watershed. Gregory and Walling (1973) classified Dd in low, medium and high classes on the basis of values which are $> 2\,km^{-1}$, $2\text{--}4\,km^{-1}$, $4\text{--}6\,km^{-1}$ and $< 6\,km^{-1}$, respectively (Fig. 8.4e). The values computed for the four subwatersheds and DG watershed are thus placed in moderate category which implies that the subwatersheds and watershed have gentle-steep inclining slopes with medium to sparse vegetation cover; moderate permeability and high erosion in lower order streams having high surface runoff.

8.5.2.2 Drainage Texture (Dt)

The Dt is the product of Dd and Fs which is a measure to calculate space between stream segments in a fluvialy dissected terrain influenced by factors like precipitation, vegetation, lithology, and alluvium (Smith 1950). On the basis of Dt values the watershed is classified into very coarse ($<2\,km^{-1}$), coarse ($2\text{--}4\,km^{-1}$), moderate ($4\text{--}6\,km^{-1}$), fine ($6\text{--}8\,km^{-1}$) and very fine ($>8\,km^{-1}$) (Horton 1945; Smith 1950). The Dt for 4 subwatersheds ranges between 4.6 and 9.3, while that of DG watershed is 12.07 (Table 8.3). The results suggest that SW2 has moderate, SW1 has fine, while SW3 and SW4 have very fine texture; whereas DG watershed has very fine texture. The majorities of subwatersheds have fine to very fine texture which infers peak flows of short duration and are associated with the presence of weak rocks prone to erosion. The subwatersheds with fine to very fine texture indicate high soil erosion and high susceptibility for floods.

8.5.2.3 Stream Frequency (Fs)

Horton (1945) defined Fs as "total number of stream segment in all orders per unit area in a watershed." The Fs has a direct relationship with the drainage density as

Table 8.3 Table showing results calculated for areal aspect perimeters

S. No	Basin shape (Bs)	Area (in km²)	Perimeter (in km)	Drainage intensity (Di)	Stream frequency (Fs)	Drainage texture (Dt)	Circularity ratio (Rc)	Elongation ratio (Re)	Form factor (Ff)	Compactness coefficient (Cc)	Texture ratio (Rt)	Infiltration number (If)	Drainage density (Dd)	Length of overland flow (Lg)
SW1	2.54	6	12	3.2	13	6.5	0.52	0.7	0.39	1.39	4.75	52.78	4.06	0.12
SW2	2.4	6	11	2.5	8.5	4.6	0.62	0.72	0.41	1.27	3.18	28.98	3.41	0.14
SW3	2.6	21	21	2.01	7.52	7.9	0.59	0.68	0.37	1.33	5.57	28.12	3.75	0.13
SW4	4.05	17	21	2.80	11.52	9.3	0.48	0.57	0.23	1.44	6.9	47.23	4.11	0.12
DG	4.6	50	40	2.46	9.66	12.08	0.39	0.52	0.21	1.63	8.85	36.67	3.9	0.12

Fs increases with increasing Dd. The stream frequency is influenced by lithology, structure, vegetation cover and precipitation. The Fs value for DG watershed is 9.66 km^{-2} and for the four subwatersheds it ranges between 7.52 and 11.52 km^{-2}. The high value of Fs in watershed suggests the lithology, and structure has major role in formation of streams (Table 8.3). The results indicate that the watershed is susceptible to soil erosion and flooding.

8.5.2.4   Drainage Intensity (Di)

The Di is expressed as "the ratio of the Fs and Dd" (Faniran 1968). The Di values of watershed helps in assessing the impact of denudation in lowering land surface. The Di for DG watershed is 2.46 whereas it ranges from 2.01 to 3.2 for all four subwatersheds (Table 8.3). The high Di values for the subwatersheds and watershed indicates that Dd and Fs have dominant role in lowering the land surface. It also implies that surface runoff in the watershed is fast and lag time is low which results in high denudation from the watershed. The lithology has important role in whole process, the weak and jointed metamorphic rocks in the watershed provides a favorable condition for removal of landmass from the watershed.

8.5.2.5   Form Factor (Ff)

The Ff is "the ratio of the watershed area to the square of the watershed length" (Horton 1932). Smaller the Ff value, more elongated is the shape and longer it takes for water to escape watershed; whereas, higher values indicate a circular shape and shorter exit time. In DG watershed, the value of Ff is low (0.21) which suggests elongated shape of watershed; the value for SW1, SW2 and SW3 are moderate suggesting the watershed is advancing toward elongated shape, in case of SW4, the values are low which suggests an elongated shape. The lower value of Ff in all subwatersheds suggests control of lithology and precipitation on the topography rather than structure.

8.5.2.6   Elongation Ratio (Re)

The Re is "the ratio of diameter of a circle of the same area as the basin to the maximum basin length" (Schumm 1956). The shape of a basin is categorized on the basis of Re values ranging 0.9–1.0, 0.8–0.9, 0.7–0.8, 0.5–0.7, and < 0.5 suggesting circular, oval, less elongated and more elongated shapes, respectively (Rai et al. 2018). The higher Re values indicate high infiltration and less runoff; while lower values are associated with low infiltration and high surface runoff. The four subwatersheds have Re values from 0.57 to 0.7 (Table 8.3), for DG watershed it is 0.52 (Table 8.3) which infers elongated shape of subwatersheds and watershed.

8.5.2.7 Circularity Ratio (Rc)

The circularity ratio is a dimensionless unit and is used as a quantitative method to express the external form of watershed (Miller 1953, Strahler 1964). The Rc is "the ratio of watershed area to the area of a circle having the same perimeter" (Rawat et al. 2021). The Rc of watershed is influenced by stream density, lithology, structure, slope, climate, relief and LULC. The Rc values varies from 0 to 1 where the values close to 1 suggests circular watershed, while lower values or close to 0 indicate an elongated basin. Low, medium and high values of Rc are associated with young, mature and old stage of landform development, respectively. The Rc for DG watershed is 0.39, whereas in four subwatersheds it ranges between 0.48 and 0.62 which infers elongated shape of the watershed and subwatersheds (Table 8.3). The interpretation suggests the watershed has homogenous lithology, permeable rocks and the watershed is progressing from youth stage to mature stage of development. The youth stage of watershed suggests that it is highly susceptible to denudation because of high surface runoff.

8.5.2.8 Compactness Coefficient (Cc)

The compactness coefficient is slope dependent unit which is independent of dimension. The Cc of a watershed is the ratio of watershed perimeter to the circumference of a circular area having equal area as watershed (Gravelius 1914). The value for the four subwatersheds ranges from 1.27 to 1.44 and for DG watershed is 1.63 (Table 8.3). The values of Cc for subwatersheds and watershed indicate elongated shape and minor side flow for shorter duration and the high main flow for longer duration, which makes the watershed prone to erosion.

8.5.2.9 Infiltration Number (if)

The ability of rocks to transmit water is computed by infiltration number for a watershed. It is the product of Dd and Fs having direct relation with the dissection and runoff potential. The "If" is inversely proportional to infiltration capability of watershed. Higher the value of If lower will be infiltration and higher the runoff, whereas lower value suggests high infiltration and low surface runoff. In case of DG watershed the If is moderate (37.67) while the value for 4 subwatersheds ranges from 28.12 to 52.78 (Table 8.3), which suggest moderate to high If. The results of "If" in the watershed and subwatersheds imply moderate infiltration and high risk of surface runoff.

8.5.2.10 Length of Overland Flow (Lg)

Length of overland flow is the length of the stream segment over the surface before it gets concentrated in a channel (Horton 1945). The Lg is equal to one half reciprocal of Di (Rai et al. 2018). The Lg has inverse relationship with slope and relief; higher the value of Lo gentler the slope and lower will be the relief, whereas lower value suggests a high relief and steeply inclined rock in the watershed (Yadav et al. 2014). The subwatersheds have Lg values varying from 0.12 to 0.14, whereas for DG watershed it is 0.12. The value of Lg suggests a high relief and steeply inclined slopes indicating high surface runoff and moderate to low permeability.

8.5.2.11 Constant of Channel Maintenance (C)

The parameter is expressed as "a unit surface area required to develop or sustain a channel 1 km long" (Schumm 1956). The C is "the ratio between the drainage basin area and the sum of stream length of all channels in watershed" (Schumm 1956). The inverse parameter of Dd is the constant of channel maintenance which is a significant aspect of landform analysis (Schumm 1956). The C of the DG watershed is 0.25. while for 4 subwatersheds it varies from 0.24 to 0.29. The values computed for the parameters are low suggesting low permeability, rapid discharge of water from the watershed. The results indicate that the watershed has high susceptibility for erosion.

8.5.3 Relief Aspects

8.5.3.1 Basin Relief (R)

The altitudinal difference between highest and lowest elevation point in the watershed is termed as Basin relief. The R value of watershed influences gradient of streams, flood patterns and transport of sediment from the watershed (Hadley and Schumm 1961). The R value for DG watershed is 1.75 km whereas for four subwatersheds it ranges from 1.04 to 1.26 km (Table 8.4). The relief of the watershed suggests a mountainous terrain with dissected hills and steep slopes (Fig. 8.4b, d).

8.5.3.2 Relief Ratio (Rr)

Rr is defined as "the ratio of total relief of the watershed to basin length (Schumm 1956). The low values of Rr are associated with resistant basement rocks and low angled slopes; while the higher value indicate high relief and fragile basement rocks establishing relationship between relief and erosion. The Rr values for all four watershed is between 0.13 and 0.33 (Table 8.4) and for DG watershed it is 0.11 (Table 8.4), which suggest high relief and erosion in the watershed.

Table 8.4 Quantified result for Relief aspects

S. No	Maximum elevation (H) In Km	Minimum elevation (H) In Km	Relief (R)	Relief ratio (Rr)	Ruggedness number (Rn)	Gradient ratio (Rg)	Dissection index (Di)
SW1	2.05	0.98	1.07	0.28	8.3	0.26	0.67
SW2	2.26	0.98	1.27	0.33	7.35	0.31	0.69
SW3	2.32	0.96	1.35	0.18	8.57	0.16	0.70
SW4	1.66	0.54	1.12	0.13	6.65	0.13	0.75
DG	2.32	0.54	1.78	0.11	8.8	0.12	0.81

8.5.3.3 Ruggedness Number (Rn)

Ruggedness number of a watershed emphasizes upon the surface unevenness of watershed (Selvan et al. 2011). Strahler (1958) defined Rn as "a product of basin relief (H) and Dd and combined the qualities of slope steepness and its length;" it has implication to understand the structural complexity and erosion potential. The Rn value for DG watershed is 8.8, while for all 4 subwatersheds it is very high ranging from 6.65 to 8.57 (Table 8.4). Thus, the Rn for present study suggests that area is highly prone to erosion and have direct relationship with Dd and relief.

8.5.3.4 Dissection Index (Dis)

To understand the dissection of terrain Dis is used which infers "degree of vertical erosion in a watershed and phase of landform development" (Schumm 1956; Singh and Dubey 1994). The value of Dis lies between 0 and 1 where 0 being the lowest value and 1 is the highest. The lower Dis values are indicator of lack of vertical erosion and horizontal topography; while higher value are associated with vertical erosion and steep slopes. The value for Dis in DG watershed is 0.81, whereas for subwatersheds it lies between 0.67 and 0.75; it indicates the watershed is highly dissected with high erosion.

8.5.3.5 Gradient Ratio (Rg)

Gradient ratio is a parameter of channel inclination enabling analysis of the runoff volume (Sreedevi et al. 2005). The DG watershed has Rg value 0.11, whereas the value for four subwatersheds ranges between 0.13 and 0.30 which suggests a moderate to high value. The value of Rg indicates a mountainous terrain with moderate to steep slopes and high-soil erosion potential.

8.5.4 Hypsometric Curve (HC) and Hypsometric Integral (HI)

The hypsometric analysis was first introduced by Langbein (1947) to indicate overall slope and form of watershed. Strahler (1952) defined HC as distribution of area at different elevation of watershed, in which upward convex shape of the HC indicates youth stage, S-shaped HC represents mature stage and upward concave suggests old stage of landscape evolution of watershed. Pike and Wilson (1971) have introduced the expression to calculate the HI by using elevation-relief ratio relationship. The HI is classified into three classes viz. youth stage (>0.5–1), mature stage (>0.35–0.5) and old stage (<0.35) suggesting the geomorphic stage of landscape evolution of the watershed.

The hypsometric analysis for DG is calculated to assess the total landmass removed by erosion and geomorphic stage of development of watershed. the HC and HI of the DG watershed suggests S-shaped curve and early mature stage of development; for all 4 subwatersheds the value of HI ranges from 0.41 to 0.51 where SW1 and SW2 are in their early mature stage, SW4 is in late mature stage and SW4 is in late youth stage (Fig. 8.5). The HC and HI value ascertain that the watershed and subwatersheds are in early mature to late youth stage which indicates that dissection and erosion activity are prevalent in the watershed. The application of adequate

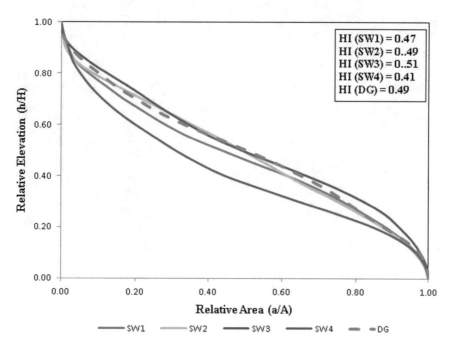

Fig. 8.5 Hypsometric curve developed for the subwatersheds (SW) and Dhundsir Gad (DG) watershed. The hypsometric integral (HI) value of the SW and DG are given on top right corner

conservation method in the watershed may help in reducing or stopping the erosion and increasing the efficiency of groundwater.

Fig. 8.6 Showing **a** jointed and fractured quartzite with quartz veins; **b** lithocontact with sheared material and seasonal nala is flowing along contact; **c** refolded crushed quartz vein in sheared and crushed phyllite suggesting tectonic influence; **d** crushed phyllite in margin of fold crest quartzite; **e** folding seen from afar along newly road-cut section near fault zone; and **f** closely placed lithocontact with a few meters of shear zone

Table 8.5 Priority table for subwatersheds in Dhundsir Gad watershed

Parameters	Subwatersheds	SW1	SW2	SW3	SW4
Drainage density (Dd)		4.06 (2)	3.41 (4)	3.74 (3)	4.10 (1)
Stream frequency (Fs)		13 (1)	8.5 (3)	7.52 (4)	11.52 (2)
Mean bifurcation ratio (Rbm)		3.85 (2)	3.31 (4)	5.39 (1)	3.5 (3)
Drainage texture (Dt)		6.5 (3)	4.6 (4)	7.9 (2)	9.3 (1)
Length of overland flow (Lg)		0.12 (3)	0.14 (1)	0.13 (2)	0.12 (3)
Circularity ratio (Rc)		0.52 (3)	0.62 (4)	0.59 (2)	0.48 (1)
Elongation ratio (Re)		0.7 (2)	0.72 (4)	0.68 (3)	0.57 (1)
Compactness coefficient (Cc)		0.52 (3)	0.62 (1)	0.59 (2)	0.48 (4)
Form factor (Ff)		0.39 (3)	0.41 (4)	0.37 (2)	0.23 (1)
Basin shape (Bs)		2.5 (2)	2.4 (1)	2.6 (3)	4.05 (4)
Compound priority		2.4	3	2.4	2.1
Final priority		Medium	Low	Medium	High

8.5.5 Prioritization

Quantitative analysis followed by ranking of geomorphic indices gives C_p of 4 subwatersheds in DG watershed. The Cp of subwatersheds ranges from 2.1–3 where the lowest value is shown by SW4 and highest by SW2 (Table 8.5). The SW4 has the lowest Cp thus highest priority is given to it; SW1 and SW3 are placed in medium priority, while lowest priority is given to SW2 (Table 8.5). The impact of thrust and faults (Fig. 8.2a, b) in the SW4 is clearly evident, mass wasting activity near thrust and fault zones has added to denudation; also, the phyllitic rocks in the area are weak and fragile which also adds to the erosion. In case of SW1 and SW3, the presence of homogenous and compact rock has restricted the denudation to a certain degree; still the denudation occurs at certain locality restricted to shear and lithocontact zones as the rocks are highly jointed and fractured near these areas. The lowest priority is of SW2 suggesting lowest denudation or erosion area in DG; as the gneiss and quartzitic rocks in the area are less jointed or fractured and are hard and compact when tested by hammering two to three time. Therefore, the priority map of the DG watershed is prepared showing priority distribution for planning in subwatersheds (Fig. 8.4f). The highest and lowest value of Cp suggest the severity of erosion and the conservation measures are suggested to be applied first in the subwatersheds with highest priority.

8.5.6 Geological Investigation

The primary objective of the study is to assess the influence of drainage network on soil erosion. Erosion of rocks and soil has a direct relationship with the lithology and structure as they influence the process in watershed. A detailed field based geological investigation is carried out along the Kirtinagar-Syalsaur road which runs in both flanks of Dhundsir Gad. The structural data is taken from different lithology to assess their relationship in landscape development and geomorphic processes; also, the impact of geological antecedents on hydrological characteristics is identified.

In detailed geological study, several lithological contact was encountered which are evident by presence of either sharp alteration in lithology or the presence of sheared material between alternate lithology (Fig. 8.6b, c, f). The weak planes of contact are of significant importance as most of the channels flows through these contacts as they are easy to erode (Fig. 8.6b). The identified contacts in the field indicates the importance and influence of these weak shear zones in controlling the hydrological activities and role of tectonic activities in developing the hydrological characteristics. The field study suggests that the rocks in the watershed have undergone deformation due to folding and thrusting (Fig. 8.6c–e) repetitively. The folding in phyllite suggests that area has undergone several phase of deformation; similarly sheared material near thrust and fault vicinity, folded and jointed quartzite suggests that area has gone through intense deformation due to tectonic activities; the reason. Alteration of several lithologies in a small terrain (Fig. 8.2a) indicates that the area has witnessed several phase of deformation. Several large scale structures like fold, faults and lithocontact were encountered in the watershed; several other mesoscale structures like foliation, lineation, joints, boudins, ripple marks and kink folding, etc., were encountered in the metamorphic rocks of DG watershed. The North Almora Thrust or Srinagar thrust (Sati et al. 2007) is the major structure in the south of DG watershed which is separating south dipping phyllite from quartzite in the north. Another major structure in the watershed is Kirtinagar fault which is encountered south of the Srinagar thrust (Fig. 8.2a). The dextral strike-slip Kirtinagar fault is evident by change in the direction of main channel flow which shifts perpendicularly left from its original course and regains its original course afterward (Fig. 8.2a).

The study suggests that due to deformation the rocks in the watershed are weak and jointed which elevates the erosion and infiltration processes. Also the weakly deformed rocks in at several places in DG watershed are resistant to erosion and increases the surface runoff with minimum infiltration increase the flood susceptibility of the region which could result in greater reduction of mass in lower elevations. The availability of loose material due to landslides, thrusting and faulting also escalates erosion processes from the watersheds. The geological investigation suggests the lithology and structure plays significant role in denudational processes in the DG watershed and should also be kept in mind during management and planning.

8.6   Conclusions

The four subwatersheds of DG watershed were analyzed for natural resources conservation and planning using geomorphic indices coupled with RS and GIS. Also, the influence of hydrological processes, lithology and structure in soil erosion from subwatersheds is studied by quantitative analysis of geomorphic aspects viz. areal, linear and relief. The study of morphometric parameters suggests that the drainage characteristics influence the dynamics of subwatersheds. The soil erosion in the subwatersheds is quantified on the basis of their relative characteristics. The hypsometric analysis suggests signifies high rate of erosion in the subwatersheds and late youth to early mature stage of development. The prioritization of the subwatersheds suggests that SW4 has the highest priority and thus any conservation and planning should start with it.

The geological field study reveals that hydrological processes coupled with lithological and structural response can accelerate the erosion and influence the hydrological response of the subwatersheds as in the case of SW4. The lithology and structure control the drainage characteristics of the watershed and hence contributes in shaping the topography of the watershed. The soil erosion in all four watersheds reaches its peak during monsoons, while all year round, it is slow and steady sheet erosion which gives elongated shape to DG watershed. The study ascertains that the morphometric parameters incorporated with GIS techniques, remote sensing data and field observations can help in understanding the impact of hydrology, climate, geology and vegetation in development of watershed. The *modus operandi* used in the present study will be of great use in any sort of watershed conservation, planning and mitigation measures around Himalaya to prevent soil erosion.

Acknowledgements Authors are thankful to head Department of geology, HNB Garhwal University; Director, Uttarakhand Space Application Center; All the research scholars of Department of Geology, HNB Garhwal University for their help during the study.

Conflict of Interest Authors have no conflict of interest.

References

Biswas A, Majumdar DD, Banerjee S (2014) Morphometry governs the dynamics of a drainage basin: analysis and implications. Geo Jour 2014:1–14. https://doi.org/10.1155/2014/927176

Chaudhary BS, Kumar S (2018) Identification of groundwater potential zones using remote sensing and GIS of K-J Watershed, India. J Geol Soc India 91:717–721. https://doi.org/10.1007/s12594-018-0929-3

Chaudhary BS, Kumar S (2017) Use of RS and GIS for land use/landcover mapping of K-J Watershed India. Int Jour Adv Remote Sens GIS 5(1):85–92

Chaudhary BS, Rani R, Kumar S, Sundriyal YP, Kumar P (2021) Analysis of land use/land cover mapping for sustainable land resources development of Hisar District, Haryana, India. In: Kumar P, Sajjad H, Chaudhary BS, Rawat JS, Rani M (eds) Remote sensing and GIScience. Springer, Cham. https://doi.org/10.1007/978-3-030-55092-9_9

Chorley RJ (1969) Introduction to physical hydrology. Methuen and Co., Ltd, Suffolk, p 211

Dowling TI, Richardson DP, O'Sullivan A, Summerell GK, Walker J, (1998) Application of the hypsometric integral and other terrain based metrices as indicators of the catchment health: a preliminary analysis. In: Technical report 20/98, CSIRO Land and Water, Canberra

Faniran A (1968) The index of drainage intensity-a provisional new drainage factor. Aust J Sci 31:328–330

Gravelius H (1914) Grundrifi der gesamten Gewcisserkunde. In: Band I: Flufikunde vol 1. Compendium of Hydrology, Rivers, in German, Goschen, Berlin

Gregory, KJ, Walling, DE (1973) Drainage basin form and process-a geomorphological approach. Edward Arnold Pub. Ltd., London, p 321. https://doi.org/10.1080/02626666809493583

Hadley RF, Schumm SA (1961) Sediment sources and drainage basin characteristics in upper cheyenne river basin. US geological survey water-supply paper 1531-B, 198

Horton RE (1932) Drainage basin characteristics. Trans Am Geophys Union 13:350–361

Horton RE (1945) Erosional development of streams and their drainage basins-hydrophysical approach to quantitative morphology. Geol Soc Am Bull 56(3):275–370

Javed A, Khanday MY, Ahmed R (2009) Prioritization of sub-watersheds based on morphometric and land use analysis using remote sensing and GIS techniques. J Indian Soc Remote Sens 37:261. https://doi.org/10.1007/s12524-009-0016-8

Keller EA, Pinter N (1996) Active tectonics, earthquake uplift and landscape. Prentice Hall, Upper Saddle River, New Jersey

Kumar G, Agarwal NC (1975) Geology of the Srinagar–Nandprayag area (Alaknanda Valley), Chamoli, Garhwal and Tehri Garhwal districts, Kumaun Himalaya, Uttar Pradesh. Himalayan Geol 5:29–59

Kumar S, Chaudhary BS (2016) GIS applications in morphometric analysis of Koshalya-Jhajhara Watershed in Northwestern India. J Geol Soc India 88:585–592

Langbein WB (1947) Topographic characteristics of drainage basins. USGS Water Supply Paper, 947-C p 157

Melton MA (1957) An analysis of the relations among elements of climate, surface properties and geomorphology. ColumbiaUniversity, New York, NY, USA

Meshram SG, Alvandi E, Meshram C, Kahya E, Al- AMF (2020) Application of SAW and TOPSIS in prioritizing watersheds. Water Res Manag 34:715–732. https://doi.org/10.1007/s11269-019-02470-x

Miller VC (1953) A quantitative geomorphologic study of drainage basin characteristics in the clinch mountain area. Virginia and Tennessee Columbia University, Department of Geology, Technical Report, No. 3, Contract N6 ONR 271–300

Nooka RK, Srivastava YK, Venkateswara V et al (2005) Check dam positioning by prioritization of micro-watersheds using SYI model and morphometric analysis—Remote sensing and GIS perspective. J Indian Soc Remote Sens 33:25. https://doi.org/10.1007/BF02989988

Pike RJ, Wilson SE (1971) Elevation–relief ratio, hypsometric integral and geomorphic area–altitude analysis. Geol Soc Am Bull 82:1079–1084

Rai PK, Chandel RS, Mishra VN, Singh P (2018) Hydrological inferences through morphometric analysis of lower Kosi river basin of India for water resource management based on remote sensing data. Appl Water Sci 8:15

Rawat A, Bisht MPS, Sundriyal YP, Banerjee S, Singh V (2021) Assessment of soil erosion, flood risk and groundwater potential of Dhanari watershed using remote sensing and geographic information system, district Uttarkashi, Uttarakhand India. App Water Sci 11:119. https://doi.org/10.1007/s13201-021-01450-0

Ritter DF, Kochel RC, Miller JR (2002) Process geomorphology. McGraw Hill, Boston

Sandilya AK, Prasad C (1982) Geomorphic studies of lineament in Garhwal Himalaya. Himalayan Geol 12

Sati SP, Sundriyal YP, Rawat GS (2007) Geomorphic indicators of neotectonic activity around Srinagar (Alaknanda basin) Uttarakhand. Curr Sci 92(6):824–829

Schumm SA (1956) Evolution of drainage systems and slopes in badlands at Perth Amboy, New Jersey. Geol Soc Am Bulln 67:597–646

Selvan MT, Ahmad S, Rashid SM (2011) Analysis of the geomorphometric parameters in high altitude Glacierised Terrain using SRTM DEM data in Central Himalaya, India. ARPN J Sci Technol 1(1):22–27

Singh S, Dubey A (1994) Geoenvironmental planning of watershed in India. Chugh Publications, Allahabad, pp 28–69

Singh O, Sarangi A, Sharma MC (2008) Hypsometric integral estimation methods and its relevance on erosion status of North-Western lesser Himalayan Watersheds. Water Res Mng 22:1545–1560

Singh P, Gupta A, Singh M (2014) Hydrological Inferences from Watershed analysis for water resource management using remote sensing and GIS techniques. Egypt J Remote Sens Space Sci 17:111–121

Smith K (1950) Standards for grading textures of erosional topography. Am J Sci 248:655–668. https://doi.org/10.2475/ajs.248.9.655

Soni S (2017) Assessment of morphometric characteristics of Chakrar watershed in Madhya Pradesh India using geospatial technique. Appl Water Sci 7:2089–2102. https://doi.org/10.1007/s13201-016-0395-2

Sreedevi PD, Subrahmanyam K et al (2005) Integrated approach for delineating potential zones to explore for groundwater in the Pageru River basin. Cuddapah District, Andhra Pradesh, India. Hyd J 13:534–545

Strahler AN (1952) Hypsometric (area-altitude) analysis of erosional topography. Bull Geol Soc Am 63:1117–1142

Strahler AN (1958) Dimensional analysis applied to fluvially eroded landforms. Geol Soc Am Bull 69:279–300

Strahler AN (1964) Quantitative geomorphology of drainage basin and channel networks. In: Chow VT (ed) Handbook of appld hydr. McGraw Hill Book, New York, pp 4–76

Weissel JK, Pratson LF, Malinverno A (1994) The length-scaling properties of topography. J Geophys Res 99:13997–14012

Yadav SK, Singh SK, Gupta M, Srivastava PK (2014) Morphometric analysis of Upper Tons basin from Northern Foreland of Peninsular India using CARTOSAT satellite and GIS. Geocarto Int 29(8):895–914. https://doi.org/10.1080/10106049.2013.868043

Chapter 9
Assessment of Groundwater Potential Zones and Resource Sustainability Through Geospatial Techniques: A Case Study of Kamina Sub-Watershed of Bhima River Basin, Maharashtra, India

Ratnaprabha Jadhav, Bhavana Umrikar, Nilima Tikone, and Brototi Biswas

Abstract Groundwater potential zones perform a crucial function in hard rock terrain. With this view, the study has been carried out to identify the groundwater prolific zones using remote sensing (RS) and geographical information system (GIS) in Kamina sub-watershed of Bhima River Basin, Shirur Taluka of Pune District, Maharashtra, India. The thematic maps such as geology, geomorphology, DEM, slope (%), rainfall, soil, drainage density, dug well density, borewell density, land use/land cover mapswere generated to identify the potential zones. Saaty's Analytical Hierarchy Process (AHP) is an efficient tool for the delineation of groundwater potential zones. The AHP proposes a weight for each evaluation criteria according to the decision maker's pairwise comparisons of the criteria. The hierarchy is built depending on the degree of influence made by each factor on groundwater potentiality. Finally, the AHP combines the criteria weight and the option scores. The ground water potential zone map so generated is divided into four classes (very low, low, moderate and high) depending on the possibility of groundwater potential. The resultant map depicts that 15.91, 23.3, 23.96, and 36.83% of the area represents "poor," "moderate," "high," and "very high" groundwater favorable zones, respectively. The highest potential area is located toward eastern and southern region because of flood plains facilitating high infiltration, drainage of Ghodriver and thick soil cover. The areas having low potential is toward the basin boundary and northern parts of the study area due to highly dissected plateau, poor soil depth, steep slopes

R. Jadhav (✉)
Department of Geography, S. N. D. T. Women's University Pune Campus, Pune, Maharashtra, India

B. Umrikar
Department of Geology, Savitribai Phule Pune University, Pune, Maharashtra 411007, India

N. Tikone · B. Biswas
Department of Geography and RM, Central University of Mizoram, Aizawl, India

© The Author(s), under exclusive license to Springer Nature Singapore Pte Ltd. 2022   191
P. K. Rai et al. (eds.), *Geospatial Technology for Landscape and Environmental Management*, Advances in Geographical and Environmental Sciences,
https://doi.org/10.1007/978-981-16-7373-3_9

and low infiltration rate. The findings of the studycan helpin the formulation of an efficient management plan for sustainable development of the area.

Keywords Groundwater · Groundwater potential zone · AHP · Sustainability

9.1 Introduction

In most parts of the world, groundwater acts as a prime source for drinking and domestic supply, hence the source strengthening becomes utmost important. Besides, the search for new sources has become essential due to the increasing demand for water. Currently, about 34% of the total annual drinking water supply is fulfilled by groundwater resources. Thus, an identification of potential areas is extremely significant, in future to have a non-contaminated source of drinking water and groundwater would also play a key role in sustainable water resource management. In India, especially in rural regions, groundwater is largely drawn from dug wells, mainly for agricultural and domestic purposes wherein about 65% area represents hard rock aquifers with poor primary porosity (<5%) and very low permeability (10^{-1}–10^{-5} m/day) (Saraf and Choudhary, 1998). Hence a very detail and extensive hydrogeological investigationsare required for thorough understanding of the groundwater status.

   The RS and GIS tools have opened new paths in water resources studies. RS-GIS techniques are widely used for the management of various natural resources (Maggirwar and Umrikar 2009; Magesh et al. 2011; Magesh and Chandrasekar 2014; Yadav et al. 2014; Samal et al. 2015). RS techniques help providing multispectral, high resolution and multi-sensor data with high periodicity (Choudhary et al. 2003). In past few decades, application of integrated remotely sensed data with conventional maps and field data, has made it easy to decipher the groundwater potential zones. Exploration and delineation of groundwater potential zones using RS-GIS techniques are performed by various researchers and the derived results are found to be reliableby verifying with field surveys (Jankowski 1995; Krishnamurthy et al. 1996; Murthy 2000; Shahid et al. 2000; Sener et al. 2005; Solomon and Quiel 2006; Nagarajan and Singh 2009; Dar et al. 2010; Maggirwar and Umrikar 2011; Dutta and Karmakar 2016). In this study a special focus is given to the identification of groundwater potential zones in Kaminasub-watershed, Shirur district, Pune using the advanced remote sensing and GIS tools for the development, policy making and effective administration of groundwater resources.

9.2 Study Area

Kamina sub-watershed lies between 18–19 ° 02″N latitude and 74–74 ° 57″ E longitudes, covering an area of 150.80 sq. km and has been included in Survey of India.

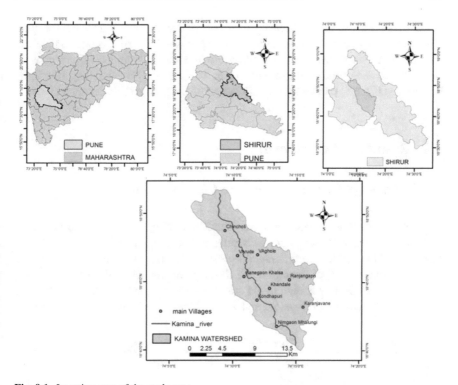

Fig. 9.1 Location map of the study area

Topographic Sheet Numbers *47J/1, 47J/2, 47J/5, 47J/6, 47J/10 and 47J/11* on the numerical scale of 1: 50,000 (Fig. 9.1). The sub-watershed includes the part of Pabal and Shirur circle along with 14 villages. The maximum length of the sub-watershed is 25 km and maximum width of about 12 km. The area prominently shows basaltic lithology comprising different lava flow units. The area is divided into two major physical divisions namely plateau region and flood plain region. It has the elevation range of 600–720 m which represents a middle level plateau. The study area experiences a hot, dry and tropical monsoon climate. The average annual rainfall in this area ranges between 600 and 700 mm.

9.3   Materials and Methods

This research is basically based on the secondary data. Topographical maps, geological map (Geological Survey of India), and Soil map (National Bureau of Soil Survey Maharashtra state) was used to prepare various thematic maps. SRTM data have been used for generating the Digital Elevation Model and slope map with the help of ArcGIS Spatial Analyst module. Density tool in GIS environment is used to prepare the drainage density, dug well density and tubewell density maps. Landsat ETM (path 147, row 47) image of 30 m resolution dated February 24th, 2015 downloaded from Bhuvan. Supervised classification method and maximum likelihood Algorithm were employed to detect the land use land cover. Thomas Satty's Analytical Hierarchy Process (AHP) method was used to allocation of weights to each parameter.

9.4 Results and Discussion

9.4.1 Geology and Morphometric Characteristics

The types of geological features were found to be simple basaltic lava flow. The exposures of vesicular/amygdaloidal basalts as well as fractured/jointed basalts are observed during the field work. This sub-watershed area consists of various lava flow units of Indrayani, Upper Ratangirh and Karla Formation with different thickness.

The drainage density is high in the middle and lower parts of the basin showing dendritic drainage pattern (Figs. 9.2 and 9.3). Drainage system is one of the important

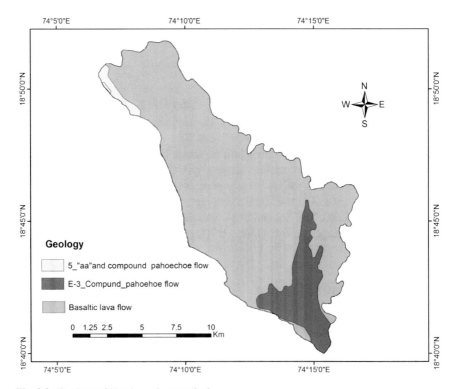

Fig. 9.2 Geology of Kamina sub-watershed

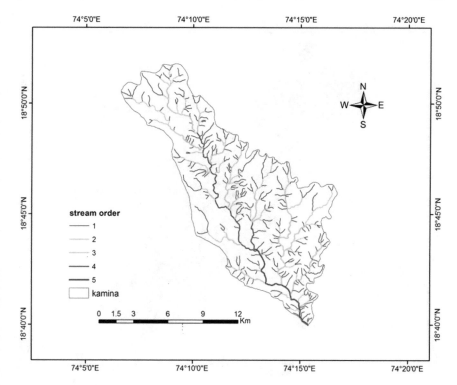

Fig. 9.3 Drainage network with stream orders. *Source* Digitized from SOI Toposheet

components of the physical environment. Kamina is the tributary of Bhima river. It originates at Gholapwadi of Kanhur Mesai village at 720 m elevation above mean sea level. Kamina is a fifth order river and the length of longest stream is 30.10 km. The total length of first order streams is 171.77 km and the fifth order stream is 9.07 km. The minimum and maximum bifurcation ratio values are 2 and 4.29.Stream frequency of Kamina sub-watershed is 2.26775.

Drainage density, according to Horton (1932), is defined as the length of streams per unit area. The drainage density value is **2.005** which indicates the study area is underlain by impermeable lithology with moderate relief. The drainage texture value of the present study area is **4.54** which indicates the intermediate drainage texture of the study area. The form factor value (0.241) indicates the elongated shape of the basin. The circulatory ratio value of 0.3544 indicates the low circularity stage of the basin (Fig. 9.4).

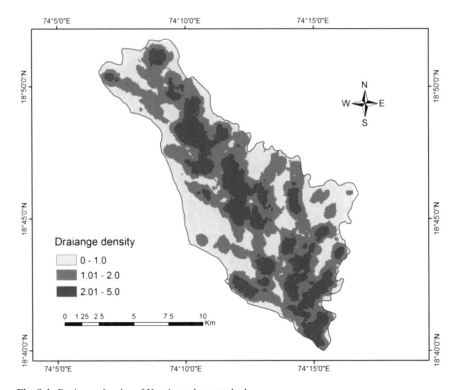

Fig. 9.4 Drainage density of Kamina sub-watershed

9.4.2 DEM and Slope (%)

Watershed relief is key factor in understanding the denudational characteristics of the watershed and plays a significant role in landform development, drainage development, surface runoff and recharge, permeability and erosional properties of the terrain (Magesh et al. 2014). The DEM was generated through 20 m contours lines, which were obtained from SOI topomaps (1:50,000). The slope (%) was derived from DEM (Figs. 9.5 and 9.6). The height is ranging between 520 to 720 m above the mean sea level. Northwestern part has maximum height (720 m) that is at the source Kamina river whereas the minimum height is at the confluence of Kamina river and Bhima river.

Slope is considered as an important criterion for selection and execution of soil water conservation sites. (Agarwal et al. 2013). Slope plays a key role in groundwater occurrence as infiltration is inversely related to slope. High basin slopes are observed at the periphery of watershed. Disposition of slopes within a watershed influence the topographical conditions and drainage network. The peripheral areas of watersheds in present study show steep slopes where first and second order streams are concentrated.

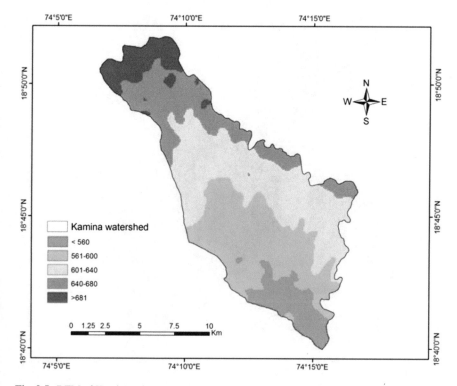

Fig. 9.5 DEM of Kamina sub-watershed

9.4.3 Soil

Soil texture, depth and composition imparts and facilitates surface water infiltration, hence it is the part of majority of the groundwater potential and recharge studies. The majority part of this sub-watershed depicts well drained, moderate to thick clayey and calcareous soil covering moderate to gentle slopes that undergoes low to moderate erosion process (Fig. 9.7). However, the middle part of the basin is covered with clayey soil with moderate salinity and moderate erosion. The soil occupying gently sloping plains and valley/valley fills undergoes slight erosion. Very shallow, well drained, loamy soil is observed at very few locations where the rate of erosion is moderate.

9.4.4 Dug Well and Bore Well Density

Wells are commonly used as a irrigation source for agriculture purpose. The well location map is prepared by collecting well point from Toposheet. Highest dug well

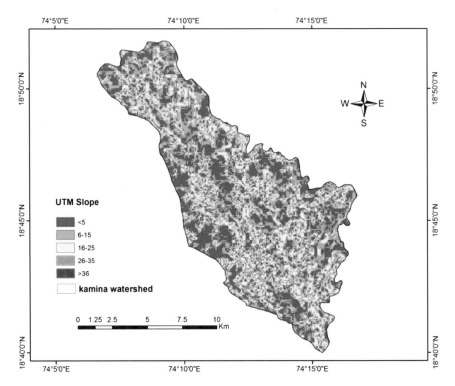

Fig. 9.6 Slope Map of Kamina sub-watershed

density is observed in the central part where as the lowest density is seen in southern part of the study area. Recently instead of large diameter dug wells, bore wells are more popular especially due to the low cost. The maximum density is mainly observed in the southern part (Figs. 9.8 and 9.9).

9.4.5 Land Use Land Cover (LULC)

Land use/land cover analysis is one of the important phenomena which has been dealt with a great emphasis in the recent past (Ambashetty et al. 2005). The LULC has higher impact on runoff-recharge processes vis-à-vis the groundwater availability. The study area is categorized into 8 classes such as barren land, fallow land, forest, pasture land, waterbody, agriculture, settlement and saline soil. Table 9.1 shows the area covered under each land use class. The data depict that 28.59% area was utilized for the agriculture purpose (Fig. 9.10).

Despite of the Chaskaman canal being the major source of irrigation, the area under agriculture is comparatively less. Agriculture is mainly observed in the central and southern part of the study area. The percentage of fallow and pasture land ranges

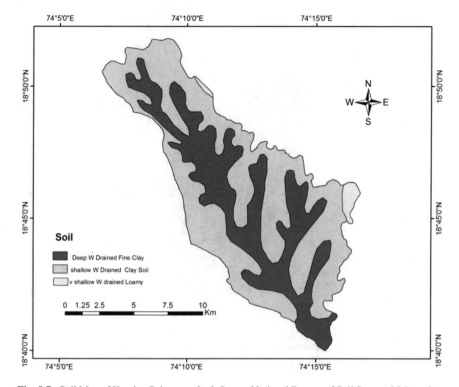

Fig. 9.7 Soil Map of Kamina Sub-watershed. *Source* National Bureau of Soil Survey, Maharashtra state

between 20 and 23% and it occupies mainly the north-western part of the sub-watershed. Saline soil is observed on a minor scale due to existing cropping pattern and over irrigation/water logging. The percentage of natural forest and waterbody is negligible (less than 1%).

9.4.6 Groundwater Potential Zones

Thomas Satty's (1980) AHP is an effective tool for dealing with complex decision making. The features of the study area such as topography, soil type, land use pattern, drainage pattern, geology, geohydrology and geomorphology were obtained and thematic maps were prepared and generated the attributes for each thematic maps. The AHP generates a weight for each evaluation criteria according to the decision maker's pairwise comparisons of the criteria. The higher the weight the more important is the considering criterion. Finally, the AHP combines the criteria weight and the option scores. The global score for given option is a weighted sum of the scores obtained with respect to all criteria.

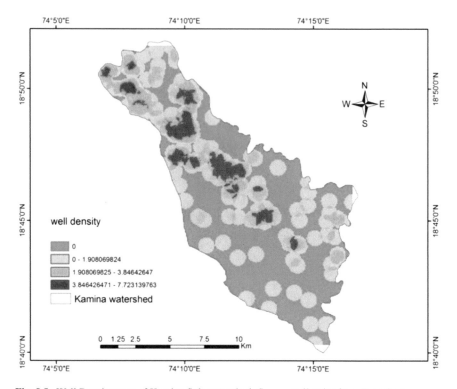

Fig. 9.8 Well Density map of Kamina Sub-watershed. *Source* well point from Toposheet

AHP involves pairwise comparisons to create a ratio matrix. In this diagonal matrix, the pairwise comparisons (using the scale proposed by Saaty comprising 1 to 9 digits) of the parameters are taken as input data and after normalization the priority matrix produces the relative weights of various parameters as an output. The sub-criteria under each parametric class have been allotted the suitable rank up to 4 depending on the degree of influence on groundwater potentiality (Table 9.2). All the reclassified raster thematic layers of various parameters have been overlaid using weighted overlay analysis in GIS environment. The resultant output map was divided into four classes by equal interval criteria and converted to vector format for computing the area covered by each class. The final map depicts that about 36% of the total area falls under very high groundwater potential category and 15.91% area shows poor groundwater availability (Table 9.3) (Fig. 9.11).

The villages from very good groundwater potential have both surface and groundwater sources for irrigation and as a result the tendency for the farmer to cultivate cash crops has been found during the field surveys. Hence, there is a need of awareness program regarding the judicial use of water so that this precious resource acquires a sustainable state. Apart from this, the village falling in poor groundwater potential may has a project of surface water harvesting structures, inclusion in piped water

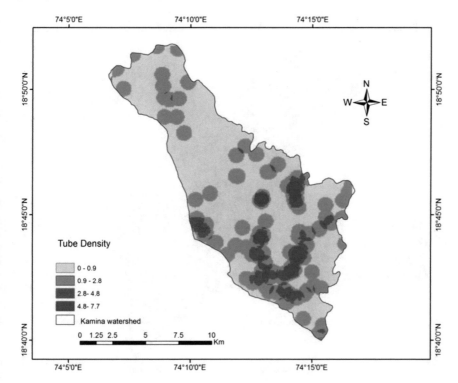

Fig. 9.9 Bore Well Density map of Kamina Sub-watershed. *Source* GPS locations of Bore wells

Table 9.1 LULC Classification (Area in Percent)

LULC Classes	Area in Percent	LULC Classes	Area in Percent
Pasture Land	22.82	Barren Land	13.74
Settlement	11.32	Forest	0.42
Fallow Land	21.59	Saline	0.15
Agriculture	28.59	Waterbody	0.41

Source Computed by researcher

supply schemes or conservation of spring water where ever applicable so as to resolve the water scarcity issues.

9.5 Conclusion

Groundwater is one of the natural prime sources of drinking water and with increase in demands there is an urgent need of sustainable management of this precious

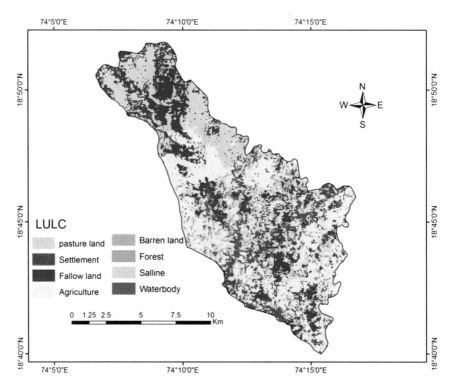

Fig. 9.10 LULC of KaminaSub-watershed. *Source* satellite image from Landsat ETM-Feb-2015

Table 9.2 Rank and weights of different Parameter for ground water potential zone

SN	Criteria	Classes	Rank	Weights(%)
1	Slope	< 5	4	*13.26*
		6–15	3	
		16–25	2	
		> 26	1	
2	DEM (m)	560–640	4	*6.15*
		640–700	3	
		700–740	2	
		< 760	1	

(continued)

resource. Remote Sensing and GIS are important tools to delineate the groundwater potentiality zone. Geology, geomorphology, elevation, slope, rainfall, Soil, drainage network, drainage density, dug well density, tubewell density and land use land cover themeswere used as input data in AHP. The study area is classified into very high, followed by high, moderate and poor groundwater potentiality by applying the AHP

Table 9.2 (continued)

SN	Criteria	Classes	Rank	Weights(%)
3	Drainage Density (km/km^2)	Very high	1	4.61
		High	2	
		Moderate	3	
		Low	4	
4	Rainfall (cm)	15–30	1	10.21
		30–45	2	
		45–60	3	
		60–75	4	
5	LULC	Agriculture	4	13.31
		Forest	3	
		Saline belt	1	
		Settlement	1	
		Waterbody	3	
6	Geology	E-3 pahoehoe flow	3	13.76
		Basaltic lava	2	
		5 "'aa" compound	1	
7	Soil	Shallow	2	7.55
		Deep well	3	
		Very shallow	1	
8	Dug Well Density	Very high	4	6.14
		High	3	
		Moderate	2	
		Low	1	
9	Tube Well Density	Very high	4	8.17
		High	3	
		Moderate	2	
		Low	1	

Table 9.3 Groundwater Potential Zones (Area in percentage)

Sr.no	Zone	Area in sq.km	Area in Percent
1	Poor	22.49	15.91
2	Moderate	29.36	23.3
3	High	36.09	23.96
4	Very High	55.54	36.83

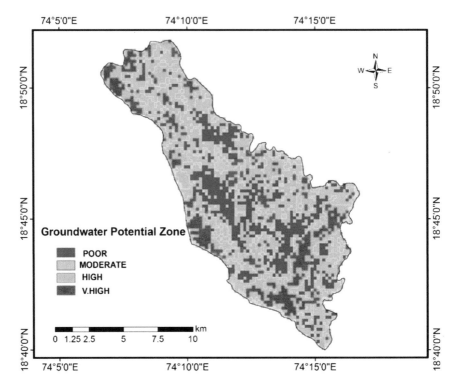

Fig. 9.11 Groundwater Potential in Kamina Sub-watershed

method. In this study, 36.83% area under falls under very high ground water potential and 15.91% area covered poor ground water potential zone. Thus, the resultant groundwater potential map would be useful for further development and management of water resources of this region. The study also recommends the use of GIS technology with remotely sensed data for the further study of groundwater potential zones thatcould minimize the cost, time, human resource with higher accuracy.

References

Agarwal R, Garg PK, Garg RD (2013) Remote sensing and GIS based approach for identification of artificial recharge sites. Water Resour Manage 27:2671–2689. https://doi.org/10.1007/s11269-013-0310-7

Chowdhary VM, Rao NH, Sarma PBS (2003) GIS based decision support system for groundwater assessment in large irrigation project areas. Agric Water Manag 62:229–252

Dar IA, Sankar K, Dar MA (2010) Deciphering groundwater potential zones in hard rock terrain using geospatial technology. Environ Monit Assess 173:597610

Dutta H, Goswami D, Karmakar R (2016) Geo-environmental assessment for land capability classification of selected subwatersheds of the Siang river of Arunachal Pradesh, India using remote sensing and GIS techniques. J Environ Res Develop 10(3):407–422

Horton RE (1932) Drainage basin characteristics. Trans Am Geophys Union 13:350–361

Jankowski P (1995) Integrating geographic information systems and multiple criteria decision making methods. Int J Geogr Inform Syst 9(3):251–273

Krishnamurthy J, Kumar NV, Jayaraman V, Manivel M (1996) An approach to demarcate groundwater potential zones through remote sensing and a geographic information system. Int J Remote Sens 17(10):1867–1884

Magesh N, Chandrasekar N (2014) GIS model-based morphometric evaluation of Tamiraparanisubbasin, Tirunelveli district, Tamil Nadu. India Arab J Geosci 7:131–141

Magesh N, Chandrasekar N, Soundranayagam J (2011) Morphometric evaluation of Papanasam and Manimuthar watersheds, parts of Western Ghats, Tirunelveli district, Tamil Nadu, India: a GIS approach. Environ Earth Sci 64(2):373–381

Maggirwar BC, Umrikar BN (2009) Possibility of artificial recharge in overdeveloped miniwatersheds: RS-GIS approach., e-Journal, Earth Science India., ISSN No. 0974-8350, vol 2 (II), pp. 101–110 (http://www.earthscienceindia.info)

Maggirwar BC, Umrikar BN (2011) Influence of various factors on the fluctuation of groundwater level in hard rock terrain and its importance in the assessment of groundwater. Int J Geol Mining Res 3(11):305–317

Massam BH (1988) Multi-criteria decision making techniques in planning. Programme Planning 30:184

Murthy KSR (2000) Groundwater potential in a semi-arid region of Andhra Pradeshda geographical information system approach. Int J Remote Sens 21:1867–1884; Nagarajan M, Singh S (2009) Assessment of groundwater potential zones using GIS technique. J Indian Soc Remote Sens 37:69–77

Nagarajan M (2009) Assessment of groundwater potential zones using GIS technique. J Indian Soc Remote Sens 37:69–77

Samal D, Gedam S, Nagarajan R (2015) GIS based drainage morphometry and its influence on hydrology in parts of Western Ghats region, Maharashtra India. Geocarto Int 30(7):755–778

Saraf AK, Choudhary PR (1998) Integrated remote sensing and GIS for groundwater exploration and indentification of artificial recharges sites. Int J Remote Sens 19(10):1825–1841

Sarma B, Saraf AK (2002) Study of land use ground water relationship using an integrated remote sensing and GIS approach. http://gisdevelopment.net

Sener E, Davraz A, Ozcelik M (2005) An integration of GIS and remote sensing in groundwater investigations: A case study in Burdur, Turkey. Hydrogeol J 13:826–834

Shahid S, Nath SK, Roy J (2000) Groundwater potential modeling in a softrock area using a GIS. Int J Remote Sens 21(9):1919–1924. https://doi.org/10.1080/014311600209823

Shetty A, Nandagiri L, Thokchom S, Rajesh MVS (2005) Land use—land cover mapping using satellite data for a forested watershed, Udupi district, Karnataka state, India. J Indian Soc Remote Sens 33:233–238

Solomon S, Quiel F (2006) Groundwater study using remote sensing and geographic information system (GIS) in the central highlands of Eritrea. Hydrogeol J 14(5):729–741. https://doi.org/10.1007/s10040-005-0477-y

Yadav S, Singh S, Gupta M, Srivastava P (2014) Morphometric analysis of upper tons basin from Northern Foreland of Peninsular India using CARTOSAT satellite and GIS. Geocarto Int 29:895–914

Chapter 10
Morphometric Analysis of Damodar River Sub-watershed, Jharkhand, India, Using Remote Sensing and GIS Techniques

Akshay Kumar, Anamika Shalini Tirkey, Rahul Ratnam, and Akhouri Pramod Krishna

Abstract Morphometric analysis is a mathematical examination of the shape and dimension of the earth's surface that illustrates the interrelationship between hydraulic parameters and geomorphologic characteristics of a drainage basin. The present study indicates the effectiveness of remote sensing (RS) and geographic information system (GIS)-based morphometric analysis of a sub-watershed of Damodar River basin in Ramgarh district, Jharkhand, India. The sub-watershed and drainage texture of the study area is extracted by ASTER DEM and topographical map in the GIS environment. Morphometric parameters such as stream order, stream length, bifurcation ratio, drainage density, stream frequency, form factor, and circulatory ratio are calculated. The sub-watershed's total drainage area is 46.71 km^2 and shows a dendritic drainage pattern that designates homogeneous lithology, gentle regional slope, and lack of structural control. The study area is designated as the fifth-order basin with a drainage density (D_d) value ranges 9.91 km/km^2. An extensive field survey supports the results. Results of the study have immense significance for engineers, managers, and planners for management of soil, water and provide watershed prioritizing management activities in the area.

Keywords Morphometric analysis · Sub-watershed · Drainage density · ASTER DEM · GIS

A. Kumar (✉) · A. P. Krishna
Department of Remote Sensing, Birla Institute of Technology, Mesra, Ranchi, Jharkhand 835215, India

A. S. Tirkey
Department of Geoinformatics, School of Natural Resource Management, Central University of Jharkhand, Brambe, Jharkhand, India

R. Ratnam
Department of Geography, Panjab University, Chandigarh 160014, India

10.1 Introduction

Watershed is the basic unit of water supply that allows the surface run-off to a defined channel, drain, stream, or river (Chopra et al. 2005). Characterization of watershed provides a reliable foundation to understand the behavior of the area that further uses for developing suitable land and water resources management practices. The dynamic ecosystem of water is directly linked with the geology, elevation, and climate of any area (Mesa 2006). Therefore, accurate quantification of the morphological, geomorphic, and topographic features of a watershed becomes imperative for evaluating the hydrologic response of watersheds (Abdulkareem et al. 2018).

There is a need for evaluation of watershed for identifying health and classification of the watershed. Hence, morphometric analysis can be used for the numerical description of a drainage basin. The morphometric analysis also helps to evaluate the hydrological parameters such as the formation of a watershed, its surface characteristics, topography, groundwater, and underlying geology (Krishnamurthy and Srinivas 1995; Nag 1998; Vittala et al. 2004; Rawat et al. 2011; Jasmin and Mallikarjuna 2013). It provides an accurate quantitative description of basin geometry, the slope, and inequalities in rock hardness, geological controls, and drainage (Strahler 1964). Morphometric analysis requires measurement of drainage, slope of channel network, relief, etc. (Nautiyal 1994). Drainage analysis is very significant for watershed development and planning. It also provides knowledge about the basin characteristics in terms of topographical and soil attributes, runoff behavior, surface water potential, etc. (Astras and Soulankellis 1992). Systematics study of drainage morphometric provides a quantitative description of the basin geometry to understand its geological and geomorphic history of drainage basin and provide helpful evidence about the hydrological behavior of the rocks exposed within the river basin or watershed (Strahler 1964).

Remote sensing (RS) and geographic information system (GIS) are a convenient tool for morphometric analysis. A satellite image provides a synoptic view of a large area and is very useful in updating the drainage networks. Various morphometric parameters such as drainage pattern, stream order, bifurcation ratio, drainage density, and other linear aspects have been studied using RS techniques and topographical maps (Zolekar and Bhagat 2015; Mesa 2006). However, before emerging GIS and remote sensing technologies, identifying drainage systems within basins or sub-basins can be attained using conventional approaches and topographic maps (Panda et al. 2019). Several researchers have done morphometric analysis coupled with RS and GIS for studying the land and water resources of an area (Saxena and Prasad 2008; Mishra et al. 2011; Bagyaraj et al. 2011; Al-Daghastani and Al-Maitah 2006; Zaidi 2011). Sarkar et al. (2020) applied morphometric analysis to assess and understand the hydrological characteristics of Nagar River basin, situated at the Indo-Bangladesh trans-boundary region. They also prepared a flood vulnerability map using the analytical hierarchy process (AHP) and flood influencing morphometric parameters. The study revealed that the lower portion of the Nagar River basin is highly vulnerable to flooding. Rai et al. (2017) utilized satellite remote sensing and morphometric analysis to study surface morphological features and their correlation with groundwater management prospects at the Kosi River basin. They also prepared land use/land cover (LU/LC) of the basin to assess the change in the dynamic of the basin for further watershed prioritization. The study revealed that maximum morphometric parameters were showing good prospects for the water management program. Markose et al. (2014) used the bAd calculator to assess the Kali River basin's morphometric parameters situated at the southwest coast of India. Their result indicated that most area sub-basins have less constant values (~3 to 4) of mean bifurcation ratio showing the uniform and systematic branching pattern of streams. Furthermore, the high drainage density is located in the Western Ghats region, whereas low drainage density zones are observed in the eastern part of the basin and near the estuary. Pareta and Pareta (2011) studied the morphometric characteristics of the Karawan watershed in the Dhasan basin, which itself is part of the mega Yamuna basin in Sagar district, Madhya Pradesh, India. They utilized ASTER DEM and GIS to compute 85 morphometric parameters of the study area. The results indicated that the erosional development of the site by the streams has progressed well beyond maturity. Moreover, that lithology has influenced drainage development. These studies are beneficial for planning rainwater harvesting and watershed management of the study area.

Coal mining practices, both surface and underground, damage the complex system of flora, fauna, and hydrological regimes of the region (Kumar and Krishna 2021). The study area is a coal mining impacted region where the groundwater remains at deeper levels and chances of contamination are much higher. In addition, mining activity, mainly opencast, changes the natural topography, which further impacts the drainage system of the area. It also makes inadequate groundwater potential capacity due to which the people of the region face severe water scarcity both for irrigation and drinking purposes. Therefore, the present study investigates the morphometric changes in the sub-watershed of Damodar River, Jharkhand, India. Further, the study will help to understand the hydrological changes that notably raised our concern about the deep groundwater levels in the area.

10.2   Study Area

The Damador River sub-watershed is located in Ramgarh district, Jharkhand state, India. The study area lies between longitudes of 85° 30' to 85° 36' E and latitudes of 23° 39' to 23° 45' N (Fig. 10.1). It covers an area of 39 km^2 and exhibits undulating topography with elevation ranging between 286 and 471 m above sea level (ASL). The physiography of the watershed is gently sloping with industrial and settlement areas followed by agricultural regions dominating the majority of the sub-watershed except for forest cover over the hilly terrain present in the north-eastern part of the watershed. Climate of the area is subtropical, with average minimum and maximum temperatures varying between 25 °C in winters and 45 °C in summers (Kumar and Krishna 2018). The area exhibits a significant seasonal variation and receives the average annual precipitation in the form of a rain of about 1050 mm, a maximum in

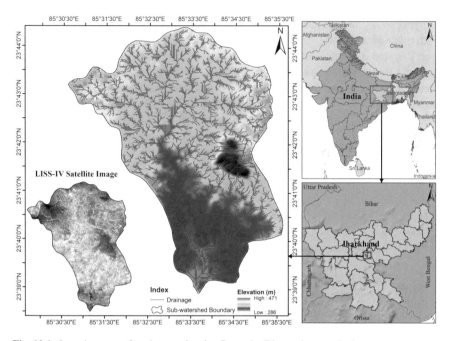

Fig. 10.1 Location map of study area showing Damodar River sub-watershed

the rainy season (June–October) (Pandey and Kumar 2014). Total volume of water accumulated within the area drains into the Damodar River. Geology of the area belongs to igneous and metamorphic terrain of the Precambrian age, whereas soil texture is characterized by fine-loamy to fine (Kumar and Pramod Krishna 2020). The Damodar River sub-watershed is shown in Fig. 10.1.

10.3 Methodology and Data Used

Multiple satellite data, conventional map, and field data were utilized in this study to analyze the morphometric parameters of the area.

The Advanced Spaceborne Thermal Emission and Reflection Radiometer Digital Elevation Model (ASTER DEM) data were downloaded from ASTER GDEM Web site (https://ssl.jspace-systems.or.jp/ersdac/GDEM/E/). Landsat-8 Operational Land Imager (OLI) data were downloaded from the USGS Web site (https://earthexplorer.usgs.gov/) and LISS-IV satellite image was procured from National Remote Sensing Agency (NRSC), Hyderabad. Survey of India (SOI) topographical map pertaining to sheet number 73 E/6 provides the details of the study area at the scale of 1:50,000 used for ground reference. Remote sensing and GIS-based morphometric analysis are a time saver and cost-effective and provide higher accuracy. The Damodar River sub-watershed, stream network, and morphometric-related data have been extracted from

Table 10.1 List of satellite data and topographical map with their specifications

Data used	Path/row	Date of acquisition	Spectral resolution (μm)	Spatial resolution (m)	Swath (km)
Landsat 8 ETM OLI/TIRS	140/44	31 January, 2019	Band 1 = 0.433–0.453	30	185*180
			Band 2 = 0.450–0.515	30	
			Band 3 = 0.525–0.600	30	
			Band 4 = 0.630–0.680	30	
			Band 5 = 0.845–0.885	30	
			Band 6 = 1.560–1.660	30	
			Band 7 = 2.100–2.300	120	
			Band 8 = 0.500–0.680	15	
			Band 9 = 1.360–1.390	30	
			Band 10 = 10.30–11.30	100	
			Band 11 = 11.50–12.50	100	
IRS-R2 LISS-IV	105/55	23 November, 2013	Band 1 = 0.52–0.59	5.8	70
			Band 2 = 0.62–0.68		
			Band 3 = 0.77–0.86		
Topographical Map	No. 73 E/6	Surveyed 1962	1:50,000 Scale	–	–

ASTER DEM in the GIS environment. Various primary topographic variables such as slope, aspect, and relief were also prepared by ASTER DEM. The specifications of satellite and ancillary data are presented in Table 10.1.

10.3.1   Delineation of Sub-watershed

The Damodar River sub-watershed is delineated from ASTER DEM using ArcGIS spatial analyst module. A pour point with DEM is required as an input parameter

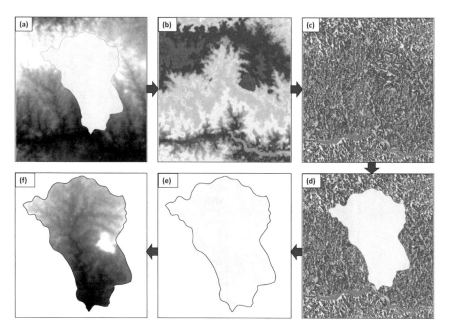

Fig. 10.2 Figure showing delineation of Damodar River sub-watershed boundary through, **a** ASTER DEM, **b** filled DEM, **c** flow direction, **d** sub-watershed raster boundary, **e** sub-watershed vector boundary, **f** clipped DEM through sub-watershed boundary

to delineate the drainage area for morphometric analysis. The pour point is a user-provided point to the cells of the highest flow accumulation in any basin (Magesh and Chandrasekar 2014; Rai et al. 2018). The result of this process will create a watershed boundary polygon from the flow direction raster data. The whole procedure to run the model for sub-watershed delineation is shown in Fig. 10.2.

10.3.2 Extraction of Drainage Network

Drainage network was extracted using ASTER DEM and validated with SOI topographical map in the GIS environment. Firstly, DEM preparation was done by filling the sinks then flow direction was calculated for each pixel to generate the drainage network. Finally, flow accumulation was also taken into account based on the flow direction of each cell (Avinash et al. 2011; Kudnar and Rajasekhar 2020). The systematic process required for the automatic extraction of the drainage networks is shown in Fig. 10.3. Further, the extracted drainage line was ordered based on Strahler's (1957) system of classification.

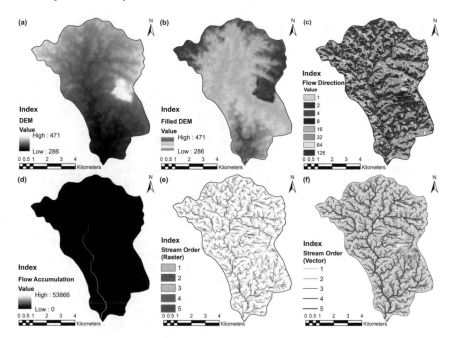

Fig. 10.3 Figure showing extraction of drainage network through, **a** ASTER DEM, **b** filled DEM, **c** flow direction, **d** flow accumulation, **e** stream order raster, **f** stream order vector

10.3.3   Morphometric Parameter Calculation

Morphometric parameters such as drainage density (D_d), bifurcation ratio (R_b), relief ratio (R_h), stream frequency (F_s), drainage texture (R_t), form factor (R_f), circulatory ratio (R_c), length of overland flow (L_g), and constant of channel maintenance (C) have been done on the basis of well-established standard mathematical formulae as described in Table 10.2.

The software ArcGIS (version 10.5) and ERDAS Imagine (version 2015) were, respectively, used for data preparation and morphometric analysis of Damodar River sub-watershed. Methodology adopted for entire study is shown in Fig. 10.4.

10.4   Result and Discussion

Morphometric characteristics is a quantitative analysis for understanding the geometry, rock structure, present diastrophism, morphological and geological outline of the river basin (Strahler 1964). The results of morphometric analysis for Damodar River sub-watershed are shown in Table 10.3. The drainage pattern of the study area is dendritic and mainly controlled by the general topography, geology, and rainfall. Topographic parameters such as slope and aspects were generated by ASTER DEM.

Table 10.2 Standard formulae for morphometric analysis

Sl. No	Morphometric parameters	Formulae	References
1	Stream order	Hierarchical rank	Strahler (1964)
2	Stream length	Length of the stream	Horton (1945)
3	Mean stream length (L_{sm})	$L_{sm} = Lu/Nu$, where L_{sm} = mean stream length, L_u = total stream length of order "u", and Nu = total number of stream segments of order u	Strahler (1964)
4	Bifurcation ratio (R_b)	$Rb = Nu/Nu + 1$, where Rb = bifurcation ratio, Nu = total no. of stream segments of order u, and N_{u+1} = number of stream segments of the next higher order	Schumn (1956)
5	Relief ratio (R_h)	$R_h = H/Lb$, where R_h = relief ratio, H = total relief (relative relief) of the basin in km, and L_b = basin length	Schumn (1956)
6	Drainage density (D_d)	D_d-Lu/A, where D_d = drainage density, Lu = total stream length of all orders, and A = area of the basin (km^2)	Horton (1932)
7	Stream frequency (F_s)	$Fs = Nu/A$, where F_s = stream frequency, Nu = total no. of streams of all orders, and A = area of the basin (km^2)	Horton (1932)
8	Drainage texture (R_t)	$R_t = Nu/p$, where R_t = drainage texture	Horton (1945)
9	Form factor (R_f)	$R_f = A/Lb^2$, where R_f = form factor, A = area of the basin (km^2), and L_b^2 = square of basin length	Horton (1932)
10	Circularity ratio (R_c)	$Rc = 4*\pi*A/P^2$, where R_c = circularity ratio; π = "π" value, i.e., 3.14; and L_b = basin length (km)	Miller (1953)
11	Length of overland flow (L_g)	$L_g = 1/D*2$, where L_g = length of overland flow and D = drainage density	Horton (1945)
12	Constant of channel maintenance (C)	$C = 1/D$, where C = constant channel maintenance and D = drainage density	Schumn (1956)

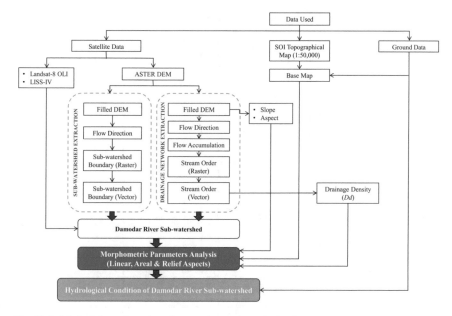

Fig. 10.4 Methodology flow chart for morphometric analysis of Damodar River sub-watershed

10.4.1 Aspect

Aspect generally refers to the direction to which a mountain slope faces. The aspect of a slope can make very significant influences on its local climate because the sun's rays are in the west at the hottest time of day in the afternoon, and so in most cases a west-facing slope will be warmer than sheltered east-facing slope. The value of the output raster dataset represents the compass direction of the aspect (Magesh et al. 2012a). The aspect map of Damodar River sub-watershed is gently sloping toward north to south direction as shown in Fig. 10.5a.

10.4.2 Slope

Slope analysis is an important parameter in geomorphological studies for watershed development and important for morphometric analysis. The slope elements, in turn, are controlled by the climato-morphogenic processes in areas having rock of varying resistance (Magesh et al. 2012a; Gayen et al. 2013).

The slope map of the study area is calculated based on ASTER DEM data using the spatial analysis tool in ArcGIS v10.5. The degree of slope in Damodar River sub-watershed varies from 2° to 27° (Fig. 10.5b). The slope map of study area is shown in Fig. 10.5b. Higher slope degree results in rapid runoff and increased erosion rate (potential soil loss) with less ground water recharge potential. Lowest slope (>2°)

Table 10.3 Results of morphometric analysis of Damodar River sub-watershed

Stream order	Stream segments	Stream length (km)	Basin area (km^2)	Perimeter (km)	Basin length (km)	Mean stream length (km)	Bifurcation ratio	Relative relief	Relief ratio
I	4388	275.55	46.71	31.95	13.16	0.063	2.62	8.95	21.73
II	1677	90.86				0.054	2.01		
III	834	46.32				0.056	1.13		
IV	739	39.14				0.053	3.57		
V	207	10.96				0.053	–		
Total	7845	462.83				0.059			
Drainage density (km/km^2)	Stream frequency	Drainage texture	Form factor	Circularity ratio	Length of overland flow	Constant of channel maintenance	Stream segment		
9.91	167.95	6.95	0.26	0.57	4.95	0.10	245.54		

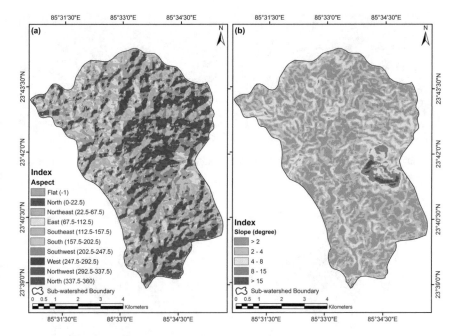

Fig. 10.5 Figures showing, **a** Aspect, **b** slope map of study area

indicating flat topography is present in the South-western region of the study area
and the slopes >15° is observed over the hilly region present in the eastern part of
study area.

10.4.3   Stream Order

It is based on the hierarchic ranking of streams proposed by Strahler (1964). The
first-order streams have no tributaries. The second-order streams have only first-
order streams as tributaries. Similarly, third-order streams have first and second-
order streams as tributaries and so on. A perusal of Table 10.3 indicates that the
catchment is designated up to the fifth-order having a total of 7845 stream segments
of different orders (Fig. 10.7a). The maximum stream order frequency is observed
in the case of first-order streams and followed with second order. Hence, it is noticed
that there is a decrease in stream frequency as the stream order increases and vice
versa (Fig. 10.6a).

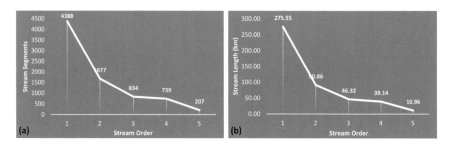

Fig. 10.6 Figures showing relationship between, **a** stream segments and stream order, **b** stream length and stream order

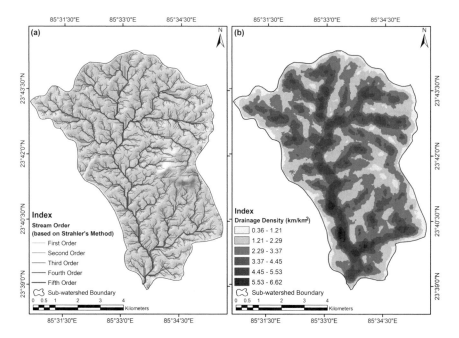

Fig. 10.7 Figures showing, **a** stream order map, **b** drainage density map of the study area

10.4.4 Stream Number

The total number of streams in a particular order is known as stream number. Horton's (1945) law of stream numbers states that the number of stream segments of each order forms an inverse geometric series with the order number. It is noted that the total number of stream decreases with an increase in stream order and hence shows compliance with Horton's law of stream ordering. This points that the catchment has no structural control or regional upliftment.

10.4.5 Stream Length (L_u)

Stream length (L_u) is characterized as the sum of the lengths of all streams in a given order. *Lu* is determined by counting the total number of streams of various orders and measuring their length. Generally, with stream order increases, the total length of stream segments decreases (Chandniha and Kansal 2017). When plotted against the order, logarithms of the number of streams of a particular order generally lie on a straight line (Horton 1945). The landscape is characterized by high relief and relatively steep slopes, underlain by varied lithology, and likely uplift across the basin, based on deviations from its general behavior (Singh and Singh 1997; Chandniha and Kansal 2017). In the present study, the first order L_u is 275.55 km, the second order is 90.86 km, the third order is 46.32 km, the fourth order is 39.14 km, and the fifth order is 10.96 km. Figure 10.6b conforming to Horton's law (1945) as the stream order increases, the length of the stream decreases.

10.4.6 Mean Stream Length (L_{sm})

The mean stream length of a channel is a dimensional property and reveals the characteristic size of drainage network components and its contributing basin surfaces (Strahler 1964). The mean stream length (L_{sm}) has been calculated by dividing the total stream length of order by the number of streams. It can be expressed by the equation given by Strahler (1964):

$$L_{sm} = \frac{L_u}{N_u} \tag{10.1}$$

where L_{sm} = mean stream length; L_u = total stream length of order 'u'; N_u = total no. of stream segments of order 'u'.

Table 10.3 indicates that L_{sm} values of the sub-watershed ranges from 0.063 km to 0.053 km. L_{sm} of any order is higher than the lower order in the sub-watershed except for the orders IV. This anomaly is due to the variation in slope due to local structural control.

10.4.7 Bifurcation Ratio (R_b)

It is the ratio of the number of streams of a given order to the number of streams of the next higher order (Schumm 1956). It is expressed by the equation (Schumm 1956)

$$R_b = \frac{N_u}{N_{u+1}} \tag{10.2}$$

where R_b = Bifurcation Ratio; N_u = Total no. of stream segments of order 'u'; N_{u+1} = Number of stream segments of the next higher order.

According to Horton (1945), the bifurcation ratio (R_b) is an index for watershed relief and dissections, whereas Strahler (1957) verified that the R_b exhibits a slight variation for different regions in different environments except at geologically controlled areas. Low R_b values mean structurally less disturbed watersheds without any distortion in drainage (Nag 1998). Table 10.1 shows that mean bifurcation ratio (R_b) of the catchment is between 2.62 and 3.57 which indicates that the sub-watershed is structurally less disturbed without any distortion in its drainage.

10.4.8 Relative Relief (R)

The highest relief of the area is 471 m, while the lowest value is recorded as 286 m. The relative relief (R) is calculated as the maximum drop in the basin relief divided by perimeter of the basin ($R = H/P$). Structurally controlled river flows along with straight courses and generally have more relative relief. The sinuosity of the river increases the channel length. The relative relief of the sub-watershed was calculated as 8.95, which is very low and indicates less structural control on drainage in the study area.

10.4.9 Relief Ratio (R_h)

Elevation difference between the highest and lowest points on the valley floor of a sub-watershed is termed its total relief (Rai et al. 2017). The maximum relief to horizontal distance ratio along the longest dimension of the basin parallel to the principal drainage line is the relief ratio (R_h) (Schumm 1956; Rai et al. 2017). It assesses the total steepness of a drainage basin and serves as an indicator of the severity of erosion processes occurring on the basin's slopes. The R_h usually increases with decreasing drainage area and size of a specific drainage basin (Gottschalk 1964). In the present study, the R_h value for the sub-watershed is 21.73, indicating that the watershed has a low topographic variation (Table 10.3).

10.4.9.1 Drainage Density (D_d)

Drainage density (D_d) is the nearness of drainage networks. It is the total length of the stream segment of all orders per unit area. It can be expressed by the equation given by Smith (1950).

$$D_d = \frac{L_u}{A} \qquad (10.3)$$

where D_d = Drainage Density; L_u = Total stream length of all orders; A = area of the basin (km^2).

The drainage density is affected by factors which control the characteristic length of the stream like resistance to weathering, permeability of rock formation, climate, vegetation, etc. Langbein (1947) recognized the significance of D_d as a factor determining the time of travel by water within the basin and suggested that it varies between 0.36 and 6.62 km/km^2 in humid region (Fig. 10.7b). In general, low value of D_d is observed in regions underlain by highly resistant permeable material with vegetative cover and low relief. High drainage density is observed in the regions of weak and impermeable subsurface material and sparse vegetation and mountainous relief. Drainage density values of more than 2.09 km/km^2 indicate that the watershed has excellent type of drainage (Deju 1971). The catchment with low drainage density of 2.1 km/km^2 indicates that the study area is underlain by impermeable subsurface material, with moderately high relief, excellent drainage, and therefore more prone to erosion.

10.4.9.2 Stream Frequency (F_S)

Stream frequency/channel frequency (F_S) is the total number of stream segments of all orders per unit area. It can be expressed by the equation given by Horton (1932).

$$F_s = \frac{N_u}{A} \qquad (10.4)$$

where F_s = Stream Frequency; N_u = Total no. of streams of all orders; A = Area of the Basin (km^2).

The stream frequency of the sub-watershed was found to be 167.95. A close correlation exists between stream frequency and drainage density value of the watershed, indicating the increase in stream population with respect to increased drainage density. Stream frequency is an essential parameter along with drainage density. Generally, it is used as an additional measure of the fineness of texture of surface runoff. A high stream frequency means higher runoff and vice versa. The present sub-watershed therefore reveals less runoff.

10.4.9.3 Drainage Texture (R_t)

It is the total number of stream segments of all orders per perimeter of that area (Horton 1945). Horton recognized infiltration capacity as the single important factor which influenced drainage texture (R_t) and considered the drainage texture to include drainage density and stream frequency. Thus, higher R_t values indicate lower infiltration capacity.

Table 10.4 Drainage texture values and their description (Smith 1950)

Values	Description
<4.0	Coarse
4.0 to 10.0	Intermediate
>10.0	Fine
>15.0	Ultra-fine (Bad land topography)

The drainage texture depends upon a number of natural factors such as climate, rainfall, vegetation, rock and soil type, infiltration capacity, relief and stage of development. The soft or weak rocks unprotected by vegetation produce a fine texture, whereas massive and resistant rocks produce a coarse texture. The texture of a rock is commonly dependent upon vegetation type and climate (Dornkamp and King 1971). In simple terms, drainage texture is a product of D_d and F_s. It is expressed by the equation (Smith 1950):

$$R_t = D_d * F_s \tag{10.5}$$

where R_t = Drainage texture.

Based on the values of R_t, Smith (1950) classified the drainage texture of any area (Table 10.4). In the present study, the sub-watershed has low drainage texture (6.95) and thereby having low infiltration capacity.

10.4.9.4   Form Factor (F_f)

It is defined as the ratio of basin area to square of the basin length (Horton 1932). It can be expressed by the equation (Horton 1932):

$$F_f = \frac{A}{L_b^2} \tag{10.6}$$

where F_f = Form factor; A = Area of the basin (km^2); L_b^2 = square of basin length.

Form factors are generally less than 0.7854 for a perfectly circular basin. Less form factors represent elongated basins, whereas basins with high form factors represent high peak flows of short duration. The elongated sub-watershed with low form factors on the other hand has a low peak flow of long duration. In the present study, the sub-watershed has a low F_f value of 0.26 which indicates its elongated shape and less intensity of flood flows for long duration. Therefore, flood flows of elongated basins are easy to manage than circular basins.

10.4.9.5   Circulatory Ratio (R_c)

The ratio of basin area to the area of the circle having the same circumference as its perimeter is circularity ratio (Miller 1953). It is subjective to the length and frequency of streams, geological structures, LU/LC, climate, relief, and slope of the basin. In the present study, the circulatory ratio of the sub-watershed is 0.57, which confirms its elongated shape and less structural control (Table 10.3). It can be expressed by the equation (Miller 1953):

$$R_c = \frac{4*\pi*A}{P^2} \qquad (10.7)$$

where R_c = Circularity Ratio; π = 'π' value, i.e., 3.14; L_b = basin length (km).

10.4.9.6   Length of Overland Flow (L_g)

It is the length of water over the ground before it gets concentrated into definite stream channels (Horton 1945). This factor relates inversely to the average slope of the channel and is quite synonymous with the length of sheet flow to a large degree. It approximately equals to half of the reciprocal drainage density (Horton 1945). It can be expressed by the equation Horton (1945).

$$L_g = \frac{1}{D_d*2} \qquad (10.8)$$

where L_g = Length of overland flow; D_d = drainage density.

The value of L_g was calculated to be 4.95. The medium length of overland flow indicates that the runoff will get sufficient time to enter the streams in the catchment thereby providing more time for infiltration. This will provide for good groundwater recharge techniques and thereby increasing of groundwater level in the study area.

10.4.9.7   Constant of Channel Maintenance (C)

Schumm (1956) has used the inverse of drainage density as a property termed constant of channel maintenance (C). It can be expressed by the equation.

$$C = \frac{1}{D_d} \qquad (10.9)$$

where C = Constant Channel Maintenance; D_d = Drainage Density.

It tells the number of square feet of watershed surface required to sustain one linear feet of channel. The value of C for the watershed is 0.10, which means that on an average 0.38 sq.ft. surface is needed in the sub-watershed to support each linear foot of the channel.

10.5   Conclusions

Morphometric analysis of drainage systems is a prerequisite to any hydrological study. Thus, the determination of stream networks' behavior and their interrelation is of great importance in many water resources studies. Implementation of morphometric analysis in a coal mining region may provide helpful insight for any hydrological investigation like an assessment of groundwater potential, groundwater management, basin management, and environmental assessment. The present study demonstrated that the remote sensing satellite data and GIS techniques had been valuable tools in drainage delineation and calculation of various morphometric parameters. The Damodar River sub-watershed is well-drained in nature, with stream order ranging from 1 to 5. The drainage density for the study area has been calculated as 9.91 km/km^2, revealing that the subsurface area is permeable, a characteristic feature of coarse drainage. Morphometric analysis of the present study will be essential for field-based observation and computation strategies to be adopted for managing purposes. The results obtained in the present study also provide valuable insight for planning groundwater recharge techniques and increasing the area's water level. The used approaches in this study include a comprehensive morphometric analysis that can apply to any drainage system elsewhere.

Acknowledgements   The authors are grateful to the Department of Remote Sensing, Birla Institute of Technology, Mesra, Ranchi to provide all the necessary facilities for carrying out this work.

References

Abdulkareem JH, Pradhan B, Sulaiman WNA, Jamil NR (2018) Quantification of runoff as influenced by morphometric characteristics in a rural complex catchment. Earth Syst Environ 2(1):145–162

Al-Daghastani NS, Al-Maitah KJ (2006) Water harvesting using morphometric analysis and GIS techniques: a case study of the HRH Tasneem Bint Ghazi for Technology Research Station. In: Proceedings of ISPRS mid-term symposium 'remote sensing: from pixels to processes', WG VII/7, Enschede, Netherlands, pp 8–11

Astras T, Soulankellis N (1992) Contribution of digital image analysis techniques on Landsat-5 TM imageries for drainage delineation. A case study from the Olympus mountain, West Macedonia, Greece. In Proceedings of the 18th Annual Conference Remote Sensing Soc., Univ of Dundee, pp 163–172

Avinash K, Jayappa KS, Deepika B (2011) Prioritization of sub-basins based on geomorphology and morphometricanalysis using remote sensing and geographic informationsystem (GIS) techniques. Geocarto Int 26(7):569–592

Bagyaraj M, Gurugnanam B, Nagar A (2011) Significance of morphometry studies, soil characteristics, erosion phenomena and landform processes using remote sensing and GIS for Kodaikanal Hills, a global biodiversity hotpot in Western Ghats, Dindigul District, Tamil Nadu, South India. Res J Environ Earth Sci 3(3):221–233

Chopra R, Dhiman RD, Sharma PK (2005) Morphometric analysis of sub-watersheds in Gurdaspur district, Punjab using remote sensing and GIS techniques. J Indian Soc Remote Sens 33(4):531–539

Chandniha SK, Kansal ML (2017) Prioritization of sub-watersheds based on morphometric analysis using geospatial technique in Piperiya watershed, India. Appl Water Sci 7(1):329–338

Deju R (1971) Regional hydrology fundamentals. Gordon and Breach Science Publishers, Newark. https://doi.org/10.1007/500254-006-0297-Y

Dornkamp JC, King M (1971) Numerical analyses in geomorphology: an introduction. St. Martins, New York, 372p

Gayen S, Bhunia GS, Shit PK (2013) Morphometric analysis of Kangshabati-Darkeswar Interfluves area in West Bengal, India using ASTER DEM and GIS techniques. J Geol Geosci 2(4):1–10

Gottschalk LC (1964) Reservoir sedimentation. In: Chow VT (ed) Handbook of applied hydrology. McGraw Hill Book Company, New York, Section 7-1

Horton RE (1932) Drainage-basin characteristics. Trans Am Geophys Union 13:350–361

Horton RE (1945) Erosional development of streams and their drainage basins: hydrophysical approach to quantitative morphology. Geol Soc Am Bull 56:275–370

Jasmin I, Mallikarjuna P (2013) Morphometric analysis of Araniar river basin using remote sensing and geographical information system in the assessment of groundwater potential. Arab J Geosci 6(10):3683–3692

Krishnamurthy J, Srinivas G (1995) Role of geological and geomorphological factors in ground water exploration: a study using IRS LISS data. Int J Remote Sens 16(14):2595–2618

Kudnar NS, Rajasekhar M (2020) A study of the morphometric analysis and cycle of erosion in Waingangā Basin, India. Modeling Earth Syst Environ 6(1):311–327

Kumar A, Krishna AP (2018) Assessment of groundwater potential zones in coal mining impacted hard-rock terrain of India by integrating geospatial and analytic hierarchy process (AHP) approach. Geocarto Int 33(2):105–129

Kumar A, Krishna AP (2021) Groundwater quality assessment using geospatial technique based water quality index (WQI) approach in a coal mining region of India. Arab J Geosci 14(12):1–26

Kumar A, Pramod Krishna A (2020) Groundwater vulnerability and contamination risk assessment using GIS-based modified DRASTIC-LU model in hard rock aquifer system in India. Geocarto Int 35(11):1149–1178

Langbein WB (1947) Topographic characteristics of drainage basins. US Geol Surv Water-Supply Paper 986(C):157–159

Magesh NS, Chandrasekar N (2014) GIS model-based morphometric evaluation of Tamiraparani subbasin, Tirunelveli district, Tamil Nadu, India. Arab J Geosci 7(1):131–141

Magesh NS, Chandrasekar N, Soundranayagam JP (2012) Delineation of groundwater potential zones in Theni district, Tamil Nadu, using remote sensing GIS and MIF Techniques. Geosci Frontiers 3(2):189–196

Magesh NS, Jitheshlal KV, Chandrasekar N, Jini KV (2012) GIS based morphometric evaluation of Chimmini and Mupily watersheds, parts of Western Ghats, Thrissur District, Kerala, India. Earth Sci Informatics 5(2):111–121

Markose VJ, Dinesh AC, Jayappa KS (2014) Quantitative analysis of morphometric parameters of Kali River basin, southern India, using bearing azimuth and drainage (bAd) calculator and GIS. Environ Earth Sci 72(8):2887–2903

Mesa LM (2006) Morphometric analysis of a subtropical Andean basin (Tucuman, Argentina). Environ Geol 50(8):1235–1242

Miller VC (1953) A quantitative geomorphologic study of drainage basin characteristics in the Clinch Mountain area, Virginia and Tennessee. Project NR 389042, Tech Report 3. Columbia University Department of Geology, ONR Geography Branch, New York

Mishra A, Dubey DP, Tiwari RN (2011) Morphometric analysis of Tons basin, Rewa District, Madhya Pradesh, based on watershed approach. Earth Science India, p 4

Nag SK (1998) Morphometric analysis using remote sensing techniques in the Chaka sub-basin, Purulia district, West Bengal. J Indian Soc Remote Sens 26(1):69–76

Nautiyal MD (1994) Morphometric analysis of a drainage basin using aerial photographs: a case study of Khairkuli Basin, District Dehradun, UP. J Indian Soc Remote Sens 22(4):251–261

Panda B, Venkatesh M, Kumar B (2019) A GIS-based approach in drainage and morphometric analysis of Ken River basin and sub-basins, Central India. J Geol Soc India 93(1):75–84

Pandey AC, Kumar A (2014) Analysing topographical changes in open cast coal-mining region of Patratu, Jharkhand using CARTOSAT-I Stereopair satellite images. Geocarto Int 29(7):731–744

Pareta K, Pareta U (2011) Quantitative morphometric analysis of a watershed of Yamuna basin, India using ASTER (DEM) data and GIS. Int J Geomatics Geosci 2(1):248

Rai PK, Chandel RS, Mishra VN, Singh P (2018) Hydrological inferences through morphometric analysis of lower Kosi river basin of India for water resource management based on remote sensing data. Appl Water Sci 8(1):1–16

Rai PK, Mishra VN, Mohan K (2017) A study of morphometric evaluation of the Son basin, India using geospatial approach. Remote Sens Appl: Soc Environ 7:9–20

Rawat PK, Tiwari PC, Pant CC, Sharama AK, Pant PD (2011) Morphometric analysis of third order river basins using high resolution satellite imagery and GIS technology: special reference to natural hazard vulnerability assessment. E-Int Sci Res J 3(2):70–87

Sarkar D, Mondal P, Sutradhar S, Sarkar P (2020) Morphometric analysis using SRTM-DEM and GIS of Nagar River Basin, Indo-Bangladesh Barind Tract. J Indian Soc Remote Sens 48(4):597–614

Saxena PR, Prasad NSR (2008) Integrated land and water resources conservation and management–development plan using remote sensing and GIS of Chenvella sub-watershed, RR District, Andhra Pradesh, India. The international archives of the photogrammetry. Remote Sens Spatial Information Sci 37:729–732

Schumm SA (1956) Evolution of drainage systems and slopes in badlands at Perth Amboy, New Jersey. Geol Soc Am Bull 67(5):597–646

Singh S, Singh MC (1997) Morphometric analysis of Kanhar river basin. Natl Geogr J India 43(1):31–43

Smith KG (1950) Standards for grading texture of erosional topography. Am J Sci 248(9):655–668

Strahler AN (1964) Quantitative geomorphology of drainage basins and channel networks. In: Chow VT (ed) Handbook of applied hydrology. McGraw Hill Book Company, New York, pp 4–11

Strahler AN (1957) Quantitative analysis of watershed geomorphology. EOS Trans Am Geophys Union 38(6):913–920

Vittala SS, Govindaiah S, Gowda HH (2004) Morphometric analysis of sub-watersheds in the Pavagada area of Tumkur district, South India using remote sensing and GIS techniques. J Indian Soc Remote Sens 32(4):351–362

Zaidi FK (2011) Drainage basin morphometry for identifying zones for artificial recharge: a case study from the Gagas River Basin, India. J Geol Soc India 77(2):160–166

Zolekar RB, Bhagat VS (2015) Multi-criteria land suitability analysis for agriculture in hilly zone: Remote sensing and GIS approach. Comput Electron Agric 118:300–321

Chapter 11
The Increasing Inevitability of IoT in Remote Disaster Monitoring Applications

Vishal Barot, Srishti Sharma, and Prashant Gupta

Abstract Disaster is a term used for an unusual event that has a negative impact on life, environment, and materials. This event is more unwanted than unusual as most of the time this is caused due to reasons, we remain unaware of. Disasters could be natural or man-made like floods, earthquakes, volcanic eruptions, accidents, gas leakage, forest fire, etc. Since the disaster occurs due to reasons that are unknown and is an untimely phenomenon, we have a lesser scope of avoiding it. The use of emerging technologies, early warnings, immediate incidence response, and post-recovery activities can well be employed to mitigate the losses that the disaster would lead to. One such emerging technology is the Internet of things (IoT). The sensor nodes used in an IoT architecture independently are capable of sensing the environment and gathering data. This data can be further sent to the sink node/base station which has the responsibility of reporting the data ahead of the network, for initiating actuation on the basis of the data gathered and analyzing the data. This can help in monitoring for the purpose of early warning of disasters. The actuations can help provide immediate incidence response. Furthermore, upon the integration of data analytics with IoT, the data generated with the help of sensors can be used for predictive analysis and inference drawing in order to enable early disaster prediction and mitigation of losses that might occur due to the same. Thus, the potential of IoT is to provide rescue, response, mitigation, and preparedness to manage a disaster. This chapter explains the role of IoT in disaster management along with proposing a generic model having distinct layers through well-defined functionality. The chapter also explains the integration of cloud and IoT that could improve the efficiency of IoT applications in disaster management.

V. Barot
LDRP Institute of Technology and Research, Gandhinagar, India

S. Sharma
Ahmedabad University, Ahmedabad, India

P. Gupta (✉)
Maharashtra Institute of Technology, Aurangabad, India

© The Author(s), under exclusive license to Springer Nature Singapore Pte Ltd. 2022   227
P. K. Rai et al. (eds.), *Geospatial Technology for Landscape and Environmental Management*, Advances in Geographical and Environmental Sciences,
https://doi.org/10.1007/978-981-16-7373-3_11

Keywords Disaster management · Natural disaster · IoT architecture · Disaster technology · Industrial IoT

11.1 Introduction

Disaster, in general, can be termed as a sudden accident or a natural catastrophe that leads to harm of life, environment, and materials. Some of the natural disasters include but is not limited to tsunamis, floods, earthquakes, landslides, hurricanes, etc. (Perry 2007; Onagh and Kumra 2012; Singh et al. 2020; Sur et al. 2021). The downside to such an event is the inability of the society at large to manage such a calamity on their own, without the use of any external assistance. These have been responsible for unpleasantly affecting the community since the early presence of civilizations with one of the oldest examples of note being Mediterranean earthquake in Egypt and Syria in the year 1201 almost taking a million human lives. The large-scale impact of such events on the society has in turn made the individuals in the community respond, leading to the development of measures to cope up with the initial impact along with post-event and recovery demands. A comprehensive effort toward the goal in the picture here can be termed as disaster management. The aim of disaster management is to minimize the harm to lives, environment, and materials across the world. The non-uniformity to execute the preparation and response measures to a disaster across the globe due to differences in societal, economic, political reasons, etc., makes some countries more accomplished than others to handle the issue at hand (Coppola 2006) (Table 11.1).

Although the practices vary from nation to nation, none of them is fully insusceptible from the harmful and damaging effects of any disaster and thus calls for having good disaster management plans in place. In today's global economy, the influence of one's economy directly or indirectly affects the global perspective of the same, and even the developed nations are not immune in such a scenario. The role of government is very crucial in formulating a centralized, synchronous action plan to address the people's needs in dire situations post the disaster event. Also, it is important to ascertain the degree of risk involved during an emergency. The national forces can also be called upon in case of an acute need to maintain law and order during the chaos and help to rebuild communities post the event (Jeanty 2017). The Federal Emergency Management Agency of the USA and National Disaster Management Authority in India are some examples of national agencies responsible for managing a disaster situation on the behalf of the government. Along with the same, some non-government, non-profit organizations such as All hands Volunteers, Direct Relief International, The World Vision, The React International, The Salvation Army, and The American Red Cross complete an exhaustive list of disaster management agencies working round the clock across the globe (Galaxy Digital 2019).

The focus till now has been on post-ops to the disaster event. With the development of applied technologies, it is possible to detect disasters early and try to be ready with

Table 11.1 Impact of major occurrences of natural disasters in last decade (Meyers 2019)

Sr. No	Year of occurrence	Disaster event and place	Effects to the nation (material and economic loss)	Human lives lost	References
1	2010	Haiti Earthquake at Haiti (magnitude 7.0)	Excess of 1,500,000 people displaced	220,000	("Fast Facts on the U.S. Government's Work in Haiti" 2013)
2	2011	Tohuku Earthquake (magnitude 9.0) and Tsunami (133 ft high wave) in Tohoku; Fukushima Daiichi Nuclear Disaster	45,700 destroyed buildings and 144,300 damaged ones with 100,000 affected children	19,575	McCurry (2011)
3	2012	Hurricane Sandy (Category 2 storm with winds around 185 km/h) starting from the Caribbean	Loss worth 70 Billion USD across 8 countries from Caribbean to Canada (including USA)	233	Diakakis et al. (2015)
5	2013	Typhoon Haiyan; (Category 5 storm equivalent with winds around 200 km/h) Philippines	11 million people affected with a loss of around 2.2 billion USD in property (1.08 million houses)	6352	(Del Rosario and Usec 2014; "Tacloban: City at the Centre of the Storm—BBC News" 2013)
6	2015	Gorkha Earthquake, (Magnitude 7.8) Nepal	22,000 injured people with 500,000 destroyed houses and an overall loss of around 7 Billion USD; around 35% of Nepal's GDP	9000	(Nepal 2015; *Bloomberg Business* 2015)
7	2017	Hurricane Harvey; (category 4 storm with winds at 215 km/hr) US coast starting from Texas	Trillions of gallons of rain in southern coast resulted in floods leading to loss of 125 billion USD and rescue of 17,000 people with around 30,000 displaced	107	(McCausland et al. 2017; NOAA's National Centers for Environmental Information (NCEI) 2018)

(continued)

Table 11.1 (continued)

Sr. No	Year of occurrence	Disaster event and place	Effects to the nation (material and economic loss)	Human lives lost	References
8	2017	Hurricane Maria; Dominica (category 5 storm with winds at 220 km/hr	3.4 million affected people with an estimated loss of 90 billion USD	3000	(NOAA's National Centers for Environmental Information (NCEI) 2018)
9	2019	Cyclone Idai; Southern Africa (winds at 195 km/hr);	Flooding of agricultural land with salty water in Mozambique, Zimbabwe, and Malawi affecting 3 million human lives	1300	(Funes 2019)
10	2019	Global wildfires	50,477 wildfires burning away 4.7 million acres of land	Unknown	("Facts + Statistics: Wildfires I III" 2017)

the challenge at hand prior to the occurrence of such an event. The use of emerging disruptive technologies such as big data, artificial intelligence (AI), and the Internet of things (IoT) can do a world of good in terms of disaster risk reduction and management with the assistance of robotics and drones. IoT is a concept that represents a virtual network of physical objects (Awasthi and Mohammed 2019). When electronic equipment such as sensors in connection with controller boards are embedded in real-world objects or are placed in the real-world environment/scenarios, they can sense the environment and measure the parameters they are designed for such as temperature, location, light, weight, pulse, and moisture. These parameters can be communicated using wireless communication technology such as GSM, GPRS, and IEEE 802.11 to a remote entity. The data can be sent over the cloud or may be sent as an SMS to a contact number or as an HTTP response and displayed to the user. On the basis of the data received, the analysis takes place over this entity present at the receiving end and an actuation takes place in response whenever necessary. For instance, if a building is well equipped with fire detection sensors, there is a notification along with the geo-coordinates of the building relayed to the nearest fire station, in case the sensors sense high temperatures that may accumulate due to fire. This can be done using a GSM module that places an automated call and also drops a message providing the required information on an urgent basis. In addition to the same, the simultaneous impact is brought about by automatic unlocking of the local fire extinguishers and the actuation of shower sprinklers resulting in water pouring at the site of the fire. The response for this case in particular is in the form of fire alarm, sprinkling of water, and intimation of the incident to the nearest fire station.

This results in an amalgamation of an embedded system, a wireless communication technology-based network, and a respondent system representing information technology usage in the form of IoT.

WSN is an important aspect of IoT-based applications. It is a cluster of sensors that are placed at a distance from each other such that they are in the range [18]. These sensors independently sense the environment and gather data that they then send to the sink node or the base station which is the authority responsible for reporting the data ahead of the network or responsible for initiating a certain actuation on the basis of the data gathered or for analysis of the data. There is a need to constantly monitor the relevant parameters to limit the damage that can be caused by natural disasters. The monitoring of these parameters at adverse locations such as in the ocean/sea, forest, earth's crust is not possible without the use of technology. One of the salient features of WSNs makes them suitable for such monitoring activities. These sensors can easily configure their network dynamically once dropped in land, air, or sea. In addition to the same, they can keep sensing the environmental parameters constantly and do not require to be attended. The data collected by these WSNs is sent over the cloud by making use of a local network architecture extended with wireless communication technology like the Internet (Renold and Ganesh 2019). The use of real-time monitoring of environmental parameters has helped researchers analyze the patterns of the changes taking place in the environmental conditions and thus predict the occurrence of a disaster in time, so as to minimize the losses of life, environment, and property. The wide coverage of WSNs has made IoT inevitable for applications such as detection of forest fires, monitoring of nuclear power stations prediction of earthquakes, landslides, flood and epidemics like dengue; monitoring of railway track health, etc.

This chapter aims at providing a holistic view of the usage of IoT in mitigating the risks posed by natural disasters on life, environment, and infrastructure to the extent possible. The authors aim to present a structured illustration of various applications to exhibit the use of IoT in remotely monitoring and mitigating conditions that may lead to natural disasters. Furthermore, it will include the scope of data analysis to enable the use of data gathered by the IoT architecture in predicting the disasters before they impact lives.

11.2   Disaster Management

Various governments have combined with non-government organizations and private companies to strategize new tactics for managing disasters occurring partially due to a change in climate and rapid development in urban areas across the globe (Olorun-toba et al. 2018). The term disaster management can be termed as the methodology of management and organization of resources/responsibilities for coping up with the readiness, response, relief, and recovery aspects of disasters or emergencies

(Toribiong 2010). The framework of disaster management in any nation is a multi-hierarchical one with state and local governments, federal bodies, civil society organizations, and the private sector involved at different jurisdiction levels. It ensures the alignment of key indicators that are measurable such as roles and responsibilities of different actors and stakeholders at various levels across the nation. Also, it designates the best practices for disaster management which considers other aspects as well, along with recovery and response, thereby providing a holistic approach which is in the form of cohesive and inclusive policy of the nation (Daly et al. 2017; Chan et al. 2020).

The disaster management cycle consists of four phases namely: Mitigation, Preparation, Response, and Recovery ("Phases of Disaster" n.d.; Backfried et al. 2013; Rolland et al. 2010; Lettieri et al. 2009) as shown in Fig. 11.1 wherein.

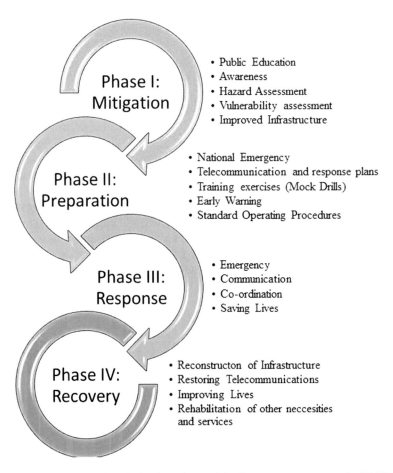

Fig. 11.1 Activities carried out during four phases of the disaster management cycle (DMC)

- Phase I i.e., Mitigation deals with making plans to decrease the likelihood and/or eliminate the sequence of events that may eventually lead to a hazard. It also seeks to deal with the hazard so as to lighten its impact on society.
- Phase II i.e., Preparedness deals with gaining an understanding of the impact of a disaster on the community and the effect of education, training, and outreach on capacity building to respond and recover from an unfortunate event. It helps in preparing who might be impacted and who might help those who are impacted by a disaster for improvement of survival rate along with the minimization of material and financial losses.
- Phase III i.e., Response involves addressing immediate threats put forth by the disaster by taking action to save lives, ensuring necessities for living such as food, clothing, shelter, safety, and public health; damage and assessment. It eventually leads to conducting repairs, public services, and restoring utilities after the emergencies are tended to.
- Phase IV i.e., Recovery deals with return of normalcy in the life of the victims following the consequences of the disaster. It follows immediately after response and carries on until months or years till the community impacted by the disaster reaches some sort of social, physical, environmental, and economic stability.

11.3   Role of IoT in Disaster Management

Disaster, man-made or natural, pose a great threat to civilizations due to the adverse effect they have on lives, property, economies, businesses, homes, etc. The advancements in technology plays a pivotal role while managing emergency situations such as disasters, thereby leading to avoiding/mitigating the losses incurred due to adverse conditions (Ray et al. 2017). The necessary components that make up IoT technology are devices consisting of sensors to pick up data, wireless network, cloud, data processing, and the resulting analytics transmitted to the user via a user interface. The monitoring of various conditions such as climate, weather, and rainfall levels can be made possible with the use of various sensors. This in combination with developments in broadband wireless networks, cloud computing, and data analytics has potentially led to the rise of real time and integrated systems. It is very difficult for IoT to stop disasters but it can prove its usefulness in the preparedness phase as it can predict and serve as an early warning system. It is capable enough to provide a timely warning system, immediate rescue operation steps, effective necessary assistance to victims, and other post-disaster follow-up mechanism and regulations techniques along with data analytics for future prevention of disaster (Kamruzzaman et al. 2017; Sakhardande et al. 2016). It can be employed for managing disasters such as forest fires, volcanic eruptions, nuclear leaks, landslides, earthquakes, floods, and road accidents (Boukerche and Coutinho 2018; Sharma and Kaur 2019). A sensor can alert a potentially dangerous situation such as a tree sensor detecting a spike in temperature, moisture, and CO_2 levels indicating nearby fire, ground sensor detecting movements in ground indicating a possible earthquake and river sensors monitoring for water level to indicate the possibility of flood situations. It has been reported (Max et al. 2005) that an excess of 95% of all disaster deaths across the globe occurred in developing countries. IoT can be a boon in developing countries as they are vulnerable to such natural calamities possibly due to poor healthcare, infrastructure, insurance, and lack of preparation to handle response and recovery from disasters. IoT can be considered as an efficient technology in the context of managing disasters and a summary of various phases in which it can be employed is given below.

1. Minimization of disaster risk and its prevention: There could be WSNs deployed at various locations to sense, monitor, and record parameters such as the temperature, amount of gases, water level, light intensity, sound, seismographic readings, and motion (Poslad et al. 2015). A closed-circuit television (CCTV) surveillance and global positioning system (GPS) can also be implemented for activity and location tracking, respectively. Also, a suitable wireless communication technology could help report an unusual monitored reading identified by the WSN to the concerned authorities like a warning. This type of a setup could help design an early warning system for disasters and can help to spread awareness about it among the masses so as implement protocols for precautionary measures.

2. Responding to the emergency—It deals with instant relief and rescue measures to be taken after the disaster has occurred. As shown in Fig. 11.2, in case of fire in a building with an IoT-based fire safety management system, the WSN deployed in the building will be able to identify the fire condition immediately with the help of temperature and smoke detection sensors (Poslad et al. 2015). After, the detection of fire, a command will be sent by the controller board, which is a component of IoT system, to the actuators that will start the sprinkler system deployed across the whole building. In addition to that, the fire extinguisher board will get unlocked. Apart from these two activities, the controller board has been programmed with the logic of notifying the nearest fire station about the fire incident via a text message or a phone call or over a control panel monitored by the fire safety team via making use of a web service. This can alert the fire extinguishing authorities and they can immediately be informed about the incident. In this IoT architecture, initial activities such as prompt action over the incident, intimating the authorities, are better than traditional emergency response and not dependent upon the availability of human resources (Zambrano et al. 2017). This kind of response to an emergency condition with the use of IoT could reduce the amount of losses incurred during the disaster to a larger extent.

3. Recovery from Disaster—IoT can be employed for post-disaster management activities, e.g., as an interface that the people could use to report if there was a person missing and CCTV surveillance and GPS tracking could be used for identifying the person during search operations. Another example could be in the use of water level identification sensors for identification of any water logging in the low-lying areas due to floods so that rescue measures could be taken up in such areas (Zambrano et al. 2017). Also, animal tracking is yet another

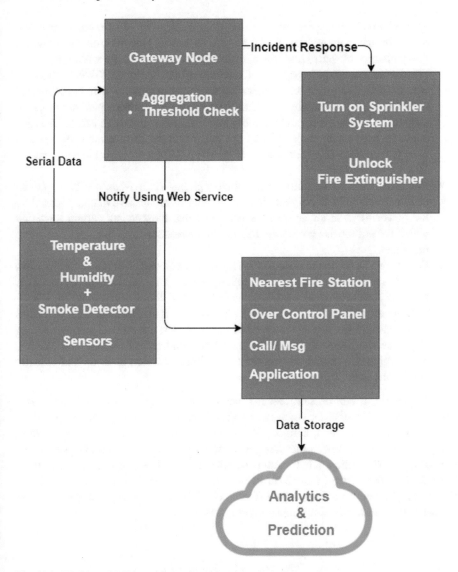

Fig. 11.2 Working of IoT-based fire safety management system

application that could help identify if there were animals trapped at a certain region in the forest amidst the forest fire.

An account of managing potential disasters with the use of IoT as the key enabling technology in regard to monitoring, mitigation, and prevention is given as under:

- Early Earthquake Warning System—When an earthquake occurs, there is a small shock wave emitted first, which is then followed by a dangerous wave. Also, the span of time between the two waves is only as large as a few minutes and may not be sufficient enough to take large scale measures. However, certain activities like halting of elevators and trains, machines, etc., can be done within this small time frame to limit damage prevention (Ghasemi and Karimian 2020). Yet another way of identification is the acceleration sensor in the phone. If the acceleration sensor in phones of nearby people in the same region shows a particular type of a signal generating a common pattern that resembles that of an earthquake, this could be used as an early warning system. An application named MyShake is used for detection of an earthquake of up to 4.5 Richter scale within a distance of 10 km (Wald 2019).

- Release of radioactive matter from a nuclear plant—An IoT-based WSN deployed at a nuclear plant can help monitor the amount of radioactive matter released by the power plant in air or water. When this release exceeds a certain threshold value, the authorities are notified about the same so as to implement corrective measures.
- Forest Fire Detection and Remedy: The WSN deployed at the forest regions has sensors such as temperature and humidity sensor, CO_2, and CO gases detection sensor. These sensors also provide the GPS coordinates of their location. Whenever the sensed value of these sensors exceeds a certain threshold value, the GPS coordinates of the region are reported to the nearest operational station that can then take the necessary measures for control and rescue (Dubey et al. 2019). The GPS system could also report congestion of traffic condition and climatic conditions such as thunderstorms.

In general, it can be said that IoT has a role to play in almost all the phases in dealing with a disaster condition ranging from its mitigation to recovery. Be it an early warning system, notifying the concerned authorities immediately, alerting the people, taking control measures, analyzing the type of disaster and parameters related to it (recording data for future avoidance and prevention), rescue operations (notifying about escape paths), statistics update (water level indication in low-lying areas) and other mitigation steps, IoT can help fight disasters in an effective manner ensuring the least possible damages and losses.

11.4 IoT Model for Disaster Management

The flow of data and subsequent operations in IoT, a key technology in the disaster management process is more or less similar for various disasters. However, the context of information sensed and recorded, followed by actuations or the countermeasures performed, often technically called response measures or post-disaster recovery methods can vary depending on the context of the application (Mouradian et al. 2018).

Figure 11.3 shows the general architecture or model for IoT that can be employed in disaster management. There are four distinct layers identified in the model namely infrastructure, event identification, incident response, and analysis as discussed below.

- Infrastructure comprises of WSN deployed in a building or any location such as in the forest or in the sea or over-bridge or in a nuclear reactor power plant, etc. The reason for opting for a WSN and not just a single sensor for reporting the sensed data is the accuracy and reliability of the readings reported by it. Hence, the aggregated recordings of various sensors in the network are more accurate and reliable compared to readings reported through single sensors (Shah et al. 2019). The nodes in WSN are capable of routing its packet containing the sensed data to the gateway node via intermediate nodes. This layer also consists of actuators

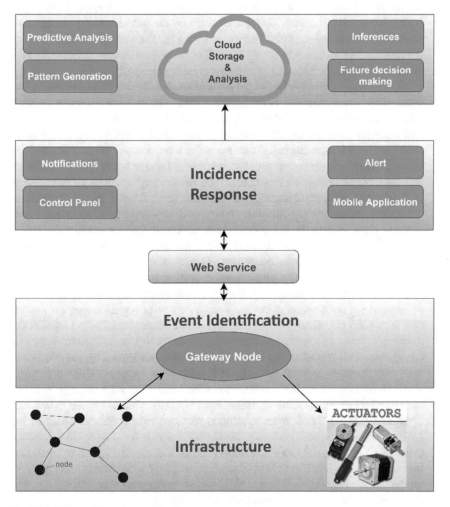

Fig. 11.3 IoT application model for disaster management

that help to control or regulate the damage done as a response to identification of the disaster. These actuators are either programmed to operate upon detection of an anomaly condition on the basis of the sensor readings or are controlled via human intervention, thus making use of a control panel or an application interface as and when required.

- Event identification occurs when the data from all the sensors in the WSN reaches the gateway node, a node closest to the edge of the network, and the aggregate reading is then tested against the threshold parameters for the sensed data. If the aggregate sensed data crosses the threshold value or falls below the threshold value (programmed logically over the controller board on the basis of previous observations), a disaster event or a pre-disaster occurrence is identified. The gateway node can gather the data from the sensor nodes of WSN in two possible ways, i.e., either via the nodes routing their packets to the gateway node in the regular manner or polling/requesting the particular sensor node by the gateway node (coordinator node) for sensed value in the particular instance. The second methodology of data gathering is prominent in case of need of the sensed environment reading for any region (Greco et al., n.d.). The gateway node is responsible for directing the actuators to perform actuation on the basis of an identified disaster or pre-disaster event. The communication between WSN and actuators via the gateway node takes place using Wi-Fi, Zigbee, Bluetooth, etc. (Furquim et al. 2018)

- Incidence response means the countermeasures implemented upon detection of an unusual event referring to a disaster event or a pre-disaster event. The response to any detected disaster event is of two types with the first being direct automated, i.e., autonomous actuation that needs to be performed for damage control so as to reduce the amount of damage and loss and second being the notification to concerned authorities for decision making and implementation of countermeasures in place. When the gateway node detects a disaster or pre-disaster event, it passes a command to the actuators in the infrastructure directly or automatically (logically programmed command as per threshold) for performing counter actuations according to the autonomous actuation response. This is believed to be an immediate measure of response toward an emergency. In notification type of response, a notification about the event along with any an important parameter, e.g., GPS coordinates of the location or any other identification number for a region is sent to the concerned authority for them to tend to the emergency as a response. This response can also be an actuation performed by the concerned authority remotely or with the help of reaching the location. This notification could be sent over a mobile application in the form of an alert message over the control panel along with the statistics. For the notification to be sent over a call or message, GSM/GPRS module setup is required with controller board of the gateway node along with a web service implementation for playing over a control panel or displaying over an application (Furquim et al. 2018). The communication technologies used for data communication in IoT for a disaster management model are Wi-Fi, 2G, 3G, 4G, and WiMAX, etc. (Furquim et al. 2018).

- Analysis in the context of disaster management means analyzing the nature of the disaster via the data gathered through sensors. As the data reported by the gateway nodes are displayed on the application interface or control panel, it can be stored over an external device, e.g., cloud by the use of web service for the purpose of analytics. It is efficient in drawing patterns or inferences out of the cloud storage such as information over the impact of the disaster on lives, property, efficiency of adopted countermeasures, rate of success or failure of these measures. If analyzed, all the above-mentioned information will help to fight against the disaster in the future (Sinha et al. 2019).

Furthermore, the concept of Industrial IoT (IIoT) is pertinent to be mentioned here as it can prove valuable in both natural and man-made disasters in the age of radical technological transformation. It refers to the concept of IoT in the industrial environment and can be defined as the equipment, machinery, humans, and computers empowering tasks of industries smartly with the help of artificial intelligence and data analytics for effective business solutions. Let us understand the transformation that can be brought upon a nuclear power plant with the implementation of IIoT. The conventional way of operating in a nuclear plant would make one find the operators moving the gauges and levers manually for any control system. The fourth industrial revolution has basically improvised the way operations are carried out in a nuclear power plant. IIoT is capable to establish a connection between the machines and devices using web services that can be deployed on the cloud, thereby enabling quick and easy access to all the machines and thus, a better way to control the operations. The setup of IIoT for nuclear plant comprises of the radioactive gas level detection sensor network for radioactive gases and water level regulator switches. The recorded data is sent to a gateway node which examines the radioactive gas levels with respect to the threshold values and induces a response. The employees are equipped with devices that have sensors and controllers can send an immediate notification to emergency responders with specific details of the incident. The location trackers save precious time of the emergency responders by intimating them for reaching to the location as soon as possible. In the worst-case scenario, depending on the extent of damage that may be occurred, the controllers may directly send a command to turn on the water switch so as to let water pour in the plant. It allows the plant to prioritize the lives of the employees. The information is instantaneously relayed through email, alarms, and even text messages. A notification is sent making use of wireless communication technology to the concerned authorities. The placement of a siren or a buzzer too may result in ringing in the form of an alarm for the employees to vacant the premise of safety and security purposes. Lastly, the real-time environment monitoring is done by the sensors. If the entire incident-related data is stored over the cloud, this disaster could be dealt with in time with more effectiveness in case of reoccurrence in the future (Wellington and Ramesh 2017).

Thus, with the integration of the fundamentals of IoT, the efficiency of the operations is improved and the system can have centralized control. The operators can take quick and effective decision based on the historical data and with the help of quick and accurate decisions, unnecessary energy expenditure can be avoided. It also contributes toward an increase in the lifespan of the machines and reduce the maintenance cost.

11.5 IoT and Cloud Integration

The role of the cloud in analytics layer of the IoT model for disaster management is explained earlier wherein the data from the gateway node is sent over the cloud via a web service to store and retrieve it as per its use. The application of machine learning algorithms for predictive analysis or drawing inferences over a large amount of gathered data takes place over the cloud due to its versatile storage capabilities and facilitation of remote access which could result in managing the disasters better if a similar kind of disaster event is encountered in the future (Chung and Park 2016).

IoT-based applications, used for managing disasters, are mostly based on WSNs that are deployed at remote locations which as often unattended. These sensors are battery-operated devices and excessive computation and communication over-heads deplete their energy level. This generally results in higher battery usage and subsequent lifespan of the sensor, thereby hampering its application by limiting the longevity of the WSN. The scope of losing out sensor nodes in WSN could ultimately lead to inaccuracy or incompetency of the application resulting in major losses of lives, properties, economies, etc. It is reported that the highest amount of energy consumption by a sensor node is caused due to the computational and communication overhead imposed upon the nodes, which is considered as the core (Gaire et al. 2018). The application of virtually unlimited computation and storage capabilities of cloud could reduce the energy consumption of IoT application nodes by moving a major part of computation over the cloud.

The techniques such as data compression can be deployed on the gateway node which serves as an edge layer introduced between the IoT architecture and cloud. The data before being sent to cloud is compressed at the edge layer and then sent to the cloud to reduce the communication overhead incurred by this gateway node, a battery-operated device working as a coordinator in this application (Azar et al. 2019). This compressed data can be decompressed over the cloud and can then be used for the purpose of analytics. Caching of data is another such technique that can be deployed over this Cloud IoT Integrated Architecture. The detection of most disasters prior to the occurrence is on the basis of a pattern extracted accurately from larger data size. If sensor readings are sent to the cloud over a periodic cycle of one week, for instance, any reading requested by a third party for analysis can be sent to them on the basis of the data already stored on the cloud. In this context, data analytics plays a major role as pattern matching needs to be performed for deciding that the previous week's data would match the current week data on basis of past data stored. Thus, data caching helps to reduce the number of communication cycles between the IoT architecture and the cloud. A load-balancing algorithm running over the cloud can represent another such technique, which would only request data from the gateway node that is not heavily loaded, making sure that a gateway node does not completely drain the sensor battery. The aggregation of data at the gateway node is yet another technique of reducing the communication overhead as the total amount of data sent over the cloud goes down (Gill et al. 2019). This gateway node forms the edge layer, a subset of cloud, in close proximity to the network. The convergence of

IoT, cloud, and edge can help effectively deploy disaster management mechanisms as explained above.

11.6 Integration of IoT with Various Technologies for Disaster Management Applications

The growth of technology has tremendously contributed to form new strategies in the enhancement of creating disaster resilience and reduction of risks which in our sense can be the multiplying factor of likelihood and consequence (Bostrom 1996). Some groundbreaking work in Information and Communication Technology (ICT) has led to assisting disaster prevention, mitigation, response, and recovery. Although the term "disruptive technology" was introduced a couple of decades ago (Christensen et al. 2015) with a possible groundbreaking impact on business, it can be truly said that the term due to innovative uses has touched upon in various aspects of life including the decrease of disaster risk and its management.

The current digital revolution due to IOT along with big data and AI has allowed its widespread use in all the four phases of disaster management. Also, associated technologies such as 5G, blockchain, social media, crowdsourcing along with the use of drones and robots have been used for disaster management activities. For simplicity in understanding, let us take all the associated technologies one by one:

11.6.1 Big Data

The concept of crisis analytics enables data analysis for large datasets regarding disaster management activities. The analysis of large user and sensor-generated data can attribute to betterment in disaster response. Some examples include analysis of social media communications to have more emphasis and filtering false information by understanding information types and creators of the content. There has been reported evidence (Frias-Martinez 2016) of population movement monitoring with the help of cellular phone data during floods. They can also be employed to analyze data generated by drones, robots, or sensors used in IoT setups.

11.6.2 Artificial Intelligence

Advancements in AI and machine learning have enabled its proficient use in classification, identification, and making predictions. This can lead to a tremendous influence on potentially predicting earthquakes along with speeding up the time for response and recovery. With the use of machine learning, there is a hope to speed up map creation in areas by removal of objects such as roads and buildings from images taken through satellites (Smith 2018). There is ongoing research work targeted toward the application of AI for earthquake prediction. As per work carried out by Kikwasi for monitoring of weather and climate in Tanzania metrological agency, the cost of technology may not necessarily be high as they employed PHP language to execute metrological observation equations and software aiding in the refinement of its calculations for effective prediction making. A user-friendly web interface in cloud system along with MySQL database management software which is free and open-source completes the setup for the above activity (Kikwasi 2016). Along with the same, AI can be used in terms of disaster management for processing large information, handling bulk calls during emergency, social media analysis and predictive analysis for analyzing past data to predict the happenings during a disaster event. One such software, Optima Protect uses processed data from emergency response systems to manage ambulance routes (Young 2017). Once integrated with online dashboards, this data can help in arranging the provision of real-time response from emergency departments and staff.

11.7 Conclusion

The potential of IoT as an enabling technology for disaster management can prove very beneficial to society as it can potentially prevent or at least minimize the losses that the community would incur due to natural disasters. IoT can help in the early identification of a disaster threat and possess the ability to provide timely incidence response, spreading awareness or notifying the people to take precautionary and rescue measures along with use in post-recovery plans. A lot of applications in various domains like flood management, volcanic eruptions, forest fires, accidents, nuclear reactor plant gas leakage, etc., are moving toward the incorporation of IoT architecture so as to minimize the scope of accidents that might happen due to human errors and for automating the process of remote monitoring the conditions and threshold-based actuations as and when necessary. This can serve as a failsafe for systems which are not capable to be fully automated due to a variety of reasons such as operation methodology, complexity of the operation, and the economics involved.

IoT model for disaster management is a generic model that explains the building blocks or the core components which can be understood as technical operations required to happen in proper sequence during the process of managing disasters. This model is universally applicable to changes only in the sensors, actuators, and threshold value based on the requirements. The integration of IoT and cloud can help with the storage and analytics of the recorded data and statistics related to a disaster. These recordings could be later analyzed to find useful patterns and draw inferences which may help to predict any possibility of the disaster occurrence in the future.

The impact of IoT in disaster management will be enormous and its integration with big data analytics, cloud computing, and artificial intelligence would make the process a lot more effective. Although IoT-based disaster management mechanism will have a lot of challenges and integration of an application in the generic model proposed will be subject to suitable parameters, the results of this adoption will definitely show a huge reduction in the losses incurred due to the disaster.

References

Awasthi Y, Mohammed AS (2019) IoT—a technological boon in natural disaster prediction. In: 2019 6th International Conference on Computing for Sustainable Global Development (INDIACom), pp 318–322

Azar J, Makhoul A, Barhamgi M, Couturier R (2019) An energy efficient IoT data compression approach for edge machine learning. Futur Gener Comput Syst 96:168–175. https://doi.org/10.1016/j.future.2019.02.005

Backfried G, Göllner J, Qirchmayr G, Rainer K, Kienast G, Thallinger G, Schmidt C, Peer A (20130 Integration of media sources for situation analysis in the different phases of disaster management: the QuOIMA project. In: 2013 European intelligence and security informatics conference, IEEE, pp 143–146

Bloomberg Business (2015) Nepal's slowing economy set for free fall without global help|Business – Gulf News, April 27, 2015. https://gulfnews.com/business/nepals-slowing-economy-set-for-free-fall-without-global-help-1.1499685

Bostrom A (1996) Book review: risk analysis, assessment and management. In: Ansell J, Wharton F (eds) John Wiley, Chichester, 1992, 220 pp, ISBN 0-471-93464-X (Pb). J Behav Decision Making 9(2): 145–146. https://doi.org/10.1002/(SICI)1099-0771(199606)9:2<145::AID-BDM 196>3.0.CO;2-N

Boukerche A, Coutinho RWL (2018) Smart disaster detection and response system for smart cities. In: IEEE symposium on computers and communications, 6

Chan C-S, Nozu K, Cheung TOL (2020) Tourism and natural disaster management process: perception of tourism stakeholders in the case of Kumamoto Earthquake in Japan. Curr Issue Tour 23(15):1864–1885

Christensen CM, Raynor M, McDonald R (2015) What is disruptive innovation? December. https://www.hbs.edu/faculty/Pages/item.aspx?num=50233

Chung K, Park RC (2016) P2P cloud network services for IoT based disaster situations information. Peer-to-Peer Netw Appl 9(3):566–577. https://doi.org/10.1007/s12083-015-0386-3

Coppola DP (2006) Introduction to international disaster management. Elsevier

Daly P, Ninglekhu S, Hollenbach P, Barenstein JD, Nguyen D (2017) Situating local stakeholders within national disaster governance structures: rebuilding urban neighbourhoods following the 2015 Nepal Earthquake. Environ Urban 29(2):403–424

Del Rosario, Usec Eduardo D (2014) Situational report re effects of Typhoon YOLANDA (HAIYAN). National Disaster Risk Reduction and Management Council, Phillipines. http://www.ndrrmc.gov.ph/attachments/article/1329/Update_on_Effects_Typhoon_YOLANDA_H aiyan_17APR2014.pdf

Diakakis M, Deligiannakis G, Katsetsiadou K, Lekkas E (2015) Hurricane Sandy Mortality in the Caribbean and Continental North America. Disaster Prev Manag 24(1):132–148. https://doi.org/10.1108/DPM-05-2014-0082

Dubey V, Prashant Kumar, Chauhan N (2019) Forest fire detection system using IoT and artificial neural network. In: Bhattacharyya S, Hassanien AE, Gupta D, Khanna A, Pan I (eds) International conference on innovative computing and communications. Lecture Notes in Networks and Systems. Springer, Singapore, pp 323–337. https://doi.org/10.1007/978-981-13-2324-9_33

Facts + Statistics: Wildfires | III (2017) April 28, 2017. https://www.iii.org/fact-statistic/facts-statistics-wildfires

Fast Facts on the U.S. Government's Work in Haiti (2013). https://web.archive.org/web/201301231
    70010/ http://haiti.usaid.gov/issues/docs/121911_shelter_fact_sheet.pdf
Frias-Martinez E (2016) How big mobile data can help manage natural disasters. Think Big (blog).
    May 13, 2016. https://en.blogthinkbig.com/2016/05/13/how-big-mobile-data-can-help-manage-
    natural-disasters/
Funes Y (2019) Cyclone Idai poised to become southern hemisphere's deadliest tropical storm,
    With More Than 1,000 Feared Dead, March 19, 2019. https://earther.gizmodo.com/cyclone-idai-
    poised-to-become-southern-hemisphere-s-dea-1833406709
Furquim G, Filho G, Jalali R, Pessin G, Pazzi R, Ueyama Jó (2018) How to improve fault tolerance
    in disaster predictions: a case study about flash floods using IoT, ML and real data. Sensors
    18(3):907. https://doi.org/10.3390/s18030907
Gaire R, Sriharsha C, Puthal D, Wijaya H, Kim J, Keshari P, Ranjan R et al (2018) Internet of
    Things (IoT) and cloud computing enabled disaster management. ArXiv:1806.07530 [Cs], June.
    http://arxiv.org/abs/1806.07530
Galaxy Digital (2019) Disaster relief organizations: the list you need to know. ReDI by Galaxy
    Digital (blog). June 28, 2019. https://www.galaxydigital.com/blog/disaster-relief-organizations/
Ghasemi P, Karimian N (2020) A qualitative study of various aspects of the application of IoT in
    disaster management. In: 2020 6th International Conference on Web Research (ICWR). IEEE,
    Tehran, Iran, pp 77–83. https://doi.org/10.1109/ICWR49608.2020.9122323
Gill SS, Garraghan P, Buyya R (2019) ROUTER: fog enabled cloud based intelligent resource
    management approach for smart home IoT devices. J Syst Softw 154:125–138. https://doi.org/
    10.1016/j.jss.2019.04.058
Greco L, Ritrovato P, Tiropanis T, Xhafa F. (n.d.) IoT and semantic web technologies for event
    detection in natural disasters, 9
Jeanty J (2017) Role of government in disaster management. The Classroom|Empowering Students
    in Their College Journey (blog). October 25, 2017. https://www.theclassroom.com/role-of-gov
    ernment-in-disaster-management-12225452.html
Kamruzzaman M, Sarkar NI, Gutierrez J, Ray SK (2017) A study of IoT-based post-disaster manage-
    ment. In: 2017 International Conference on Information Networking (ICOIN). IEEE, Da Nang,
    Vietnam, pp 406–410. https://doi.org/10.1109/ICOIN.2017.7899468
Kikwasi WK (2016) Improving efficiency and quality in weather observation and climate moni-
    toring by using artificial intelligence and Information Communication Technology (ICT) infras-
    tructure. In 16. www.wmo.int/pages/prog/www/IMOP/publications/IOM-125_TECO_2016/Ses
    sion_4/P4(20)_Kikwasi.pdf
Lettieri E, Masella C, Radaelli G (2009) Disaster management: findings from a systematic review.
    Disaster Prevention Manage: Int J
Max D, Chen R, Deichmann U, Lerner-Lam A, Arnold M (2005) National disaster hotspots: a
    global risk analysis. The World Bank, Hazard Management Unit. http://documents1.worldbank.
    org/curated/en/621711468175150317/pdf/344230PAPER0Na101official0use0only1.pdf
McCausland P, Arkin D, Chirbas K (2017) Hurricane Harvey: at least 2 dead after storm hits Texas,
    August 27, 2017. https://www.nbcnews.com/storyline/hurricane-harvey/hurricane-harvey-least-
    1-dead-after-storm-hits-texas-causing-n796316
McCurry J (2011) Japan Earthquake: 100,000 children displaced, says charity. The Guardian, March
    15, 2011, sec. World news. https://www.theguardian.com/world/2011/mar/15/japan-earthquake-
    children-displaced-charity
Meyers T (2019) 10 disasters that changed the world. Direct Relief (blog). December 29, 2019.
    https://www.directrelief.org/2019/12/10-disasters-that-changed-the-world/
Mouradian C, Jahromi NT, Glitho RH (2018) NFV and SDN-based distributed IoT gateway for
    large-scale disaster management. IEEE Internet Things J 5(5):4119–4131. https://doi.org/10.
    1109/JIOT.2018.2867255
Onagh M, Kumra VK, Rai PK (2012) Landslide susceptibility mapping in a part of Uttarkashi
    District (India) by multiple linear regression method. Int J Geol Earth Environ Sci 2(2): 102–120

Nepal, Ministry of Home Affairs (2015) Incident report of Earthquake 2015. http://drrportal.gov. np/uploads/document/14.pdf

NOAA's National Centers for Environmental Information (NCEI) (2018) Costliest U. S. tropical cyclones tables update. https://www.nhc.noaa.gov/news/UpdatedCostliest.pdf

Oloruntoba R, Sridharan R, Davison G (2018) A proposed framework of key activities and processes in the preparedness and recovery phases of disaster management. Disasters 42(3):541–570. https:// doi.org/10.1111/disa.12268

Perry M (2007) Natural disaster management planning: a study of logistics managers responding to the Tsunami. Int J Phys Distrib Logist Manag 37(5):409–433. https://doi.org/10.1108/096000 30710758455

Phases of Disaster (n.d.) Government. Restore Your Economy. Accessed October 7, 2020. https://restoreyoureconomy.org/index.php?submenu=phasesdisaster&src=gendocs&ref= 362&category=Main

Poslad S, Middleton SE, Chaves F, Tao R, Necmioglu O, Bugel U (2015) A semantic IoT early warning system for natural environment crisis management. IEEE Trans Emerg Top Comput 3(2):246–257. https://doi.org/10.1109/TETC.2015.2432742

Pravin Renold A, Balaji Ganesh A (2019) Energy efficient secure data collection with path-constrained mobile sink in duty-cycled unattended wireless sensor network. Pervasive Mob Comput 55:1–12. https://doi.org/10.1016/j.pmcj.2019.02.002

Ray PP, Mukherjee M, Shu L (2017) Internet of Things for disaster management: state-of-the-art and prospects. IEEE Access 5:18818–18835. https://doi.org/10.1109/ACCESS.2017.2752174

Rolland E, Patterson RA, Ward K, Dodin B (2010) Decision support for disaster management. Oper Manag Res 3(1–2):68–79

Sakhardande P, Hanagal S, Kulkarni S (2016) Design of disaster management system using IoT based interconnected network with smart city monitoring. In: 2016 International Conference on Internet of Things and Applications (IOTA). IEEE, Pune, India, pp 185–190. https://doi.org/10. 1109/IOTA.2016.7562719

Shah SA, Seker DZ, Hameed S, Draheim D (2019) The Rising role of big data analytics and IoT in disaster management: recent advances, taxonomy and prospects. IEEE Access 7:54595–54614. https://doi.org/10.1109/ACCESS.2019.2913340

Sharma M, Kaur J (2019) Disaster management using Internet of Things: Computer Science & IT Book Chapter|IGI Global. In: Handbook of research on big data and the IoT. https://www.igi-glo bal.com/chapter/disaster-management-using-internet-of-things/224271

Singh P, Sharma A, Sur U, Rai PK (2020) Comparative landslide susceptibility assessment using statistical information value and index of entropy model in Bhanupali Beri region, Himachal Pradesh, India. Environ Dev Sustain 1–20. https://doi.org/10.1007/s10668-020-00811-0

Sinha A, Kumar P, Rana NP, Islam R, Dwivedi YK (2019) Impact of Internet of Things (IoT) in disaster management: a task-technology fit perspective. Ann Oper Res 283(1–2):759–794. https:// doi.org/10.1007/s10479-017-2658-1

Smith N (2018) Humanitarian OpenStreetMap Team|Integrating Machine Learning into the tasking manager: notes on a direction. Non Profit Organization Site. Humanitarian Open Street Map Team. September 17, 2018. https://www.hotosm.org/updates/integrating-machine-learning-into-the-tasking-manager/

Sur U, Singh P, Rai PK (2021) Landslide probability mapping by considering fuzzy Numerical Risk Factor (FNRF) and landscape change for road corridor of Uttarakhand, India. Environ Dev Sustain (springer) 1–29. https://doi.org/10.1007/s10668-021-01226-1

Tacloban: City at the Centre of the Storm - BBC News (2013) November 12, 2013. https://www. bbc.com/news/world-asia-24891456

Toribiong J (2010) National Disaster Risk Management Framework 2010. National Emergency Management Office, Republic of Palau. https://reliefweb.int/sites/reliefweb.int/files/resources/ National%20Disaster%20Risk%20management%20Framework%202010.pdf

Wald DJ (2019) Opportunities and challenges for the delivery of a continuum of real-time earthquake information products. AGU Fall Meeting Abstracts 44 (December). http://adsabs.harvard.edu/abs/2019AGUFM.S44C..08W

Wellington JJ, Ramesh P (2017) Role of Internet of Things in disaster management. In: 2017 International Conference on Innovations in Information, Embedded and Communication Systems (ICIIECS), 1–4. IEEE, Coimbatore. https://doi.org/10.1109/ICIIECS.2017.8275928

Young A (2017) Can predictive analytics help avert Pittsburgh's next disaster? Government Technology. April 28, 2017. https://www.govtech.com/data/Can-Predictive-Analytics-Help-Avert-Pittsburghs-Next-Disaster.html

Zambrano AM, Perez I, Palau C, Esteve M (2017) Technologies of Internet of Things applied to an earthquake early warning system. Futur Gener Comput Syst 75:206–215. https://doi.org/10.1016/j.future.2016.10.009

Chapter 12
Countering Challenges of Smart Cities Mission Through Participatory Approach

Deepak Kumar and Tavishi Tewary

Abstract The application of information and communication technologies has led the reforms in the governance in Indian cities. It has also read to greater citizen participation in governance. People are important because their participation is the precondition for successfully functioning of the smart city system. The paper tries to develop a framework of systematic analysis to explain e-participation of citizens in India. The paper tries to identify and measure key indicators of e-participation in India. Survey of 200 respondents is analyzed from four smart cities across India. The paper applies regression and concludes that all the indicators have a significant impact on e-participation of citizens in smart cities. The study finds that though the government is investing a huge amount of funds in smart cities development but still a lot is to be done. Also, the concept of participatory approach is currently not a prominent research theme among scholars. Therefore, the current work will address this gap to unravel the conceptual framework of e-participation of citizens in smart cities in India.

Keywords E-participation · Smart cities · Smart living · Smart mobility · Participatory approach

12.1 Introduction

The concept of a smart city is based on six key pillars, i.e., smart city economy, smart people, smart governance, smart mobility, smart living, and smart governance (Bezai et al. 2021). These pillars are interlinked and constitute what is known as a smart city system. There are studies which give equal weight to all the components, but some researchers believe that smart people are the most important block of the

D. Kumar (✉)
Amity Institute of Geoinformatics and Remote Sensing (AIGIRS), AMITY University, Uttar Pradesh, Sector 125, Noida 201303, India

T. Tewary
Jaipuria Institute of Management, Noida, Uttar Pradesh, 32A, Sector 62, Noida 201309, India

© The Author(s), under exclusive license to Springer Nature Singapore Pte Ltd. 2022   247
P. K. Rai et al. (eds.), *Geospatial Technology for Landscape and Environmental Management*, Advances in Geographical and Environmental Sciences, https://doi.org/10.1007/978-981-16-7373-3_12

Fig. 12.1 Framework of
smart cities system

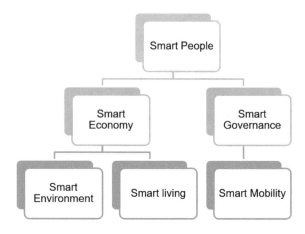

system (Vinod Kumar 2014). People are important because their participation is the precondition for successfully functioning of the smart city system (Fig. 12.1).

12.1.1 Global Scenarios

The table mentioned below portrays high scorer cities of the world. Amsterdam has 67% of all trips done by walking or cycling. Copenhagen has the most ambitious carbon reduction plan and is trying to achieve carbon neutrality by 2025 (Table 12.1).

Table 12.1 Top 10 smart cities in the world

S. No	City	Characteristics
1	Vienna	Digital governance
2	Toronto	Smart innovation and technology
3	Paris	Smart innovation and green city
4	New York	High innovation accompanied by low quality of life
5	London	Sustainability and infrastructure
6	Tokyo	Innovation and digitalization
7	Berlin	innovation and green city
8	Copenhagen	innovation and green city
9	Hong Kong	Digital governance
10	Barcelona	Efficiency in energy

Source Smart city ranking by Cohen (2012) and Kate Brown (2014)

12.1.2 Indian Scenario

India faces challenges in the form of high population, resource constraints haphazard growth of cities, environmental pollution, infrastructural crunch, and pressure on public finances (Ghisellini et al. 2016; Khan and Chatterjee 2016; Lam et al. 2020). The solution lies in the systematic coordination between the central and the state governments (Aswani et al. 2020; Rajput and Arora 2017). Good traffic management system, efficient energy usage, infrastructural growth, e-governance facilities, public safety education, and health are prerequisites for having a problem-solving city (Guan 2011; Sun and Southworth 2013). The government of India has recently launched the smart cities mission with a focus on inclusive and sustainable growth to meet the challenges of rapid urbanization.

The fact has been well established in economic literature that there exists a positive correlation between economic development and urbanization (Bassuk et al. 2015; Madhavi et al. 2016). The analysis was undertaken by World Bank and UN-habit further confirms this argument by noting that the increase in urban population from 33 to 51% between 1960 and 2010 led to growth in per capita by 152%. The core element of smart cities would comprise of assured electricity, sanitation, water supply, waste management, efficient urban mobility, affordable housing, good governance, sustainable environment, IT connectivity, health and education, safety and digitalizatio n (Adelaja et al. 2011; Ghisellini et al. 2016). The mission focuses on city improvement, renewal, and extension and pan-city initiative. For the current study, we have selected four cities covering the spatial distribution throughout the India, namely Delhi, Bhubaneswar, Jaipur, and Coimbatore, respectively (from east, west, north, and south part).

The smart city does not refer to intensive use of technology (Bassuk et al. 2015; Caragliu et al.2011; Rai and Kumra 2011). It is a city which integrates technology to cater to the needs of the citizens and to make them more active participatory in the decision-making process (Joshi et al. 2016; Kumar 2019). It is a matter of great interest in how the government and the citizen can be brought together with the help of technology.

Information and communication technologies (ICT) have facilitated implementation of new forms of governance where citizens can actively participate in the decision making after accessing the information available on public domain (Manisha et al. 2020; Moonen et al. 2012). It helps in making the process transparent and reinforcing trust in the system (Démurger and Fournier 2011). The term "e-participation" can be defined as the use of ICT tools for deepening political participation by connecting the citizens with other citizens and with the government (Bezai et al. 2021; Kossieris et al. 2014). The study explores the key determinants of e-participation in smart cities in India (Kumar 2020a; Loukanov et al. 2020; Pandey et al. 2012). The work is divided into five sections. The next section presents the theoretical framework including the main explanatory factors of e-participation. Section 3 describes the methodology used for achieving the objective. Section 4 discusses the results, and the final section draws the concluding remark.

12.2 Study Area

The study area is located in the selected cities of India, distributed throughout India. For this work, we have selected four major cities covering the spatial distribution throughout the India, namely Delhi, Bhubaneswar, Jaipur, and Coimbatore, respectively (from east, west, north, and south part). Major part of the agricultural field have crops like wheat, green gram, soybean, rice groundnut (Fig. 12.2).

Seasonal calendar are broadly distributed into three distinct seasons as monsoon, winter, and summer. The climate is subtropical, the highest temperature and lowest.

Fig. 12.2 Location of smart cities (among first 20 declared cities) for the current study

12.3   Research Methodology

The study is undertaken to discover the key indicators of e-participation in smart cities in India. It further attempts to assess the impact of each key indicator on e-participation of citizens in smart cities of India. The various categories of smart cities initiatives and the developments associated with the same are detailed in the table below. These indicators were used to develop a framework which was then used to study a sample of 200 respondents via the convenience sampling methodology to obtain their views and understanding of the importance of implementation of these indicators in context of smart cities in India. The first prerequisite for e-participation is to have an online presence. Without basic connectedness, it becomes impossible for governments to solicit ideas and participate electronically (Table 12.2).

The results of the studies showed the following results in terms of implementation of the above indicators:

12.3.1   Demographic Analysis

A sample of 50 respondents from each city was chosen for the current study. Out of the 200 respondents, 57 were females and 143 males.

Figure 12.3 illustrates the gender bifurcation of the respondents. It can be seen that 29% of the total population constitutes female population and 71% consititue male population of all respondents.

Figure 12.4 illustrates the income condition of the respondents. It can be seen that 26% of the total population constitutes high income group, 23% lies in low income group, and 51% lies in middle income group.

Figure 12.5 illustrates the educational profile of the respondents. It can be seen that 27 responsdents are 12th pass, 63 respondents are graduate, 80 respondents are post-graduate, and 30 respondents have different educational background.

Figure 12.6 illustrates the composition of different indicators for e-participation. It can be seen that it constitutes components like e-governance, ICT infrastructure, toursim, health, transporation mobility, economy development, energy sustainable development, waste management and water resources, and security.

Thus we can conclude that the country needs to work on energy sustainable development, waste management and water resources and implementation of security for e-resources.

12.3.2   Regression Analysis

For testing, the above framework regression analysis was applied. First correlation measures were calculated and the results were significant.

Table 12.2 Key parameters for research

Category	Developments
ICT infrastructure	Provision of free wi-fi in municipal buildings and public areas
	Implementation and provision of optical fiber network (MAN)
	Infrastructure (data centers) for collecting and storing data from Internet of things (IoT) sensors
	Upgradation of hardware and software in municipal departments for increased efficiency of the back office
	Electronic document work flow management systems for municipal offices—departments
	Installation of information kiosks in order to provide relevant information to visitors as well as citizens
	Installation of electronic boards to display real-time information of weather, local news, etc
Environment	Installing noise measurement sensors
	Installation of ERMS—electromagnetic radiation measurement sensors
	Installing rain level measurement sensors
	Installing air pollution measurement sensors
	Atmospheric micro particles measurement sensors and their installation
	Installation of light level measurement sensors
Transportation-mobility	Usage of specialized applications to incorporate actions for traffic management improvement in real time within/inter municipal areas
	Usage of intelligent systems for safe movement at pedestrian crossings
	Implementation of smart stops (e.g., online bus arrival marking) for public transportation
	Installation of sensors to means of transportation or roads for traffic flow monitoring
	Installation of smart information signs for traffic condition
	Installation of car parking spaces sensors to provide assistance to drivers for parking availability
Health	Installation and implementation of healthcare tele monitoring system for supporting groups of people at health risks (disabled, Alzheimer's disease, etc.)
	Implementing telemedicine system for measurements of key indicators (pressure, sugar, etc.) for citizens and medical records archive incorporating advice from doctors

(continued)

Table 12.2 (continued)

Category	Developments
	Implementation of different applications which would allow to remotely monitor patient progress in remote locations
Waste management and water resources	Drinking water online quality measurement systems
	Online monitoring system which would have appropriate sensors to detect water leaks in water mains
	Online monitoring system for instantly detecting any possible water leaks in closed irrigation canals/irrigation tanks
	Creating awareness in citizens on recycling via tele-education
	System for online monitoring and management of pumping, boring stations
	Irrigation management system with dam operation control, pumping stations control, water flow control
	A system for online waste containers management (with occupancy sensors) which would also allow for waste collection fleet management (GPS)
Energy—sustainable development	Solar cells installation in municipal buildings
	Setting up wind farms
	Smart lighting—Achievement of energy saving in municipal street lighting and public spaces
	Taking actions to create citizen awareness (via tele-education), which may aim to create energy savings
	Fuel consumption reduction on municipal transport vehicles by redesigning routes to choose most appropriate path (fleet management)
Tourism—culture	Designing and creating cultural infrastructure, agents' management system with detailed reporting and promotion via municipal website
	Development of an electronic local tourist guide
	Development of tourism content mobile applications
	Protection, promotion and enhancement of museums, galleries, monuments, etc., through virtual tours
Economy-development	Undertaking actions in order to promote entrepreneurship on MSME's website
	Actions to reinforce, promote, and sell locally made products through the municipal website
	Carrying out actions for supporting high tech farming, like precision farming in the municipal fields

(continued)

Table 12.2 (continued)

Category	Developments
	Promoting innovative technological activities
	On the government web platform, provision of interactive services for young entrepreneurs
Security	Implementation of early warning system and response to fires
	Undertaking actions for the addressable of citizens and relevant protection plans for emergencies like earthquakes and floods
	Providence to guard public buildings—facilities
	Monitoring and forecasting the weather for agricultural uses
E-government	Electronic voting applications
	Application of e-consultation for important plans like business plan, etc
	Collection of electronic signatures on important stances pertaining to citizens
	Development of applications for citizens regarding their problems and requests reporting
	Freely accessible open data for usage by public agencies or individuals
	GIS applications for urban building construction
	Implementing e-Government services provision framework

The following relationship was tested.

$$EP = \alpha + \beta 1\,EG + \beta 2\,ICT + \beta 3\,TR + \beta 4\,HT + \beta 5\,TRNS$$
$$+ \beta 6\,EN + \beta 7\,EG + \beta 8\,WM + \beta 9\,SC + \varepsilon$$

where

α	intercept and ε is the error term.
EP	represents the dependent variable e-participation.
EG	e-governance.
ICTICT	infrastructure.
TR	Tourism.
HT	Health.
TRNS	Transportation mobility.
ENE	conomy development.
EGE	nergy sustainable development.
WM	Waste management and water resources.
SC	Security.

Fig. 12.3   Gender profile

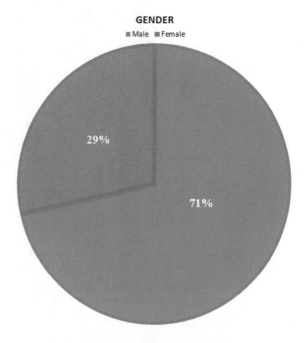

Fig. 12.4   Income profile

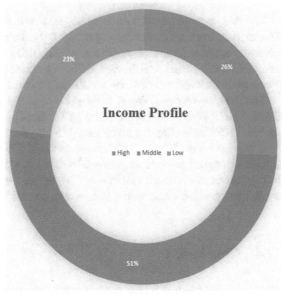

EDUCATIONAL PROFILE

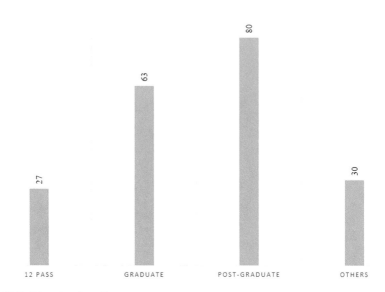

Fig. 12.5 Educational profile

The study applies partial least square regression to estimate the above relationship. This method has been chosen because it not only allows for estimation of models for relatively smaller sample sizes (Kumar 2021), it is also suitable for its prediction-oriented nature (Misra and Kumar 2020). This method also helps in avoiding the problem of multicollinearity (Kumar 2020b) (Table 12.3).

The Durban–Watson (DW) statistic was used to test the autocorrelation in the error terms. The DW value was found to be 1.96 indicating absence of autocorrelation. R square is 87% indicating the given model to be a robust model.

It can be inferred from the above-mentioned table that all the indicators have significant impact on e-participation of citizens in smart cities. The contribution of ICT infrastructure is maximum followed by e-governance, transport mobility, health, waste management, security, tourism, security, economy, and energy, respectively.

12.4 Conclusion

The concept of smart cities has drawn attention from professionals and policymakers as it manifests technological and social innovations. Three main components are core of smart cities: technology, people, and institutions. It is difficult to predict the future of smart cities. The current work makes an attempt to understand the predictors of participatory approach in smart cities in India. The study finds that though the government is investing heavily in the development of smart cities, still a lot is to

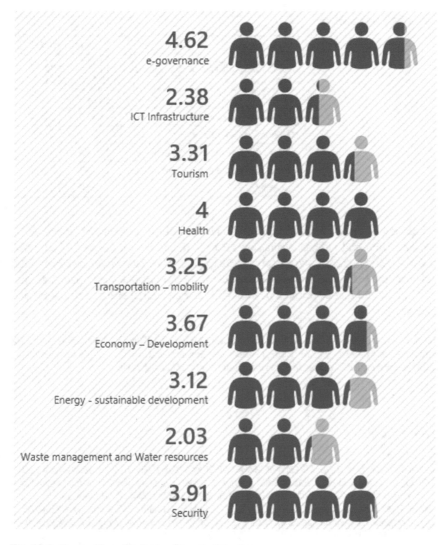

Fig. 12.6 Composition of indicators for e-participation

be done. The concept of participatory approach is currently not a research focus among scholars. To address this gap, this study unravels the conceptual framework of e-participation of citizens in smart cities in India. The study proposes and tests the impact of 9 key indicators on e-participation. It was found that all the 9 indicators have positive significant impact on the dependent variable (e-participation).

Political and administrative leadership should understand that any major transformations from the conventional methods will be challenging. They must nevertheless remain committed to seeing this transformation through to the end. Smart cities should incorporate e-participation and should promote participatory usage of

Table 12.3 Derived values

Indicators	B value	P value
e-governance	13.21	0.000
ICT infrastructure	19.81	0.000
Tourism	8.70	0.003
Health	11.90	0.004
Transportation mobility	14.15	0.015
Economy development	6.77	0.001
Energy sustainable development	5.55	0.000
Waste management and water resources	9.17	0.003
Security	7.89	0.000

technology. It should be staffed as legitimate part of the government. It should be translated into real policies scenarios. There is reluctance on the part of administration to fully digitalize the citizen's participation. They face difficulties in gaining and sustaining citizen's interest. The cities willing to create e-participatory framework might find that investment payoffs will be high both in terms of resources and legitimacy.

Acknowledgements The authors thankful to Amity University, Uttar Pradesh, Noida, India for providing the necessary infrastructure to carry out this research work. Authors also expresses gratitude for constant encouragement to carry out the present research work.

References

Adelaja S, Shaw J, W Beyea JD Charles McKeown G Agugiaro F Nex … www.imagingnotes.com (2011) Potential and prospects of solar energy in Uttara Kannada, District of Karnataka State India. Solar Energy 02(2) 1986–2000.https://doi.org/10.1080/00908319708908903

Aswani R, Ghrera SP, Chandra S, Kar AK (2020) A hybrid evolutionary approach for identifying spam websites for search engine marketing. Evol Intelle (0123456789). https://doi.org/10.1007/s12065-020-00461-1

Bassuk NL, Universite AB, Jean M, Universite C, Theoretical L, Politics U, Bibliography AA (2015) On using landscape metrics for landscape similarity search. Landsc Urban Plan 117(1):1–12.https://doi.org/10.1038/srep11160

Bezai NE, Medjdoub B, Al-Habaibeh A, Chalal ML, Fadli F (2021) Future cities and autonomous vehicles: analysis of the barriers to full adoption. Energy and Built Environ 2(1):65–81. https://doi.org/10.1016/j.enbenv.2020.05.002

Caragliu A, Bo CD, Kourtit K, Nijkamp P (2011) Comparative performance assessment of Smart Cities around the North Sea basin. Netw Industr Quart 13(3):15–17. https://doi.org/10.1017/CBO9781107415324.004

Cohen B (2011) The top 10 smart cities on the planet. Fast Company, 11 Jan 2011. Web. Last accessed 12 Feb 2014. http://www.fastcoexist.com/1679127/the-top-10-smart-cities-on-the-planet. 20 Oct 2017

Cohen B (2012) What exactly is a smart city? Fast Company Co.Exist. 19 Sep 2012. http://www.fastcoexist.com/1680538/what-exactly-is-a-smart-city. 20 Oct 2017

Cohen B (2015) The three generations of smart city. https://www.fastcompany.com/3047795/the-3-generations-of-smart-cities. 20 Oct 2017

Démurger S, Fournier M (2011) Poverty and firewood consumption: a case study of rural households in northern China. China Econ Rev 22(4):512–523. https://doi.org/10.1016/j.chieco.2010.09.009

Ghisellini P, Cialani C, Ulgiati S (2016) A review on circular economy: the expected transition to a balanced interplay of environmental and economic systems. J Clean Prod. https://doi.org/10.1016/j.jclepro.2015.09.007

Guan K (2011) Surface and ambient air temperatures associated with different ground material: a case study at the University of California, Berkeley. 14. Retrieved from http://nature.berkeley.edu/classes/es196/projects/2011final/GuanK_2011.pdf

Jain M, Dawa D, Mehta R, Pandit APDMK (2016) Monitoring land use change and its drivers in Delhi, India using multi-temporal satellite data. Model Earth Syst Environ 2(1):1–14. https://doi.org/10.1007/s40808-016-0075-0

Jain M, Korzhenevych A (2020) Urbanisation as the rise of census towns in India: an outcome of traditional master planning? Cities, 99(January 2019), 102627. https://doi.org/10.1016/j.cities.2020.102627

Joshi N, Baumann M, Ehammer A, Fensholt R, Grogan K, Hostert P, Jepsen MR, Kuemmerle T, Meyfroidt P, Mitchard ET, Reiche J (2016) A review of the application of optical and radar remote sensing data fusion to land use mapping and monitoring. Remote Sens 1–23.https://doi.org/10.3390/rs8010070

Khan A, Chatterjee S (2016) Numerical simulation of urban heat island intensity under urban—suburban surface and reference site in Kolkata. India. Model Earth Syst Environ 2(2):1–11. https://doi.org/10.1007/s40808-016-0119-5

Kossieris P, Kozanis S, Hashmi A Katsiri E, Vamvakeridou-Lyroudia LS, Farmani R, Makropoulos C, Savic D (2014) A web-based platform for water efficient households. Proc Eng 89:1128–1135.https://doi.org/10.1016/j.proeng.2014.11.234

Kumar D (2019) Hyper-temporal variability analysis of solar insolation with respect to local seasons. Remote Sens Appl: Soc Environ 15:100241. https://doi.org/10.1016/j.rsase.2019.100241

Kumar D (2020a) Urban energy system management for enhanced energy potential for upcoming smart cities. Energy Explor Exploit 014459872093752. https://doi.org/10.1177/0144598720937529

Kumar D (2020b) Satellite-based solar energy potential analysis for southern states of India. Energy Rep 6:1487–1500. ISSN: 2352-4847 (Online). https://doi.org/10.1016/j.egyr.2020.05.028, Impact Factor- 6.870, WoS/Scopus Indexed

Kumar D (2021) Urban objects detection from C-band synthetic aperture radar (SAR) satellite images through simulating filter properties. Sci Rep. ISSN 2045-2322 (Online), https://doi.org/10.1038/s41598-021-85121-9, Impact Factor- 5.133, WoS/Scopus/DOAJ Indexed

Lam DPM, Martín-López B, Wiek A, Bennett EM, Frantzeskaki N, Horcea-Milcu AI, Lang DJ (2020) Scaling the impact of sustainability initiatives: a typology of amplification processes. Urban Transformations 2(1). https://doi.org/10.1186/s42854-020-00007-9

Loukanov A, Allaoui N, Omor El A, Elmadani FZ, Bouayad K, Seiichiro N, He HS (2020) Effects of neighborhood building density, height, greenspace, and cleanliness on indoor environment and health of building occupants. Environ Res 106(February):213–222.https://doi.org/10.1016/j.buildenv.2018.06.028

Misra M, Kumar D (2020) A hybrid indexing approach for sustainable smart cities development, J Indian Soc Remote Sens 48:1639–1643. Electronic ISSN-0974-3006 / Print ISSN-0255-660X. Impact Factor- 1.563, https://doi.org/10.1007/s12524-020-01171-y, WoS/ Scopus Indexed

Moonen P, Defraeye T, Dorer V, Blocken B, Carmeliet J (2012) Urban physics: effect of the microclimate on comfort, health and energy demand. Front Architectural Res 1(3):197–228. https://doi.org/10.1016/j.foar.2012.05.002

Pandey P, Kumar D, Prakash A, Masih J, Singh M, Kumar S, Jain VK, Kumar K (2012) A study of urban heat island and its association with particulate matter during winter months over Delhi. Sci Total Environ 414:494–507https://doi.org/10.1016/j.scitotenv.2011.10.043

Rai PK, Kumra VK (2011) Role of geoinformatics in Urban Planning. J Sci Res 55:11–24

Rajput S, Arora K (2017) Sustainable smart cities in India. Sustainable Smart Cities in India: Challenges and Future Perspectives, (March), 369–382. https://doi.org/10.1007/978-3-319-471 45-7

Sun J, Southworth J (2013) Remote sensing-based fractal analysis and scale dependence associated with forest fragmentation in an amazon tri-national frontier. Remote Sens 5(2):454–472. https://doi.org/10.3390/rs5020454

Vinod Kumar TM (2014) E-governance for smart. Cities. https://doi.org/10.1007/978-981-287-287-6_1

Chapter 13
Urban Growth Modeling and Prediction of Land Use Land Cover Change Over Nagpur City, India Using Cellular Automata Approach

Farhan Khan⊙, Bhumika Das, and Pir Mohammad

Abstract The monitoring of land use land cover (LULC) change is essential to estimate the urban sprawl as the rapid growth of urban areas affects the ecology and eminence of city life. LULC forms a reference line of the spatial map for observing, managing, and planning activities for urban development. The LULC change dynamics is self-explanatory using GIS and remote sensing techniques. Thus, the present study uses these techniques to understand the spatial–temporal variability of LULC of Nagpur city, Maharashtra, from 2000 to 2020. The study area is a center for economic, education, and medical activities; therefore, changes should be analyzed to understand urban growth trends. The LULC classification is performed considering four different classes, i.e., barren land, built up, agriculture (include shrubs, urban forest, small plantation, vegetation area), and water bodies. The LULC results show that the built-up area is increased by 26.62% from 2000 (41.24%) to 2020 (67.86%), with a slight increase in water bodies 0.19% is also evident. On the other hand, the area covered with vegetation is decreased by 15.93% from 2000 (30.17%) to 2020 (14.24%), and barren land is reduced by 10.88%. The present study also includes predicting the LULC map using the artificial neural network-based (ANN) cellular automata (CA) model, using seven different driving parameters, like elevation, slope, aspect, distance to major roads, distance to water bodies, central building distance, and population. The prediction model showed an overall accuracy of 81.23% in predicting the 2025 LULC maps with the help of 2015 and 2020 LULC data. The result of the prediction model evidents a maximum growth of 30.88% in the built-up area as compared to year 2020. Therefore, the study results show that the use of LULC and CA-ANN model will be suitable to understand the

F. Khan (✉)
Department of Civil Engineering, Rungta College of Engineering and Technology, Bhilai, Chhattisgarh, India
e-mail: farhan1@rungta.ac.in

B. Das
Department of Mining, Mats University, Raipur, Chhattisgarh, India

P. Mohammad
Department of Earth Sciences, Indian Institute of Technology Roorkee, Roorkee, Uttarakhand, India

© The Author(s), under exclusive license to Springer Nature Singapore Pte Ltd. 2022   261
P. K. Rai et al. (eds.), *Geospatial Technology for Landscape and Environmental Management*, Advances in Geographical and Environmental Sciences,
https://doi.org/10.1007/978-981-16-7373-3_13

future trend, and it will help the administration and planner for the development of the sustainable city.

Keywords Cellular automata · Artificial neural network · LULC · Modeling · GIS · Remote sensing

13.1 Introduction

The growth of urban city depends on proper planning and management of utilities across the city and its neighboring suburban areas. Managers and developers interested in urban and environmentally sustainable planning are increasingly concerned about land use land cover (LULC) reform (Mohamed and Worku 2020; Mondal et al. 2019). Land use land cover transition is the result of human activities which has altered the natural surface of earth (Urgessa and Lemessa 2020). The satellite-based land classification is proven to be very helpful for analyzing the land use land cover change scenario (Mishra et al. 2018, 2020). The study of urban expansion is simplified using satellite imageries combine with different spatial techniques (Jelil Niang et al. 2020). The physical properties of the earth's surface, such as forest, water, crops, and urban development, are defined by land cover; however, land use is the alteration of land cover to meet human needs and activities (Goyal et al. 2019; Hussain et al. 2020; Liping et al. 2018). Since metropolitan centers have an effect on the earth environment that extends beyond their physical boundaries, they need careful consideration in the current situation. As the number and percentage of urban residents continue to grow, spatiotemporal shift identification in large cities and adjacent suburban areas may become increasingly significant (Dutta and Guchhait 2020; Gohain et al. 2021).The trend of urbanization is marked by urban land growth and population shifts, as well as a dramatic shift in land use patterns that affect the physical borders of cities (Ghosh et al. 2019; Vishwakarma et al. 2016; Magesh and Chandrasekar 2017). The increasing urban growth is affecting the sustainability of cities and also affecting the climatic condition; this changes should be monitored using geospatial techniques (Fonseka et al. 2019; Kaichang et al. 2012). Minta et al. (2018) states that LULC is the effect of human activities depends on various factors like population, political, and socioeconomical. Due to urbanization the population is increasing, which causes many environmental related problems; for example in turkey the population is 14.6 million, which is causing man-made deforestation (Akyürek et al. 2018). The urban sprawl is deteriorating the ecosystem resulting in decreasing water bodies, vegetation land, and forest regions (Bhat et al. 2017; Kaliraj et al. 2017; Mohammad and Goswami 2021b). Due to the land use land cover the climate change is taking place in Sundarbans biosphere, and the land surface temperature is also increasing per decade by a rate of 0.5 °C (Sahana et al. 2016). Urbanization results in the substitution of natural ground covers with impermeable construction materials, as well as changes to the biophysical climate and land surface energy cycles (Fu and Weng 2016; Mohammad and Goswami

2021a). For land use/cover mapping and transition study, remote sensing is a cost-effective alternative to ground-based surveys. Remotely sensed data time series may be used to investigate the temporal characteristics of urban attributes or processes. Furthermore, post-classification comparative approaches generate "from-to" transition knowledge between land groups, which can assist in capturing the essence of land transitions (Liu and Yang 2015). LULC changes will also have an effect on ecological diversity, habitat fragmentation, soil degradation, ecosystem resources, socio-cultural processes, and increase natural disasters including floods (Hadi et al. 2014; Kindu et al. 2013). Rapid land change is affecting the water, land, and air of the Brazil resulting in increasing temperature (Ogashawara and Bastos 2012). From the research in Atlanta city, it has observed that urban growth is adversely affecting environment, traffic condition, air pollution, and reduction in water quality (Mohammad et al. 2019; Yang and Lo 2002). Land use/cover shift identification is beneficial for a greater understanding of the environment. Comprehending the dynamics of the terrain over a specific period of time for long-term management coverage/use of land shift is a vast and ever-increasing operation, primarily because influenced by both normal and human-made events, This, in particular, triggers modifications that have an effect on ecologies (Imran Basha et al. 2018).

As the land change analysis is important, likewise, change prediction is also important to get the knowledge about the future change that can occur which helps in proper planning of urban areas. In current scenario, the cellular automata (CA) and artificial neural network (ANN) are used to generate the prediction model for LULC and drivers affecting it. Since the last two decades, several models based on simulations have been developed for LULC transition modeling on a global scale (Rimal et al. 2018; Yatoo et al. 2020). Due to the global increase in land use, it is crucial to understand the future land use pattern for proper planning and management of resources (Munthali et al. 2020; Singh et al. 2015). The proper understanding of future land use will help building a better plan to reduce future crisis (Mohamed and Worku 2020). Anand and Oinam (2020), Khan et al. (2018) state that geographic information system and remote sensing are widely used tools which is effective for determining and predicting future land use land change. By estimating the location of a pixel based on its previous condition, adjacent neighborhood impacts, and transformation laws, the CA is a popular method for simulating the LULC transition. CA model can effectively reflect nonlinear spatially stochastic LULC shift processes and produce complex correlations (Khan et al. 2021; Kumar et al. 2016; Saputra and Lee 2019). To synthesize, predict and future land use land cover simulation for Egypt cities was performed using Markov model (Mohamed and El-Raey 2019). Mahamud et al. (2019) develop a LULC prediction model to investigate the future trends in Kelantan state by integrating GIS and CA–Markov model which will benefit the urban planner and local authorities for decision and policy making. The cellular automata–Markov (CA–Markov) model was used to forecast the LULC changes. Markov chain is used in forecasting long time series with CA, which is based on spatial relationships and can more precisely simulate LULC changes in space and time (Lu et al. 2019; Tadese et al. 2021). The CA–Markov model was used to quantify the change for thirty years and to determine future change considering socioeconomic and biophysical factors (Hishe et al. 2019). Rimal et al. (2018) investigated the LULC change classes affected due to urban sprawl and urban growth expected in the future. Another literature, (Hyandye and Martz 2017), used CA–Markov model to predict the future water balance and simulate the LULC for Usangu catchment. The CA–Markov model has outperformed regression-based simulations in terms of stimulating and predicting spatial transformations in diverse land use processes (Mishra and Rai 2016; Mishra et al. 2016; Gidey et al. 2017; Wang et al. 2021).

From the above literatures, author has identified the problem statement for Nagpur city. In this study, author attempts to analyze the land use land cover changes of Nagpur city using cellular automata and artificial neural network to predict the future LULC of the study area. The specific aim of the research is as follows: (a) to determine the land use/land cover and changes during last two decades (2000–2020) at each successive five-year interval time period, (b) to determine the drivers affecting the urban growth in the study area, (c) to generate LULC prediction model for the year 2025 using CA-ANN technique. This study of urban land change from 2000–2020 and future model 2025 will definitely help the policies makers and planners for better understanding of land change in the Nagpur city in the state of Maharashtra, India.

13.2   Study Area

Nagpur is the third largest city in Maharashtra, after Mumbai and Pune. It is located at an altitude of 310 m above mean sea level on the Deccan plateau of the Indian Peninsula, between 21° 2′ 59″ N and 21° 13′ 57″ N latitudes and 78° 59′ 29″ E to 78° 12′ 13″E longitudes (Fig. 13.1). The area covered by study area is of 217 km². It has a tropical rainy and dry atmosphere, with dry weather for the majority of the year. Nagpur has an average annual precipitation of 1000–1300 mm, with 80% of it coming during the monsoon season (Sakhre et al. 2020). Summers are scorching, lasting from March to June, with May being the hottest month of the year. Winter lasts from November to January, and temperature could drop below 10 °C. The city's natural drainage pattern is formed by the Nag River, Pilli River, and drains. Nagpur is covered with natural and man-made lakes, the largest of which is Ambazari Lake. Gorewada Lake and Telangkhedi Lake are two other natural lakes. The city's historical rulers built Sonegaon Lake and Gandhi Sagar Lake, which are man-made reservoirs. The total population of Nagpur city according to census-2011 is 4,653,171 making it ninth largest urban city. Nagpur's population increased by 19.21% in 2011 relative to 2001, according to the census. It is undeniably one of central India's most urban centers. As a result, the city will have to extend its physical presence in the coming decades to meet both economic and population growth. Nagpur is Maharashtra's main commercial and political center, and it is known throughout the world as "Orange City" for being a major trading center for the region's oranges.

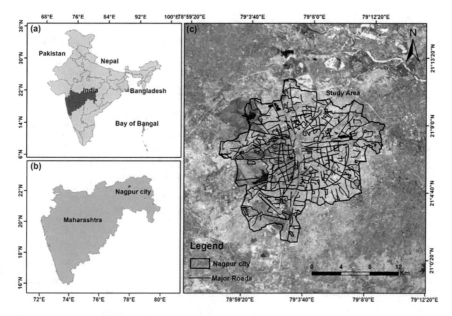

Fig. 13.1 Location of the study area

13.3   Data and Methods

13.3.1   Data Used

For land use land cover classification and prediction model, following data are used: (a) Landsat 7 ETM + imageries for the year 2000, (b) Landsat 5 images for 2005 and 2010, (c) Landsat 8 OLI-TIRS imageries for the year 2015 and 2020 (source: https:// earthexplorer.usgs.gov/), (c) ArcGIS software is used for pre- and post-classification of satellite images using maximum likelihood classification; (d) QGIS software is used for making prediction model using MOLUSCE plugin for the year 2025; (e) Nagpur metropolitan boundary, (f)The ground truth points are collected using Garmin trek 2.0 GPS for the same month corresponding to the satellite imageries used for the year 2020. (g) Topographic map of Nagpur city of scale 1:250,000 (source: https://soinakshe.uk.gov.in/), (h) AOI shape file with reference to the topographic sheet of Nagpur city. The complete details of Landsat satellite data type and time are mentioned in Table 13.1.

Table 13.1 Details of satellite imageries (Path/row: 144/045)

Landsat	Scene ID	Date and time	Cloud cover (%)
Landsat 7 ETM +	"LE71440452000417SGS00"	17–04–2000, "05:00:45.8099530Z"	0.00
Landsat 5 TM	"LT514404520050407BKT00"	07–04–2005, "04:55:35.1650380Z"	1.00
Landsat 5 TM	"LT514404520100421KHC00"	21–04–2010, "04:59:21.0040130Z"	0.00
Landsat 8 OLI/TIRS	"LC814404520150419LGN01"	19–04–2015, "05:07:51.5322230Z"	0.00
Landsat 8 OLI/TIRS	"LC814404520200416LGN00"	16–04–2020, "05:08:40.3204420Z"	0.00

Table 13.2 Description of land use classes

LULC classes	Description
Barren land	Open spaces, rocky areas, waste land, and infertile land
Built up	Industrial, commercial, residential, roads, transportation
Vegetation	Shrubs, trees, grasslands, thick vegetation, light vegetation
Water bodies	Lakes, ponds, reservoirs, permanent open water area with > 50% water

13.3.2 LULC Classification

The land use land cover classification of Nagpur city was performed to determine LULC pattern. The supervised classification technique was used to get accurate result. For LULC classification, maximum likelihood classification method was used inArcGIS 10.8 software. This method of classification is globally used for land classification. For Nagpur city, four classes were identified (1) Barren land, (2) Built-up, (3) Vegetation, and (4) Water bodies. For accuracy assessment of classified images, 100 ground control points are chosen. The description of different LULC classes is shown in Table 13.2.

13.3.3 Cellular Automata-Based (CA) Urban Growth Modeling

In this research, the change prediction simulation is performed in open-source QGIS software. The modules for land use change evaluation (MOLUSCE) plugin is used for simulation of prediction model. For prediction of future LULC, we have considered cellular automata with artificial neural network (CA-ANN) algorithm is used (Guidigan et al. 2019; Jokar Arsanjani et al. 2013; Saba et al. 2017). The process of prediction is considered as stochastic process as result depends on the random variables. To run the algorithm, past and present LULC maps are required depending on which future prediction is performed. The prediction of future LULC also depends on some drivers like distance to water bodies, distance to road network, building network, slope, and elevation. It is an important aspect of the research because it predicts future LULC for the next five years using simulation. Artificial neural networks (ANN), Markov chains, hybrid neural models, regression models, and other approaches are used to forecast LULC. The flow of process of prediction used in this study is explained in Fig. 13.2.

The MOLUSCE plugin tool implements MLP-ANN with training algorithm in a traditional way. It also performs factor variable normalization, screening, and model training. Normalization, in most cases, makes for more effective preparation and more reliable prediction results. The study was done with data from the last two

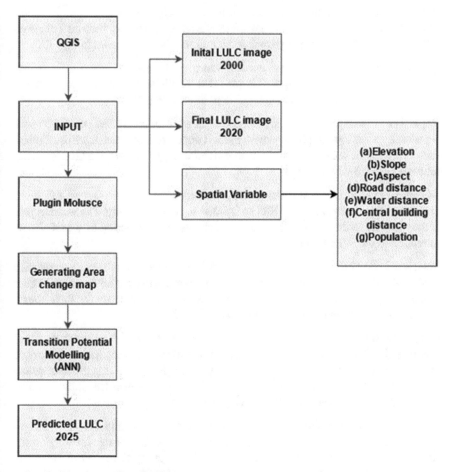

Fig. 13.2 Flow chart for generating prediction map

decades of LULC change modeling at the global level. Cellular automata is used to simulate changes in land use. To stable the learning graph, the model in this analysis was trained with a learning rate of 0.090 and momentum of 0.001. Based on the distance between the original and final year, the simulation will run once and then send the simulation for the next time frame. The simulator scans a set number of pixels with the highest confidence for each transformation belonging to the most possible transformations and then updates the pixels' class. Potential transitions are determined for each class, and the simulator creates a raster of the most possible transitions.

13.3.4 Parameters Used for Urban Growth Modeling

The study has used multiple parameters to accurately collect data from different sources to represent the urban expansion of Nagpur city. We have considered seven different parameters for urban growth modeling in this study shown in Fig. 13.3.

Elevation: In this study, we have used SRTM digital elevation model to know the elevation information of the area. It is seen that the higher elevation is witnesses in the western part of the city as compared to their eastern part.

Slope: The slope layer of the study area is prepared from the SRTM DEM. The area in generally a flat topography area with limited undulation. As seen from the figures, there is very little area with greater than 6-degree slope. Most of the area are lying between 1- and 3-degree slopes.

Aspect: It is an important influencing parameter responsible for causing urbanization. The aspect layer prepared from SRTM DEM is used in this study. The western part of the city, where there is variation in slope, shows much variation in aspect also as compared to the other part of the city.

Road Euclidian Distance: The road network is extracted from the open street map (OSM). Road is always a main central parameter responsible for urbanization. There is general agreement that everyone want a better road connectivity for their house, which would cause many benefits to their livelihood.

Water Euclidian Distance: The water body layer is obtained from the open street map (OSM). The role of water is essential in urbanization. Since ancient history, it can be seen that the most ancient civilization were developed along the river coast. Water is an important part to sustain in earth. So, this study uses Euclidian distance from water bodies as one of LULC driving parameter.

Central Building Distance (CBD): Proximity toward the central business district has been considered as an important parameter in the study for representing multi-urban functional areas along with connectivity. In this study, the center of the city is considered as central building point, and from there, a buffer of equal size was drawn,

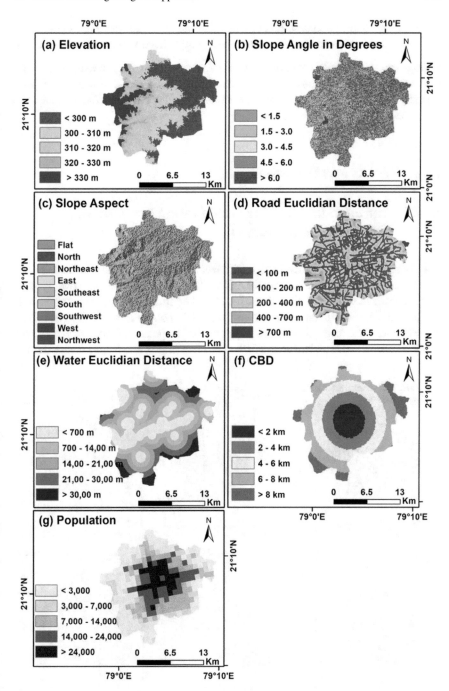

Fig. 13.3 Drivers for prediction model

up to the city outer boundary. It is obvious that the inner part of CBD is associated with rapid urban growth, while the outer part is associated with lower urbanization.

Population: The population is very essential in determining the future urban growth. In this study, the population data from land scan are considered as shown in the figure. The city center portion shown higher population, with greater than 24,000 people as compared to the surrounding areas.

13.4 Results and Discussion

13.4.1 Variation in Land Use Land Cover Maps Derived from Satellite Imagery

The spatial distribution of land use land cover map obtained from Landsat series f satellite for the year 2000, 2005, 2010, 2015, 2020, and predicted 2025 over Nagpur city is shown in Fig. 13.4. To validate the classified images, accuracy assessment is performed using confusion matrix, considering 100 points, 25 points for each class. In assessment, the original pixel is compared with the generated pixel to validate the result (Das et al. 2021; Islam et al. 2018). The error matrix and Kappa coefficient are

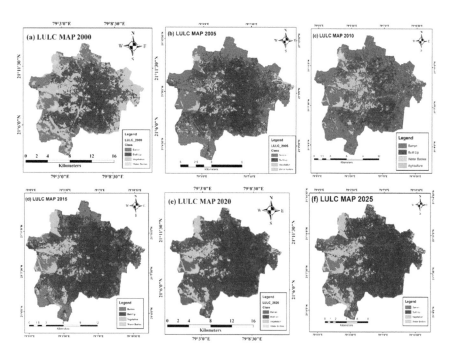

Fig. 13.4 LULC maps for different years from 2000 to 2025

generated. The result provides data of user's accuracy and producer's accuracy from which the overall accuracy is calculated. Kappa statistics is a method of comparing referenced and user-identified classified data. A Kappa value of 0.81–1.00 indicates a nearly perfect/perfect fit between the categorized and referenced data in the classification scheme. The result of assessment is generalized in Table 13.3. Results shows the accuracy of the classified LULC map prepared for the year 2000, 2005, 2010, 2015, and 2020 is 89.5%, 88.75%, 93.0%, 92.75%, and 90.25%, respectively, whereas the Kappa coefficient accuracy is 0.87, 0.86, 0.90, 0.92, and 0.89 for the subsequent years.

The built-up class has covered larger part of the study area. The migration of population residing in rural areas toward city urban part for education, careers, and better livelihood. The land demand is increased as determined from the study. The major change is seen in the built-up area as compared to other classes in last two decade.

The major affected class was built up as it shows a dramatic change in last two decades from 41.24% (92.13 km^2) in 2000 to 67.86% (151.59 km^2) in 2020 as seen in Table 13.4. The open spaces in the Nagpur urban area are decreased as a result of urbanization. The major effect of urbanization was on mix vegetation area; this land is converted to non-vegetation area for built-up purpose. The vegetation area shows decreasing trend over the study period from 67.4 km^2 in 2000 to 32.44 km^2 in 2020. The barren land also witnessed decreasing trend from 61.55 km^2 in 2000 to 28.33 km^2 in 2020. The built-up area shows increasing trend in expense of decreasing trend of barren land and vegetation. This suggests that most of the barren land and vegetation area converted to built-up land to accommodate ever-rising urban population.

The transformation of one to another LULC class is essential in understanding the changing LULC dynamics over the study area. Thus, in response to this, a change map is prepared using LULC data for the study period to analyze the change occurred from one class to another. The change data for one to another LULC classes are mentioned in Table 13.5. The spatial distribution of changes classes from 2000 to 2020 is shown in Fig. 13.5. The area change for the time period of 20 years is shown in Fig. 13.6.

The study explains that the change is occurred by conversion of one class to another. For Nagpur city, the four classes are determined, this change of land use pattern occurred between these four classes. As the maximum changes occurred between three classes: (a) barren land got converted into built up, (b) vegetation area got converted into built up and, (c) vegetation area converted to barren. The result of change is clear, for barren to build up, urban sprawl is responsible, and due to increase in population, the agricultural land inside the city boundary is utilized for urban built up. Many agricultural lands are converted to non-agriculture land for built-up purpose.

Table 13.3 Accuracy assessment

Year	User's accuracy				Producer's accuracy				Overall accuracy (%)	Kappa coefficient
	Barren	Built up	Vegetation	Water bodies	Barren	Built up	Vegetation	Water bodies		
2000	93	82	87	96	83.4	70.6	88.7	100	89.5	0.87
2005	86	84	87	98	87.4	91.5	88.8	100	88.75	0.86
2010	91	91	95	95	84.6	95.3	86.3	98.5	93	0.90
2015	92	88	95	96	95.8	87.6	91.8	92.7	92.75	0.92
2020	87	92	89	93	93.9	92.2	94.6	100	90.25	0.89

Table 13.4 LULC data from year 2000 to 2025

Land use class	2000		2005		2010		2015		2020		2025	
	Area (km²)	Area (%)	Area (km²)	Area (%)	Area (km²)	Area (%)	Area (km²)	Area (%)	Area (km²)	Area (%)	Area (km²)	Area (%)
Barren	61.55	27.55	95.51	42.75	88.1	39.43	58.42	26.15	37.23	16.67	28.33	12.71
Built-up	92.13	41.24	107.6	48.16	113.6	50.85	137.01	61.33	151.59	67.86	160.82	72.12
Vegetation	67.4	30.17	17.94	8.03	19.24	8.61	23.89	10.69	31.82	14.24	32.44	14.55
Water bodies	2.33	1.04	2.36	1.06	2.47	1.11	4.09	1.83	2.76	1.23	1.41	0.63
Total	223.4	100	223.4	100	223.4	100	223.4	100	223.4	100	223.4	100

Table 13.5 Change data from 2000 to 2020

Change	Area change (2000–2020)	Change	Area change (2000–2020)
Barren–barren	25.01	Built-up–water bodies	0.06
Barren–built-up	30.14	Vegetation–barren	11.33
Barren–vegetation	6.29	Vegetation–built-up	31.65
Barren–water bodies	0.01	Vegetation–vegetation	23.81
Built-up–barren	0.83	Vegetation–water bodies	0.5
Built-up–built-up	89.69	Water bodies–built-up	0.01
Built-up–vegetation	1.54	Water bodies–vegetation	0.13
		Water bodies–water bodies	2.18

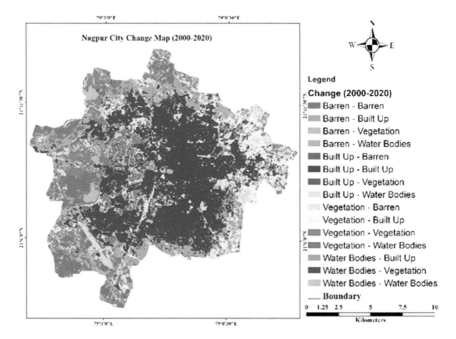

Fig. 13.5 Change map of one to another LULC classes during 2000–2020

13.4.2 Prediction Model

As the study consists of LULC classification for twenty-year time period and prediction of next five-year LULC, we have generated the prediction model in QGIS as discussed earlier using MOLUSCE plugin shown in Fig. 13.4. The result obtained from the prediction model is generalized in Table 13.6. The study of prediction model suggests that from year 2020 to 2025 the built-up area will increase to 72.12% (160.82

Fig. 13.6 Area change
graph during 2000–2020

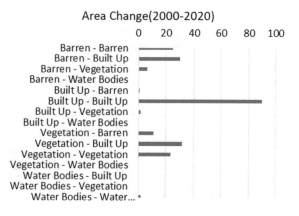

Table 13.6 LULC data of year 2025 and change in LULC classes from 2020 to 2025

Land use class	2025		Change class (2020–2025)	
	Area (km^2)	Area (%)	Area (km^2)	Percentage
Barren	28.33	12.71	−8.9	−3.96
Built up	160.82	72.12	9.23	4.26
Vegetation	32.44	14.55	0.62	0.31
Water bodies	1.41	0.63	−1.35	−0.60
Total	223.4	100.00		

km^2), and water bodies will decrease to 0.63% (1.41 km^2). The positive and negative changes are given in Table 13.6. The total area for different classes undergoes changes from 2020 to 2025, barren land decreased by 3.96%, built-up area increased by 4.26%, vegetation increased by 0.04%, and water bodies decreased by 0.60%.

13.4.3   Validation of the CA model

The result's spatial accuracy has been estimated concerning the spatial location of the built-up land in satellite-based LULC. The transition probability matrix has been computed by using the temporal LULC maps. After the study, the CA–ANN model's conclusions must be validated. The Kappa coefficient is measured using the MOLUSCE plugin in QGIS as part of the validation process. Different input parameters are required to validate the CA–ANN-based LULC prediction, like number of samples taken 1000 pixels of each land cover class in a random manner, learning rate of 0.001, hidden layers 10, and maximum iteration of 100 and momentum of 0.001. While running the artificial neural network model in MOLUSCE tool, a validation graph as shown below is generated along the train and validation data that indicated the overall validation of the input data. Three different Kappa was obtained from the

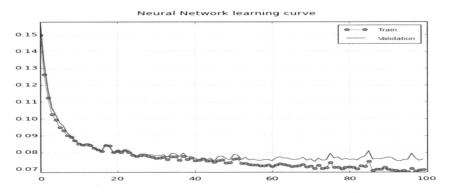

Fig. 13.7 Neural network learning curve

model as Kappa local, Kappa histogram, and Kappa overall, which vary 0.79, 0.85, and 0.81, respectively, in this present study. The percentage of accuracy, that was 81.23%, is often used to validate the CA–ANN. It means that there has been 81.23% correctness between the predicted 2020 LULC with the actual LULC. Considering, if the current land use transition continues in the same way, we predict the 2025 LULC in this case using the LULC data of 2015 and 2020, which will be discussed in next section (Fig. 13.7).

13.4.4 Urban Growth in Nagpur

Land use land cover study explains the trend of different classes consider during classification. The study of LULC for Nagpur explains the trend of urban sprawl. From the result, it is clear that built up is the most changed class from 2000 to 2020 and the in predicted model of 2025. We have discussed above the factors which are responsible for these changes. The Result shown in Table 13.4 characterized the area and percentage area change from 2000 to 2025. Figures 13.8 and 13.9 show the urban expansion for the period of twenty year of study area.

13.5 Conclusion

This study has shown the use of GIS and remote sensing for image classification of Nagpur city. The classification data show the increase in urbanization with decrease in vegetation and water bodies. This urban growth is affecting the sustainability of environment, as the agriculture land is converted to urban land. The study suggests that 67.86% of the Nagpur city area is utilized by urban population for building residential, commercial, industrial structures for the year 2020. The urban growth

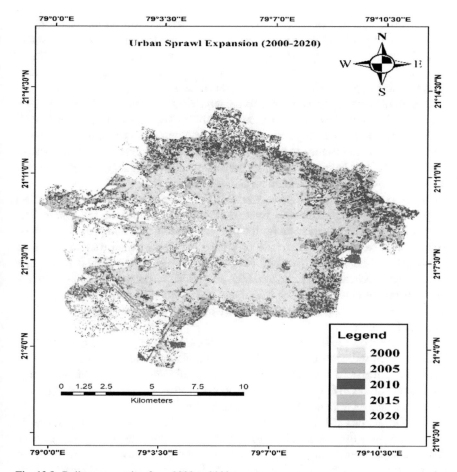

Fig. 13.8 Built-up expansion from 2000 to 2020

Fig. 13.9 Urban area
change (2000–2025)

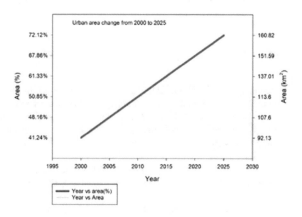

from 2000 to 2020 increased by 26.62%, which is serious issue for local authorities. In terms of area, 59.46 km^2 of urban built-up land is increases from past two decade. The Kappa index is used to assess the LULC classification of images; the value of Kappa varies from 0.86 to 0.92 which is considered as almost perfect or perfect match. The Nagpur urban sprawl cannot be constrained as it is the hub of several multinational companies which attracts the job seekers results in land demand. From the research, authors have figure out some variable drivers on which the urban growth depends, i.e., elevation, slope, aspect, central building distance, population, major road distance, and distance to water. The urban trend from previous studies shows that where there is better opportunities and resources available the urban population will follow that path for development and growth.

In Nagpur, the multi-layer perceptron (MLP) and artificial neural network (ANN) approach were used to accurately forecast the future LULC transition scenario. The prediction simulation was done for the year 2025 using past and present classified images as a base image. The result of prediction shows increase in built-up area by 4.26% as compared to year 2020; vegetation area is increased by 0.41%, water bodies and barren land decreased by 0.60% and 3.96%, respectively. The simulation's findings provide insight into the magnitude and position of potential changes. The prediction model will give idea about the urban sprawl which can be used for making policies for future.

This approach can be applied to other areas that are experiencing rapid urbanization which will save cost and time both and also provide quantified data. This technique can also be used for land management and monitoring. The findings could aid municipalities in designing suitable counter measures to protect valuable natural resources such as wetlands and agricultural land lands.

Despite the fact that this research's overall accuracy was adequate, there were certain concerns that needs to be resolved. It was attempted to classify images from the same sensor, but due to cloud cover and a lack of high-quality images and distorted images, this was not possible. This resulted in issues such as class mismatching. As a result, it is the best to use images from the same satellite for future research.

References

Akyürek D, Koç Ö, Akbaba EM, Sunar F (2018) Land use/land cover change detection using multi–temporal satellite dataset: a case study in Istanbul new airport. ISPRS Int Arch Photogramm Remote Sens Spat Inf Sci 42:17–22. https://doi.org/10.5194/isprs-archives-XLII-3-W4-17-2018

Anand V, Oinam B (2020) Future land use land cover prediction with special emphasis on urbanization and wetlands. Remote Sens Lett 11(3):225–234. https://doi.org/10.1080/2150704X.2019.1704304

Bhat PA, Shafiq MU, Mir AA, Ahmed P (2017) Urban sprawl and its impact on landuse/land cover dynamics of Dehradun City, India. Int J Sustain Built Environ 6(2):513–521. https://doi.org/10.1016/j.ijsbe.2017.10.003

Das N, Mondal P, Sutradhar S, Ghosh R (2021) Assessment of variation of land use/land cover and its impact on land surface temperature of Asansol subdivision. Egypt J Remote Sens Space Sci 24(1):131–149. https://doi.org/10.1016/j.ejrs.2020.05.001

Dutta S, Guchhait SK (2020) Assessment of land use land cover dynamics and urban growth of Kanksa Block in Paschim Barddhaman District, West Bengal. GeoJournal. https://doi.org/10. 1007/s10708-020-10292-3

Fonseka HPU, Zhang H, Sun Y, Su H, Lin H, Lin Y (2019) Urbanization and Its impacts on land surface temperature in Colombo metropolitan area, Sri Lanka, from 1988 to 2016. Remote Sens 11(8):957. https://www.mdpi.com/2072-4292/11/8/957

Fu P, Weng Q (2016) A time series analysis of urbanization induced land use and land cover change and its impact on land surface temperature with Landsat imagery. Remote Sens Environ 175:205–214. https://doi.org/10.1016/j.rse.2015.12.040

Ghosh S, Chatterjee ND, Dinda S (2019) Relation between urban biophysical composition and dynamics of land surface temperature in the Kolkata metropolitan area: a GIS and statistical based analysis for sustainable planning. Modeling Earth Syst Environ 5(1):307–329. https://doi. org/10.1007/s40808-018-0535-9

Gidey E, Dikinya O, Sebego R, Segosebe E, Zenebe A (2017) Cellular automata and Markov Chain (CA_Markov) model-based predictions of future land use and land cover scenarios (2015–2033) in Raya, northern Ethiopia. Modeling Earth Syst Environ 3(4):1245–1262. https://doi.org/10. 1007/s40808-017-0397-6

Gohain KJ, Mohammad P, Goswami A (2021) Assessing the impact of land use land cover changes on land surface temperature over Pune city, India. Quat Int 575–576:259–269. https://doi.org/10. 1016/j.quaint.2020.04.052

Goyal A, Sharma M, Singh DD (2019) Land use/land cover change detection using geoinformatics in Gurugram District, Haryana, India. Int J Recent Technol Eng 8(2):3753–3755. https://doi.org/ 10.35940/ijrte.B3388.078219

Guidigan MLG, Sanou CL, Ragatoa DS, Fafa CO, Mishra VN (2019) Assessing land use/land cover dynamic and its impact in Benin Republic using land change model and CCI-LC products. Earth Syst Environ 3(1):127–137. https://doi.org/10.1007/s41748-018-0083-5

Hadi SJ, Shafri HZM, Mahir MD (2014) Modelling LULC for the period 2010–2030 using GIS and remote sensing: a case study of Tikrit, Iraq. In: 7th IGRSM international remote sensing AND GIS conference and exhibition Kaula Lampur, Malaysia.https://doi.org/10.1088/1755-1315/20/ 1/012053

Hishe S, Bewket W, Nyssen J, Lyimo J (2019) Analysing past land use land cover change and CA-Markov-based future modelling in the Middle Suluh Valley, Northern Ethiopia. Geocarto Int 35(3):225–255. https://doi.org/10.1080/10106049.2018.1516241

Hussain S, Mubeen M, Ahmad A, Akram W, Hammad HM, Ali M, Masood N, Amin A, Farid HU, Sultana SR, Fahad S, Wang D, Nasim W (2020) Using GIS tools to detect the land use/land cover changes during forty years in Lodhran District of Pakistan. Environ Sci Pollut Res 27(32):39676–39692. https://doi.org/10.1007/s11356-019-06072-3

Hyandye C, Martz LW (2017) A Markovian and cellular automata land-use change predictive model of the Usangu Catchment. Int J Remote Sens 38(1):64–81. https://doi.org/10.1080/014 31161.2016.1259675

Imran Basha U, Suresh U, Sudarsana Raju G, Rajasekhar M, Veeraswamy G, Balaji E (2018) Landuse and landcover analysis using remote sensing and GIS: a case study in Somavathi River, Anantapur District, Andhra Pradesh, India. Nat Environ Pollut Technol 17(3):1029–1033. http:// www.neptjournal.com/upload-images/NL-65-50-(48)B-3514.pdf

Islam K, Jashimuddin M, Nath B, Nath TK (2018) Land use classification and change detection by using multi-temporal remotely sensed imagery: the case of Chunati wildlife sanctuary, Bangladesh. Egypt J Remote Sens Space Sci 21(1):37–47. https://doi.org/10.1016/j.ejrs.2016. 12.005

Jelil Niang A, Hermas E, Alharbi O, Al-Shaery A (2020) Monitoring landscape changes and spatial urban expansion using multi-source remote sensing imagery in Al-Aziziyah Valley, Makkah, KSA. Egypt J Remote Sens Space Sci 23(1):89–96. https://doi.org/10.1016/j.ejrs.2018.06.001

Jokar Arsanjani J, Helbich M, Kainz W, Darvishi Boloorani A (2013) Integration of logistic regression, Markov chain and cellular automata models to simulate urban expansion. Int J Appl Earth Obs Geoinf 21:265–275. https://doi.org/10.1016/j.jag.2011.12.014

Kaichang D, Deren L, Deyi L (2012) Remote sensing image classification with GIS data based on spatial data mining techniques. Geo-Spat Inf Sci 3(4):30–35. https://doi.org/10.1007/bf02829393

Kaliraj S, Chandrasekar N, Ramachandran KK, Srinivas Y, Saravanan S (2017) Coastal landuse and land cover change and transformations of Kanyakumari coast, India using remote sensing and GIS. Egypt J Remote Sens Space Sci 20(2):169–185. https://doi.org/10.1016/j.ejrs.2017.04.003

Khan F, Das B, Ram Krishna Mishra S, Awasthy M (2021) A review on the feasibility and application of geospatial techniques in geotechnical engineering field. Mater Today Proc. https://doi.org/10.1016/j.matpr.2021.02.108

Khan F, Rao TK, Bhave HD (2018) Classification of foundation soil: using geoinformatics(GIS). Int J Civil Eng Technol (IJCIET) 9(4):1199–1207. Article IJCIET_09_04_134. http://www.iaeme.com/MasterAdmin/Journal_uploads/IJCIET/VOLUME_9_ISSUE_4/IJCIET_09_04_134.pdf

Kindu M, Schneider T, Teketay D, Knoke T (2013) Land use/land cover change analysis using object-based classification approach in Munessa-Shashemene landscape of the Ethiopian highlands. Remote Sens 5(5):2411–2435. https://doi.org/10.3390/rs5052411

Kumar KS, Kumari KP, Bhaskar PU (2016) Application of Markov chain and cellular automata based model for prediction of urban transitions. In: 2016 international conference on electrical, electronics, and optimization techniques (ICEEOT). https://doi.org/10.1109/ICEEOT.2016.7755466

Liping C, Yujun S, Saeed S (2018) Monitoring and predicting land use and land cover changes using remote sensing and GIS techniques-A case study of a hilly area, Jiangle, China. PLoS One 13(7):e0200493. https://doi.org/10.1371/journal.pone.0200493

Liu T, Yang X (2015) Monitoring land changes in an urban area using satellite imagery, GIS and landscape metrics. Appl Geogr 56:42–54. https://doi.org/10.1016/j.apgeog.2014.10.002

Lu Y, Wu P, Ma X, Li X (2019) Detection and prediction of land use/land cover change using spatiotemporal data fusion and the Cellular Automata–Markov model. Environ Monit Assess 191(2):68. https://doi.org/10.1007/s10661-019-7200-2

Magesh NS, Chandrasekar N (2017) Driving forces behind land transformations in the Tamiraparani sub-basin, South India. Remote Sens Appl Soc Environ 8:12–19. https://doi.org/10.1016/j.rsase.2017.07.003

Mahamud MA, Samat N, Tan ML, Chan NW, Tew YL (2019) Prediction of future land use land cover changes of Kelantan, Malaysia. ISPRS Int Arch Photogrammetry Remote Sens Spat Inf Sci 42:379–384. https://doi.org/10.5194/isprs-archives-XLII-4-W16-379-2019

Minta M, Kibret K, Thorne P, Nigussie T, Nigatu L (2018) Land use and land cover dynamics in Dendi-Jeldu Hilly-mountainous areas in the central Ethiopian highlands. Geoderma 314:27–36. https://doi.org/10.1016/j.geoderma.2017.10.035

Mishra VN, Rai PK, Rajendra P, Punia M, Nistor MM (2018) Prediction of spatio-temporal land use/land cover dynamics in rapidly developing Varanasi district of Uttar Pradesh India, using geospatial approach: a comparison of hybrid models. Appl Geomat 10:257–276. https://doi.org/10.1007/s12518-018-0223-5

Mishra PK, Rai A, Rai SC (2020) Land use and land cover change detection using geospatial techniques in the Sikkim Himalaya, India. Egypt J Remote Sens Space Sci 23(2):133–143. https://doi.org/10.1016/j.ejrs.2019.02.001

Mishra VN, Rai PK (2016) A remote sensing aided multi-layer perceptron-Marcove Chain analysis for land use and land cover change prediction in Patna district (Bihar), India. Arab J Geosci 9(1):1–18. https://doi.org/10.1007/s12517-015-2138-3

Mishra VN, Rai PK, Kumar P, Prasad R (2016) Evaluation of land use/land covers classification accuracy using multi-temporal remote sensing images. Forum Geographic 15(1):45–53

Mohamed A, Worku H (2020) Simulating urban land use and cover dynamics using cellular automata and Markov chain approach in Addis Ababa and the surrounding. Urban Clim 31:100545. https://doi.org/10.1016/j.uclim.2019.100545

Mohamed SA, El-Raey ME (2019) Land cover classification and change detection analysis of Qaroun and Wadi El-Rayyan lakes using multi-temporal remotely sensed imagery. Environ Monit Assess 191(4):229. https://doi.org/10.1007/s10661-019-7339-x

Mohammad P, Goswami A (2021a) Spatial variation of surface urban heat island magnitude along the urban-rural gradient of four rapidly growing Indian cities. Geocarto Int 1–23. https://doi.org/10.1080/10106049.2021.1886338

Mohammad P, Goswami A (2021b) A spatio-temporal assessment and prediction of surface urban heat island intensity using multiple linear regression techniques over Ahmedabad City, Gujarat. J Indian Soc Remote Sens. https://doi.org/10.1007/s12524-020-01299-x

Mohammad P, Goswami A, Bonafoni S (2019) The impact of the land cover dynamics on surface urban heat island variations in semi-arid cities: a case study in Ahmedabad City, India, using multi-sensor/source data. Sensors 19(17):3701. https://www.mdpi.com/1424-8220/19/17/3701

Mondal I, Thakur S, Ghosh P, De TK, Bandyopadhyay J (2019) Land use/land cover modeling of Sagar Island, India using remote sensing and GIS techniques. In: Emerging technologies in data mining and information security, pp 771–785. https://doi.org/10.1007/978-981-13-1951-8_69

Munthali MG, Mustak S, Adeola A, Botai J, Singh SK, Davis N (2020) Modelling land use and land cover dynamics of Dedza district of Malawi using hybrid Cellular Automata and Markov model. Remote Sens Appl Soc Environ 17:100276. https://doi.org/10.1016/j.rsase.2019.100276

Ogashawara I, Bastos V (2012) A quantitative approach for analyzing the relationship between urban heat islands and land cover. Remote Sens 4(11):3596–3618. https://doi.org/10.3390/rs4113596

Rimal B, Zhang L, Keshtkar H, Haack BN, Rijal S, Zhang P (2018) Land use/land cover dynamics and modeling of urban land expansion by the integration of cellular automata and Markov Chain. ISPRS Int J Geo-Inf 7(4):154. https://www.mdpi.com/2220-9964/7/4/154

Saba T, Rehman A, AlGhamdi JS (2017) Weather forecasting based on hybrid neural model. Appl Water Sci 7(7):3869–3874. https://doi.org/10.1007/s13201-017-0538-0

Sahana M, Ahmed R, Sajjad H (2016) Analyzing land surface temperature distribution in response to land use/land cover change using split window algorithm and spectral radiance model in Sundarban Biosphere Reserve, India. Modeling Earth Syst Environ 2(2):81. https://doi.org/10.1007/s40808-016-0135-5

Sakhre S, Dey J, Vijay R, Kumar R (2020) Geospatial assessment of land surface temperature in Nagpur, India: an impact of urbanization. Environ Earth Sci 79(10):226. https://doi.org/10.1007/s12665-020-08952-1

Saputra MH, Lee HS (2019) Prediction of land use and land cover changes for North Sumatra, Indonesia, using an artificial-neural-network-based cellular automaton. Sustainability 11(11):3024. https://www.mdpi.com/2071-1050/11/11/3024

Singh SK, Mustak S, Srivastava PK, Szabó S, Islam T (2015) Predicting spatial and decadal LULC changes through cellular automata Markov chain models using earth observation datasets and geo-information. Environ Process 2(1):61–78

Tadese S, Soromessa T, Bekele T (2021) Analysis of the current and future prediction of land use/land cover change using remote sensing and the CA-Markov Model in Majang forest biosphere reserves of Gambella, Southwestern Ethiopia. Sci World J 2021:6685045. https://doi.org/10.1155/2021/6685045

Urgessa T, Lemessa D (2020) Spatiotemporal landuse land cover changes in Walmara District, Central Oromia, Ethiopia. Earth Sci 9(1). https://doi.org/10.11648/j.earth.20200901.14

Vishwakarma CS, Thakur S, Rai PK, Kamal V, Mukharjee S (2016) Changing land trajectories: a case study from India using remote sensing. Eur J Geogr 7(2):63–73

Wang SW, Munkhnasan L, Lee W-K (2021) Land use and land cover change detection and prediction in Bhutan's high altitude city of Thimphu, using cellular automata and Markov chain. Environ Challenges 2:100017. https://doi.org/10.1016/j.envc.2020.100017

Yang X, Lo CP (2002) Using a time series of satellite imagery to detect land use and land cover changes in the Atlanta, Georgia metropolitan area. Int J Remote Sens 23(9):1775–1798. https://doi.org/10.1080/01431160110075802

Yatoo SA, Sahu P, Kalubarme MH, Kansara BB (2020) Monitoring land use changes and its future prospects using cellular automata simulation and artificial neural network for Ahmedabad city. GeoJournal, India. https://doi.org/10.1007/s10708-020-10274-5

Chapter 14
Slum Categorization for Efficient Development Plan—A Case Study of Udhampur City, Jammu and Kashmir Using Remote Sensing and GIS

Majid Farooq, Gowhar Meraj, Rishabh, Shruti Kanga, Ritu Nathawat, Suraj Kumar Singh, and Vikram Ranga

Abstract Urbanization is likely to increase the rate of slum growth, as it has in the past. According to the 2011 Indian census, the population of Udhampur urban area is 91,366, with 35,507 people living in the Udhampur municipal area and 48,508 people living in Udhampur's outgrowths. Udhampur city is divided into 21 municipal wards. Out of 21 slums, 11 slums are non-notified, and ten are notified. This research attempts to categorize slums based on living standards, which will help formulate sustainable development techniques for better implementation of slum improvement projects. Data about the socioeconomic and physical condition of the slums have been collected using field surveys. For clustering slums in different categories, a 2 × 2 × 2 matrix is formed. For creating an indicative matrix, essential inputs were identified, and an overall matrix table for all the slums with their scores was prepared. A georeferenced very high-resolution satellite imagery with a ward boundary map was used to create a base map. Different maps were generated showing current slum distribution and also the spatial distribution of varying slum categories. Maps were validated with field survey and with field photographs.

M. Farooq · G. Meraj
Department of Ecology, Environment and Remote Sensing, Government of Jammu and Kashmir, Kashmir 190018, India

Rishabh · S. Kanga · R. Nathawat
Center for Climate Change & Water Research, Suresh Gyan Vihar University, Jaipur, Rajasthan 302017, India
e-mail: shruti.kanga@mygyanvihar.com

S. K. Singh (✉)
Center for Sustainable Development, Suresh Gyan Vihar University, Jaipur, Rajasthan 302017, India
e-mail: suraj.kumar@mygyanvihar.com

V. Ranga
Center for the Study of Regional Development, Jawaharlal Nehru University, New Delhi 110067, India

Keywords Slum management · Urbanization · Smart cities · Remote sensing · GIS · Jammu and Kashmir

14.1 Introduction

India is home to one-third of the world's impoverished people. According to a World Bank estimate from 2005, 41.% of India's total population lives below the global poverty level of US $1.25 a day. According to United Nations Development Program data from 2010, an estimated 37.2% of Indians live below the national poverty level. According to the most recent UNICEF data, India has one in every three malnourished children worldwide, with 42% of children under five being underweight (UN-HSP 2010). According to the 2011 Global Hunger Index (GHI) report, India is one of three countries where the GHI increased from 22.9 to 23.7 between 1996 and 2011, while 78 of the 81 developing countries studied, including Bangladesh, Pakistan, Nepal, Vietnam, Zimbabwe, Myanmar, Kenya, Nigeria, Uganda, and Malawi, saw significant improvements.

Since its first five-year plan, the Indian government has implemented several programs to reduce poverty, including subsidizing food and other essentials, expanding access to credit, enhancing farming technology and price supports, and boosting education and birth control (Baud et al. 2008; Bhagat and Mohanty 2009; Baud et al. 2010). These policies have aided in abolishing famines, reduced absolute poverty by more than half, and reduced illiteracy and malnutrition. Despite India's tremendous macroeconomic development, one-quarter of the populace receives less than the government-defined poverty line of Rs. 32 a day. According to a recent World Bank assessment, India is on track to accomplish its poverty reduction targets. However, by 2015, it is still approximated that a projected 53 million people will live in extreme poverty (Sur et al. 2004; Jain et al. 2005; No C. 2011).

Slum definition adopted for Jammu and Kashmir is a contiguous area with 10–15 households having slum-like characteristics identified as having roof material predominantly other than concrete (RBC/RCC); having the drinking water source availability, not within premises of the census house, having the availability of latrine, not within premises of the census house, and having no drainage or open drainage. Udhampur, known as the land of Dhruva, also called "Devika Nagri," is a picturesque place situated in the southern part of Jammu and Kashmir, touching its periphery in the north by Anantnag, northeast by Doda, southeast by Kathua, and southwest by Jammu. Udhampur lies between $32°\ 34'$ to $39°\ 30'$ North latitude and $74°\ 16'$ to $75°\ 38'$ East longitude (Fig. 14.1). The altitude ranges from 600 to 3000 m above mean sea level, so the temperature variation is perceptible.

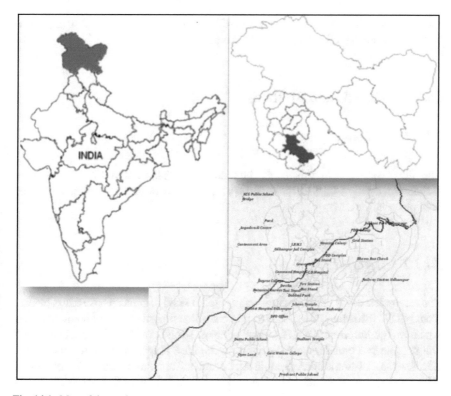

Fig. 14.1 Map of the study area

14.2 Study Area

The district has a varied topography. It is interwoven with Shivalik ranges of hills. There are, however, very few inhabited areas above the height of 1112 m that experience snowfall and severe cold in the winter. Udhampur is the fifth largest district of Jammu and Kashmir with an area of 4550 km^2 stretched across five tehsils, viz Udhampur, Ramnagar, Reasi, Chenani, and Gool Gulabgarh, with the forest area of 2343 km^2. The district's administrative headquarters are in Udhampur town at a distance of 66 km from the winter capital of the state, "Jammu," on the national highway. With a total road length of 717 km, the district is interwoven within the Web of 624 villages; out of which, 568 villages are fully electrified, and 609 villages receive a regular potable water supply. The population of the district is 555,357 as per the 2011 census. This gives it the ranking of 538th in India out of a total of 640. The forest area stands at around 1920 sq. km, and the gross area sown is 71,000 ha, with the principal crop being maize. Udhampur district has 12 community blocks in total, viz Gdhampur, Chenani, Panchari, Ghordi, Ramnagar, Majalta, Dudu-Basantgarh, Reasi, Pouni, Arenas, Mahore, and Gool with 117 panchayats. Besides the rural stretch, the district has six towns, viz Udhampur, Reasi, Katra, Ramnagar, Chenani, and Rehamble. The temperature rises to 42° in summers and dips to 1.5° in winter. Most of the rainfall takes place in July, August, and September.

14.3 Materials and Methods

14.3.1 Datasets

The database used in this study includes data gathered in the 21 slums of Udhampur through GPS surveys, surveys conducted through questionnaires, and field observations.

14.3.2 Methods

The research area's satellite picture was georeferenced to create a base map obtained from NRSC (Shekhar, 2014). The municipal office provided a ward boundary map, which was georeferenced with the satellite image (Kohli et al. 2011). Ward boundaries and notable placemarks were depicted on the base map created in a GIS system. Because the real-home borders could not be drawn from the cartosat-1 (2.5 m) image, GPS was used to determine the exact position of each family (Kuffer et al. 2016). A point theme was created, and the aggregate of these points yielded a slum boundary concept, which was then polygonized to make a slum boundary. The matrix was created using survey data acquired in the field as input (Mason and Fraser 1998).

The satellite image was used to create a detailed land use/land cover map of the town. The results of the demographic survey and matrix were visualized as maps that were linked to specific slums (Sliuzas 2008). The reports shown here were created with the use of a geographic query in a GIS. The overall methodology used in this work is shown in Fig. 14.2.

14.3.2.1 Slum Categorization – Development of Matrix

The information collected during this survey on poverty in slum households based on the actual housing and living conditions, infrastructure facilities, and tenure status has helped categorize these slums based on a $2 \times 2 \times 2$ matrix. The analysis of this indicative matrix has been instrumental in developing a systematic and transparent process for categorizing the slums as per their vulnerability and deficiency in terms of housing, infrastructure, and tenure status. This deficiency matrix would help arrive at a suitable development option to maximize the benefits by properly planning and prioritizing the resources.

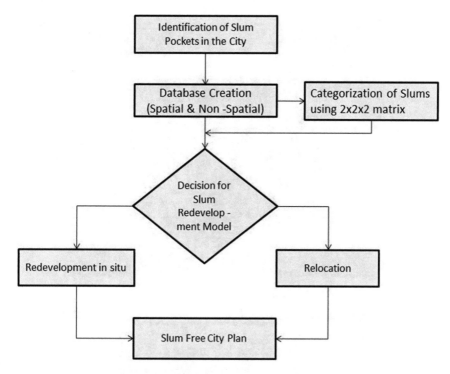

Fig. 14.2 Overall methodology used in the present study

The methodology adopted is systematic categorization of slums on a 2 × 2 × 2 basis has been strictly followed using data on the parameters of housing, infrastructure, and tenure developed on the data collected, which influence the housing, infrastructure, and tenure individually. The data collected on all the parameters of importance have been placed under the three components, and each component has been given due weight. Eventually, the scores have been generated. A uniform procedure to attach the weightage has been adopted so that each parameter gets its position as per its significance and relative importance.

The methodology in identifying the infrastructure deficiency in the slums is based on taking on board the parameters that have a bearing on infrastructure, including water supply, drainage, solid waste management, sanitation, and roads (Fig. 14.3). The scoring has been done as per the procedure adopted given below, and then, high and low level of infrastructure status is arrived at. The total score ranges from a minimum of 5 to a maximum of 10. Then, the final total score is calculated by taking less or equal to 7 as a high level of infrastructure given the code "1" and greater than seven as low levels of infrastructure given the code "2".

Water supply:

Score 1—60% or more of household connected.

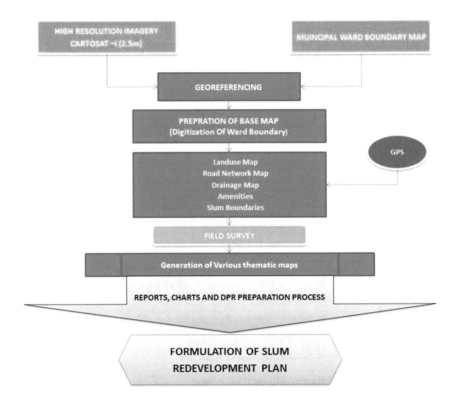

Fig. 14.3 Methodology flowchart for map generation

Score 2—Less than 60% of households connected.

Drainage:

Score 1—60% or more of in-slums connected with drains.
Score 2—Less than 60% of in-slum connected with drains.

Solid Waste Management:

Score 1—60% or more slums having garbage disposal daily, or once in 15 days + 60% or more slums having dedicated municipal staff arrangement for garbage disposal + 60% or more slums having a clearance of open drains once in 15 days.
Score 2—Less than 60% slums having garbage disposal daily or once in 15 days + Less than 60% slums having dedicated municipal staff arrangement for garbage disposal + Less than 60% slums having a clearance of open drains once in 15 days.

Sanitation:

Score 1—less than 60% going for open defecation.
Score 2—60% or more going for open defecation.

Road:

Score 1—60% or more having motorable pucca roads.
Score 2—less than 60% having motorable pucca roads.

Housing analysis is also done on the same pattern. The parameters taken into consideration for arriving at the score have been based on the structural condition.

Score 1—60% or more having pucca / semi-pucca housing.
Score 2—less than 60% having pucca / semi-pucca housing.

Tenure includes the parameters which classify the slums into secure and insecure tenure.

Score 1—Secure tenure 60% or more have possession certificate/occupancy rights.
Score 2—Insecure tenure less than 60% having possession certificate/occupancy rights, living in tented houses, or encroached public or private land.

14.3.2.2 Matrix Formation

The matrix has been developed to categorize the slums is eight categories, four in each with secure and insecure tenure placing housing on the x-axis and infrastructure on the y-axis in both. This pattern gives an idea of slums falling into categories with poor or good infrastructure and poor or good housing facilities within each secure and insecure tenure segment (Fig. 14.4).

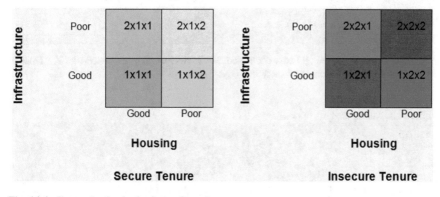

Fig. 14.4 Categorization in the form of matrix

14.4 Results and Discussion

After analyzing the parameters under infrastructure, housing, and tenure as per the dispensation described above, the slums have been ranked as per their respective position in each parameter and categorized in the matrix. For example, if a slum receives a final score of 2 for infrastructure and 2 for housing, it is assigned to the 2 × 1 × 2 matrix segment (Tables 14.1, 14.2, 14.3, 14.4 and 14.5).

The analysis of the slums indicates that 14 out of 21 slums fall in the 2 × 1 × 2 category, 3 in 1 × 1 × 2, 2 in 2 × 2 × 2, and one in 2 × 1 × 1 and 2 × 2 × 1.

Table 14.1 Matrix analysis of infrastructure

Slums		Bharat Nagar	Christain colony	Garian Talab	Idd Gah	Lower Omera	Omala Jakherni	Sail Salan	Sangoor	Upper Sail Salan	Upper Sangoor	Barrian	Dandyal	Gangera Dallah	Kaller	Kharene Omera	Nagrota	Rount	Sajalta	Sambol	Subash Nagar	Thanda Padde
House holds		45	43	73	75	88	179	104	60	34	38	17	62	102	22	32	82	33	16	56	39	70
Water supply		p	n	c	c	c	n	n	n	n	n	n	p	p	c	p	c	c	c	c	p	p
Score		2	2	1	1	1	2	2	2	2	2	2	2	2	1	2	1	1	1	1	2	2
Drainage		n	n	n	p	n	n	n	n	n	n	n	n	n	n	n	n	n	n	n	n	n
Score		2	2	2	2	2	2	2	2	2	2	2	2	2	2	2	2	2	2	2	2	2
Solid waste management	Frequency of Garbage disposal	n	n	O/15	Daily	n	n	n	n	n	n	O/w	n	n	n	n	n	n	O/w	n	n	n
	Arrangement for Garbage disposal	n	M/S	M/S	M/S	M/S	n	n	n	M/S	n	M/S	M/S	n	M/S	n	n	M/S	M/S	M/S	n	M/S
	Frequency of clearance of open drains	n	O/15	O/15	O/2	n	n	n	n	n	n	n	O/15	n	n	n	n	n	O/15	n	n	n
Score		2	1	1	1	2	2	2	2	2	2	1	2	2	2	2	2	1	2	2	2	2
Sanitation	Open defecation	45	42	73	74	88	177	103	57	33	35	17	62	100	22	32	82	33	16	54	39	69
	Shared	0	0	0	1	0	1	1	1	1	0	0	0	1	0	0	0	0	0	0	2	0
	Own dry	0	1	0	0	0	1	0	1	0	0	0	0	0	0	0	0	0	0	0	0	1
	Own Septic	0	0	0	0	0	0	1	0	1	0	3	0	0	1	0	0	0	0	0	0	0
	% open defecation	100	97.67	100	98.67	100	98.88	99.04	95.00	97.06	92.11	100	100	98.04	100	100	100	100	100	96.43	100	98.57
Score		1	1	1	1	1	1	1	1	1	1	1	1	1	1	1	1	1	1	1	1	1
Roads	Motorable Pucca	e	n	e	e	e	n	n	n	n	n	n	e	n	e	n	e	e	e	e	n	n
	Motorable Katcha	n	n	n	n	n	n	n	n	n	n	n	n	n	e	n	n	n	n	n	n	n
	Non-Motorable Katcha	n	e	n	n	n	e	n	e	n	e	e	n	n	n	n	e	n	n	n	n	e
Score		1	2	1	1	1	2	2	2	2	2	2	1	2	1	2	1	1	1	1	1	2
Net Score		2	2	1	1	2	2	2	2	2	2	2	2	2	2	2	1	2	2	2	2	2

N Not connected; *e* Existing; *p* Partly connected; *M/s* Municipal Staff; *c* Connected; *O/15* Once in 15 days; *O/w* Once in a weak; *O/2* Once in 2 days

Table 14.2 Matrix analysis of housing

Slums		Bharat Nagar	Christain colony	Garian Talab	Idd Gah	Lower Omera	Omala Jakherni	Sail Salan	Sangoor	Upper Sail Salan	Upper Sangoor	Barrian	Dandyal	Gangera Dallah	Kaller	Kharene Omera	Nagrota	Rount	Sajalta	Sambol	Subash Nagar	Thanda Padde
House holds		45	43	73	75	88	179	104	60	34	38	17	62	102	22	32	82	33	16	56	39	70
Housing	% Kaccha Houses	77.78	76.74	54.79	76.00	70.45	75.98	79.81	73.33	91.18	39.47	0.00	56.45	79.41	63.64	71.88	50.00	87.88	87.50	66.07	64.10	82.86
	% Semi-Pucca Houses	22.22	23.26	45.21	24.00	29.55	24.02	20.19	26.67	8.82	60.53	100	43.55	20.59	36.36	28.13	50.00	12.12	12.50	33.93	35.90	17.14
Score		2	2	2	2	2	2	2	2	1	1	2	2	2	2	2	2	2	2	2	2	2

Table 14.3 Matrix analysis of tenure status

Slums		Bharat Nagar	Christain colony	Garian Talab	Idd Gah	Lower Omera	Omala Jakheni	Sail Salan	Sangoor	Upper Sail Salan	Upper Sangoor	Barrion	Dandayal	Gangera Dallah	Kaller	Kharene Omera	Nagrota	Rount	Sajalta	Sambal	Subash Nagar	Thanda Padder
House holds		45	43	73	75	88	179	104	60	34	38	17	62	102	22	32	82	33	16	56	39	70
Tenure	Possession certificate/ Occupancy rights	100	88.37	73.97	64.00	43.18	93.85	96.15	43.33	97.06	44.74	100	93.55	67.65	95.45	87.50	93.90	84.85	100	100	79.49	84.29
	Living in tented houses /encroached public or private land	0.00	11.63	26.03	36.00	56.82	6.15	3.85	56.67	2.94	55.26	0.00	6.45	32.35	4.55	12.50	6.10	15.15	0.00	0.00	20.51	15.71
Score		1	1	1	1	2	1	1	2	1	2	1	1	1	1	1	1	1	1	1	1	1

Table 14.4 Overall matrix table

Slums	Bharat Nagar	Christain colony	Garian Talab	Idd Gah	Lower Omera	Omala Jakheni	Sail Salan	Sangoor	Upper Sail Salan	Upper Sangoor	Barrion	Dandayal	Gangera Dallah	Kaller	Kharene Omera	Nagrota	Rount	Sajalta	Sambal	Subash Nagar	Thanda Padder
Infrastructure	2	2	1	1	2	2	2	2	2	2	2	2	2	2	2	2	1	2	2	2	2
Tenure	1	1	1	1	2	1	1	2	1	2	1	1	1	1	1	1	1	1	1	1	1
Housing	2	2	2	2	2	2	2	2	2	1	1	2	2	2	2	2	2	2	2	2	2
Matrix position	2x1x2	2x1x2	1x1x2	1x1x2	2x2x2	2x1x2	2x1x2	2x2x2	2x1x2	2x2x1	2x1x1	2x1x2	2x1x2	2x1x2	2x1x2	2x1x2	1x1x2	2x1x2	2x1x2	2x1x2	2x1 x2

Table 14.5 Final resultant matrix

Matrix Position	No. of Slums
Secure Tenure	
$1 \times 1 \times 1$	0
$1 \times 1 \times 2$	3
$2 \times 1 \times 1$	1
$2 \times 1 \times 2$	14
Insecure Tenure	
$1 \times 2 \times 1$	0
$1 \times 2 \times 2$	0
$2 \times 2 \times 1$	1
$2 \times 2 \times 2$	2

14.5   Analysis of Slums in $2 \times 1 \times 2$ Category

Fourteen slums fall in the category of $2 \times 1 \times 2$ out of which "5" slums, namely Bharat Nagar, Christain colony, Omala Jakhen, Sail Salan, Upper Sail Salan are notified and "9," namely Dandayal, Gangera Dallah, Kaller, Kharene Omera, Nagrota, Sajalta, Sambal, Subhash Nagar, and Thanda Padder are non-notified (Fig. 14.5). These slums account for 66.66% of the total no. of slums. Out of the total population

Slums Falling in the Matrix of 2 x 1 x 2

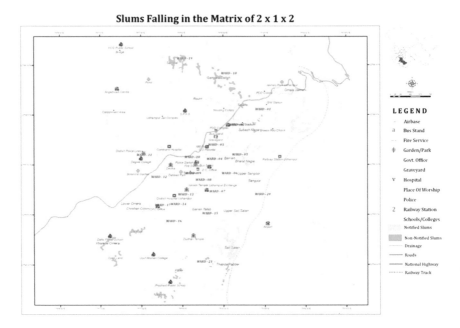

Fig. 14.5 Location map of slums falling in the matrix of $2 \times 1 \times 2$

of 5770 persons, the population of these slums in aggregate is 4017, which accounts for 69.62% of the total population living in these 14 slums. If we go individually by infrastructure facilities, except for Kaller, Nagrota, Sajalta, and Subhash Nagar, no slum has a dedicated water supply facility. None amongst these slums have a proper means of garbage disposal. Except for Sajalta and Dandhyal, where the open drains are cleared once in 15 days, none of the slums have any such facility. Only two shared, three individual dries, and one individual septic toilet facility are available out of the total of 886 households which doesn't even account for 1% of the total number of households. This means that the dwellers of 880 households are subjected to open defecation. Going by the position of roads, none of these slums have the motorable pucca road. Likewise, 73% of the households of these slums have kaccha houses to live in, going by the housing facilities. However, the "Security of tenure" parameter gives an encouraging picture that 91% of households possess the possession certificate or ownership rights.

14.6 Analysis of Slums in $1 \times 1 \times 2$

In $2 \times 1 \times 2$, only three slums out of 21 falls are Garian Talab, Idd Gah, and Ap Raunt. Only Raunt is non-notified, and the other two are notified (Fig. 14.6). These slums account for 14.28% of the total no. of slums. Out of the total population of

Slums Falling in the Matrix of 1 x 1 x 2

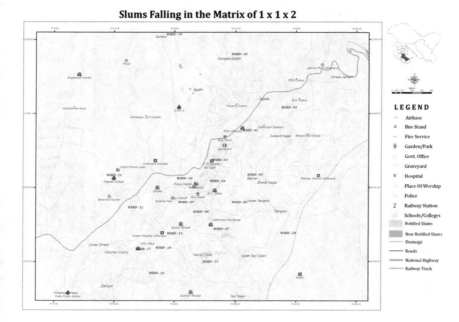

Fig. 14.6 Location map of slums falling in the matrix of 2 × 1 × 2

5770 persons, the population of these slums in aggregate is only 835, accounting for 14.47% of the total population living in these three slums. If we go individually by infrastructure facilities, all three slums have dedicated water supply facilities. Except for Rount these slums have a proper garbage disposal mechanism, which is cleared daily in Idd Gah and once in 15 days in Garian Talab. All three slums have municipal staff to look into the issues of solid waste management. In all three slums, once in 15 days, the open drains are cleared.

Only one individual dry toilet facility is available out of the total 181 households. This means that the dwellers of 180 households are subjected to open defecation. Going by the position of roads, all the three slums have access to the motorable pucca road, though. Likewise, going by the housing facilities, 72.66% of the households of these slums have kaccha houses to live in. The security of tenure' parameter shows that 74.33% of households possess the possession certificate or ownership rights.

14.7   Analysis of Slums in 2 × 2 × 2

This category reveals the abject poverty condition of these slums. In this category of 2 × 2 × 2, though only two slums out of 21 fall, Lower Omera and Sangoor, the condition in almost all the parameters is highly pathetic (Fig. 14.7). Only Rount is non-notified, and the other two are notified. These slums account for nearly 10% of

Slums Falling in the Matrix of 2 x 2 x 2

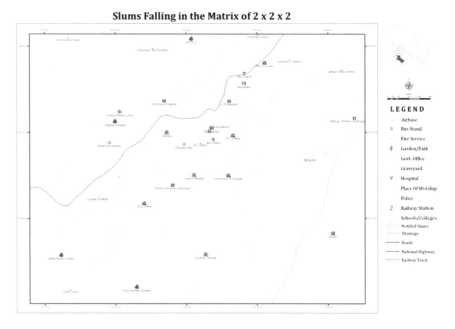

Fig. 14.7 Location map of slums falling in the matrix of $2 \times 2 \times 2$

the total no. of slums. Out of the total population of 5770 persons, the population of these slums in aggregate is 651, accounting for 11.28% of the total population living in these two slums. If we go individually by infrastructure facilities, Lower Omera is connected to the dedicated water supply, but Sangoor has no such facility. There is no drainage facility in either of the slums. None amongst them have a proper garbage disposal mechanism. Both the slums have only one shared latrine, each out of the total 148 households. This means that the dwellers of 146 households are subjected to open defecation. Going by the position of roads, Lower Omera has access to the motorable pucca road, but Sangoor is not having any such facility. Likewise, going by the housing facilities, 72% of the households of these slums have kaccha houses to live in. The security of tenure' parameter shows that 43.26% of households possess the possession certificate or ownership rights.

14.8   Analysis of Slums in 2 × 1 × 1

Only one slum, "Barrian," out of 21 slums, falls in the category of $2 \times 1 \times 1$. This slum has a population of 82 souls only (Fig. 14.8). Barrian has partial water supply connectivity. Regarding solid waste management, once a week, the waste is disposed of by the municipal staff, and once every 15 days, the drains are cleared. All 17 households are subjected to open defecation; there is neither any public lavatory

Slums Falling in the Matrix of 2 x 1 x 1

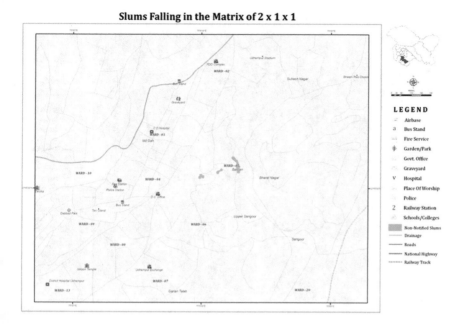

Fig. 14.8   Location map of slums falling in the matrix of 2 × 1 × 1

nor any individual arrangement of toilets. There is only a non-motorable pucca road available in the slum. 100% of the houses are semi-pucca, and all of them have possession certificates or ownership rights. So, this slum is having comparatively good housing and tenure status.

14.9   Analysis of Slums in 2 × 2 × 1

Again only one slum, "Upper Sangoor," out of 21 slums, falls in the category of 2 × 2 × 1. This slum has a population of 185 souls. Upper Sangoor has no water supply connectivity. There is no arrangement of solid waste management though there is a municipal staff (Fig. 14.9). Only three households have personal septic toilets; the other 35 households are subjected to open defecation. There is no motorable pucca road available in the slum. 60.53% of the houses are semi-pucca, and the rest are kuccha. 55.26% of the households live in tents/encroached public or private land when only 44.74% have possession certificate/ownership rights.

The United Nations passed a resolution referred to as "55/2 United Nations Millennium Declaration" during its 55th session on September 18, 2000, in which the 3rd declaration part 11 states, "We will make every attempt to liberate peoples fellow men, women, and children from the abject and dehumanizing conditions of extreme poverty, which currently affect over a billion people. We are dedicated to making

Slums Falling in the Matrix of 2 x 2 x 1

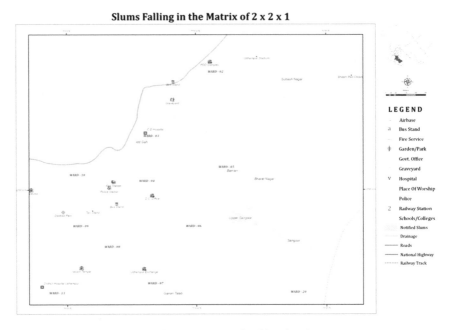

Fig. 14.9 Location map of slums falling in the matrix of $2 \times 2 \times 1$

everyone's right to growth a reality and to liberating the entire human race from day-to-day poverty."

In the year 2000, 189 countries pledged to release people from extreme poverty and various deprivations. This pledge was turned into the eight Millennium Development Goals, which must be met by 2015. The world recommitted itself to accelerating progress toward these goals in September 2010. This growing global concern has moved the international community's attention to the "Slums," as evidenced by the UN Millennium Declaration and subsequent identification of new development priorities. The growing number of slum dwellers has compelled governments to set a target for slums termed MDG 7, Target 11, to significantly improve the lives of at least 100 million slum residents by 2020. Given the expected rise of about two billion people in the next 30 years, the international community aspires for the "basic minimum" in the shape of this Millennium Development Goal. As a result, the current efforts in this regard are insufficient. To make these slums participating elements in driving away anguish, deprivation, and other characteristics of poverty, national and international policies must be directed by genuine care and compassion.

According to the Planning Commission, an estimated 26% of India's urban population lives less than Rs. 32 a day, putting them below the poverty line (Shekhar 2020). Eighty percent of their low earnings go toward covering their food and energy demands, leaving very little to cover the rising costs of living in an inflationary economy. The majority of this population, estimated to be over 75.2 million in 2001, lives in slums and squatter settlements in inhumane conditions that deprive them of dignity, housing, and access to essential civic utilities and social services. To add to their suffering and despair, these slums are frequently plagued by terrible health, high crime rates, and diseases, all of which push them farther and deeper into poverty: slums, both recognized and unnotified, house over a quarter of the Indian population. The estimated share of urban households in the next two decades is likely to climb from 28 to 50% of the overall population as urbanization continues. The number of slums may expand exponentially as a result of this. By denying essential utilities, shelter, and security, the spread of slums will cripple the creative potential of an increasing number of people.

Thus, the Government of India launched the Jawaharlal Nehru National Urban Renewal Mission, a prominent flagship program with a two-pronged aim, in December 2005. The first is the sub-mission on Basic Services to the Urban Poor, which focuses on 65 mission cities. The second is the Integrated Housing and Slum Development Program, which focuses on small and medium towns and develops slums holistically. By facilitating the building of 15 lakh dwelling units with basic facilities, the mission aims to alleviate the living conditions of slum inhabitants and the urban poor. Since the JNNURM, the BSUP, and the IHSDP have not been able to achieve the best results and have not been able to alleviate many states' concerns about the conferral of legal property rights to the urban poor, who fear that slum dwellers will sell the property and create a new encroachment, the government recently declared a change in policy with the declaration of Rajiv Awas Yojina (RAY). The President of India launched RAY in a speech to both Houses of Parliament in June 2009 and the Prime Minister on Independence Day, both of which detailed the vision of a "Slum Free India." This program intends to assist states that are prepared to grant slum dwellers property rights. RAY proposes a multi-pronged approach to addressing the problem of slums, focusing on bringing existing slums into the formal system and enabling them to access the same amount of general amenities as the rest of the town; resolving the standard system failures that lead to the creation of slums and addressing the scarcity of urban land and housing.

Udhampur city has been divided into 21 municipal wards. Out of 21 slums, 11 slums are non-notified, and ten are notified. The notified slums are located in wards 3, 7, 17, and 20. The other 6 wards, i.e., 1, 5, 11, 18, 19, 21, encompass the non-notified slums (Fig. 14.10).

GIS for the slum surveys based on spatial and non-spatial data makes the planning process more manageable. Slums in Udhampur have not previously been mapped in detail and tend therefore to be considered as chaotic and disordered masses. They contain patterns of lanes, houses, and facilities like any other part of the city. For infrastructure provision, it is essential to know the layout within a slum. GIS is useful for mapping the locations of the slums concerning the city as a whole. In other cities, this has resulted in maps that clearly show slum settlements across the urban area in a form that can be easily updated and manipulated. GIS can also be used to analyze different aspects, for example, service provision, on a city-wide level.

A view of the satellite image of the Udhampur city showing the slum pockets

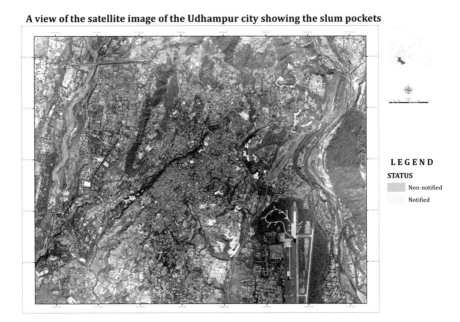

LEGEND
STATUS
Non-notified
Notified

Fig. 14.10 Location of slum pockets in Udhampur city

14.10 Conclusions

Slum characterization is critical in maximizing the efficient management of homeless people in any part of the world. In this work, slum categorization of the Udhampur district in the UT of Jammu and Kashmir was carried out using a GIS-based tenure-based methodology. Different matrices were categorized, and it was found that various areas in the district are heavily congested with slums. This work aimed to pave the way for the effective implementation of the Rajiv Awas Yojana (RAY). RAY proposes to address the problem of slums holistically and definitively, adopting a multi-pronged approach focusing on bringing existing slums within the formal system and enabling them to avail of the same level of basic amenities as the rest of the town. Moreover, it will address the faults of the primary system that have resulted in the creation of slums, as well as the scarcity of urban housing, which needs to keep shelter out of reach for the urban poor, forcing them to resort to illegal means to keep their sources of income and employment. It is envisaged that the current work will aid in the inclusion of the Udhampur district in the Government of India's smart cities program for the Jammu and Kashmir UT.

References

Baud ISA, Sridharan N, Pfeffer K (2008) Mapping urban poverty for local governance in an Indian mega-city: the case of Delhi. Urban Stud 45(7):1385–1412

Baud I, Kuffer M, Pfeffer K, Sliuzas R, Karuppannan S (2010) Understanding heterogeneity in metropolitan India: The added value of remote sensing data for analyzing sub-standard residential areas. Int J Appl Earth Obs Geoinf 12(5):359–374

Bhagat RB, Mohanty S (2009) Emerging pattern of urbanization and the contribution of migration in urban growth in India. Asian Popul Stud 5(1):5–20

Jain S, Sokhi BS, Sur U (2005) Slum identification using high-resolution satellite data. GIM Int 19(9):60

Kohli D, Sliuzas R, Kerle N, Stein A (2012) An ontology of slums for image-based classification. Comput Environ Urban Syst 36(2):154–163

Kuffer M, Pfeffer K, Sliuzas R (2016) Slums from space—15 years of slum mapping using remote sensing. Remote Sens 8(6):455

Mason SO, Fraser CS (1998) Image sources for informal settlement management. Photogram Rec 16(92):313–330

No C (2011) Census of India. Age 6(18p):13

Shekhar S (2014) Improving the slum planning through geospatial decision support system. Int Archiv Photogrammetry Remote Sens Spat Inf Sci 40(2):99

Shekhar S (2020) Effective management of slums-Case study of Kalaburagi city, Karnataka India. J Urban Manage 9(1):35–53

Sliuzas RV (2008) Improving the performance of urban planning and management with remote sensing systems. In: Rgens CJ (ed) Remote sensing: new challenges of high resolution, EARSeL, joint workshop, 5–7 Mar 2008 Bochum, Germany. Bochum: EARSeL. ISBN 978-3-925143-79-3, pp 1–13. EARSeL

Sur U, Jain S, Sokhi BS (2004) Identification and mapping of slum environment using IKONOS satellite data: a case study of Dehradun, India. In: Proceedings of the map India 2004 conference, New Delhi, India, pp 28–30

United Nations Human Settlements Programme (2010) State of the world's cities 2010/2011: bridging the urban divide. Earthscan

Chapter 15
Urban Growth Trend Analysis Using Shannon Entropy Approach—A Case Study of Dehradun City of Uttarakhand, India

Kamal Ahuja, Maya Kumari, and Shivangi Somvanshi

Abstract Urbanisation is the process of becoming urban. It is an anthropologic process which studies rising proportion of population of a region or city lives in urban area. Urbanization can be a result of demographic phenomenon, structural change in society or it can also be a result of behavioural processes. Various aspects of remote sensing technique i.e. spectral, temporal, and spatial aspect of remote sensing techniques can be effectively used in the study of such dynamic phenomenon. Remote sensing data can be effectively used in change detection mapping and processes, and therefore aiding in urban planning and management. The present study aims to study urban growth and sprawl in Dehradun city of Uttarakhand, which is one of the city in government's smart city project list. By using the approach of Shannon's Entropy, urban sprawl of Dehradun can be analysed. As per the result the entropy value obtained for the year 2008 is 0.877 and 2016 is 1.598, in which the value of 2016 is near to the value of upper limit of log n (i.e. 1.591) which depicts more urban sprawl in 2016 than in 2008. The present study effectively uses Landsat TM data of year 2008 and 2016. Urbanization have different impacts on natural, economic and social structure of any region, therefore this study can help for better planning and sustainable management of resources of a certain region and can help government officials and planners to monitor and analyse current urbanization and plan for future growth and requirements. Particularly Dehradun is selected for this study as there is a rapid increase in urbanization in the city since 2008–09 and also the city is considered by Indian government for its Smart city project. The chief goal of government is to prepare a green, clean and economically attractive city. This particular study can be helpful to understand the urbanisation pattern in the city.

K. Ahuja
Amity Institute of Geoinformatics and Remote Sensing, Amity University, Noida, Uttar Pradesh, India

M. Kumari (✉)
Amity School of Natural Resources and Sustainable Development, Amity University, Noida, Uttar Pradesh, India

S. Somvanshi
Geospatial World, Noida, Uttar Pradesh, India

© The Author(s), under exclusive license to Springer Nature Singapore Pte Ltd. 2022   301
P. K. Rai et al. (eds.), *Geospatial Technology for Landscape and Environmental Management*, Advances in Geographical and Environmental Sciences,
https://doi.org/10.1007/978-981-16-7373-3_15

Keywords Shannon entropy · Urbanization · Dehradun · Urban sprawl · Remote sensing

15.1 Introduction

Urbanisation is the process of becoming urban. It is a characteristic of economically advancing nations, where it is occurring at much faster rate than the developed world. Urbanisation is generally related to industrialization. Urbanisation is connected with the increment, concentration and dispersion of population in towns and cities (Sudhira et al. 2004).

There are three linked concepts which are associated with urbanization. These are:

i.   Demographic Phenomenon
ii. Organizational transition in society
iii. Urbanization as a deportment.

As an anthropologic phenomenon, urbanization can be analysed as a phenomena involving the total and corresponding swelling of cities and towns in specified regions. Linked with the demographic process is the structural change in the society, consequent upon the development of industrial capitalism. The third concept associated with urbanization is that urbanization is behavioural process. Urban areas, especially large cities, have been recognized centre of societal transition, outlook, morals and behaviour. Urban generally means relating to cities and towns. Towns are of many different sizes, ranging, from small country towns, sometimes smaller than village elsewhere, enormous sprawling cities will several million inhabitants. What distinguishes towns from rural settlements is not their size actually but economic activities of their inhabitants. The major economic activities in town pertains to industries, trade, commerce and services. The Level of Urbanisation in India, which is also regarded as an index of economic development, increased from 27.81% in 2001 to 31.16% in the 2011 Census of India.

In India, population is growing at a faster rate along with increase in migration, with such faster population growth urbanisation in India is not a result of deliberate forethought. More and more towns and cities are growing with a change in transportation systems across cities and towns. Such distribution of urban population across dense urban areas growing with main road networks can be described as what we call urban sprawl (Hamad 2020; Ji et al. 2006). Sprawl can be described as an unmanaged and unplanned growth or expansion of densely populated urban area which swells across the outskirts of city along main road networks and important centres or market places of city. Some of the factors of urban sprawl include—migration due to various reasons, natural growth of population, change in economic pattern, patterns of infrastructural facilities. Direct implication of urban sprawl is change in land cover, land use of that region. Today's world is largely carried forward by various spectra of science and technology (Kpienbaareh and Luginaah 2020). On one side where developed world is growing with and adapting new ways living induced by science and technology, the other side i.e. developing world is still struggling scientifically, technologically and economically to serve all the emerging urban areas with adequate sanitation, housing, health and even safe drinking water as well as transportation facilities. Urban sprawl can be taken as a possible ultimatum for such kind of development.

If the patterns and arrangement of any type of extension or increment can be analysed in advance then it will become easy of any political or social machinery to plan or develop adequate, basic infrastructural services such as transportation, housing, sewage and sanitation, electricity, water etc. such study on any urban sprawl show us about the nature and extent of urban growth or extension spread across an area or a city and also about the factors responsible for such extension. Such pieces of information make up a good pie and give a base full of information to urban planners and developers to understand growth patterns and plan various infrastructural services. Shannon's Entropy can be the GIS driven process to provide not only the information about urban sprawl, but also about change in built up area and fashion of population growth around important road networks and vital centres of any city or region.

15.1.1   Types of Urban Sprawl

Low Density Sprawl: This type sprawl grows along the margins of existing metropolitan cities. This type of urban sprawl is dependent on basic amenities such as water, sewer, power, roads etc.

Ribbon: This type of urban dispersion is developed along the major transportation routes coming out of the from urban centres.

Leapfrog Development: This is defined as a discontinuous type of urbanisation it has patches of developed lands that are widely separated from each other and also from boundaries.

15.1.2   GIS, Remote Sensing in Urban and Regional Planning

Urbanization is a very dynamic phenomenon, which is affected by various other factors. Urbanisation is a very major driver of transition in land cover land use of any particular area between a particular time period, especially in a developing country like India. Usage of Geoinformatics and remote sensing sciences in the studies of urban growth and urban model making can be a very effective science and can be a very important tool which can aid urban planners in many ways (El Garouani et al. 2017; Mishra et al. 2018).

Such studies can be further carried out for generating more information by applying more GIS models and applications and carrying tests like carrying capacity test for further analysing the suitability of site and study area for managing the increasing urban population (Farooq and Ahmad 2008). Mapping urban sprawl is a very important aspect as it provides a base for planning a proper development of any area facing such sprawl along with this it also makes the base for recognizing environmental, social and economic consequences generated due to such sprawl or growth.

GIS and remote sensing are very useful techniques in detection of any geographical changes with respect to space and time, which are vital stats while working on urban designing or planning so that any future development can be well structured in terms of society and environment (Mishra and Rai 2016; Krishnaveni and Anilkumar 2020; Rahman et al. 2011; Mishra et al. 2018). Through GIS and remote sensing, we can get spatial and temporal data and understand the arrangement of present and future urban growth this also help in tracing the pattern and type of urban sprawl (Rahman 2016; Tewolde and Cabral 2011; Mishra et al. 2019).

15.1.3   Urbanisation in Dehradun

Dehradun is witnessing increase in population, the city is rising and extending with respect to industrialisation. Total population of Dehradun city is accounted as 1,282,143 out of which 678,742 is urban population. After creation of state of Uttarakhand in the year 2000, and Dehradun as its headquarters there is a great amount of migration from hills all around, as well as high tourists' footfall. There is great population pressure on the city of Dehradun which in turn causing unavoidable burden on available infrastructural facilities which further hampers the social and economic growth of the city. This also gave rise to other social, economic and most importantly environmental consequences in the city, air quality is worsening day by day. WHO have also issued report saying, Dehradun stands 30 on the rank board of most polluted cities.

15.2   Study Area

Dehradun, is the capital as well as headquarter of the state Uttarakhand, after Uttarakhand became state in the year 2000. The city lies in the foothills of Himalayas with the drainage basin formed out of river Ganga in the east and river Yamuna in the west, with an overall estimated urban population of about 1.2 billion. This region is famous for its mild weather, British era institutions as well as Litchi plantations. This region is a very attractive tourist place during hot summers in plains especially due to its pleasant climatic conditions and hosts several institutions such as the Indian Military Academy, ITBP Academy and Indira Gandhi National Forest Academy (IGNFA),

Zoological Survey of India (ZSI), Forest Research Institute (FRI) among several others.

Dehradun district is surrounded by the Himalayas in the northern part, Shivalik in the southern part and Ganga from eastern side and Yamuna from western side. Dehradun is located between latitudes 29 58′N and 31 2′N and longitudes 77 34′E and 78 18′E. The district is divided into 6 talukas namely Dehradun, Chakrata, Vikasnagar, Raipur, Doiwala and Sahaspur that comprises of 17 towns and 764 villages (Fig. 15.1).

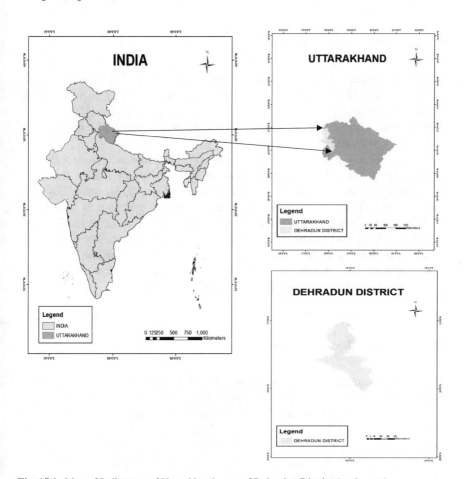

Fig. 15.1 Map of India, map of Uttarakhand, map of Dehradun District (study area)

15.3 Methodology

The images used in the following research are two Landsat images of two different years i.e. 2008 and 2016. The satellite data covering the study area were obtained from earth explorer site. The remotely sensed images of Landsat 4–5 TM and Landsat 8 OLI/TIRS are used to study the urban sprawl in Dehradun city of Uttarakhand state. Software used for the study are ERDAS imagine and Arc GIS.

15.3.1 Hybrid Classification

In order to prepare a land cover land use map of images of the study area for the year 2008 and 2016 unsupervised classification is performed. This map will help to compare the transition in land use land cover pattern of both the years. Unsupervised classification is performed in ERDAS imagine, to prepare a Land Cover Land use map of the study area (Sonde et al. 2020).

Unsupervised classification is a pixel-based classification which can be performed by grouping of pixels with common characteristics. The user specifies number of classes and spectral classes based on pixel values are automatically created. Various clustering algorithms work behind statistical and natural grouping of data. In the following study area 36 classes are classified with the process of unsupervised classification. In this classification we have obtained six classes that are, water body, agricultural area, forest area, river bed, mountainous and rocky region and settlements. Following flow chart depicts the steps for unsupervised classification (Joorabian Shooshtari et al. 2020; Meer and Mishra 2020) (Fig. 15.2).

15.3.2 Accuracy Assessment

After performing image classification other step to process the image and to reduce the errors is accuracy assessment. This process compares classified image with the other data that is contemplated to be accurate with respect to the data which is true to the ground. Ground truth data can be collected by various processes i.e. field surveys, image interpretation, or may be through existing classified imagery or data. This process helps in checking the reliability of the classified image. We have used ERDAS imagine to perform accuracy assessment of the classified image (Kumari and Sarma 2017). We have used accuracy assessment tool in which after giving input of classified image we will mark some random points on the image in the viewer and then after correcting point discrepancies we can obtain accuracy report from the same accuracy assessment tool, which proves the reliability of the image.

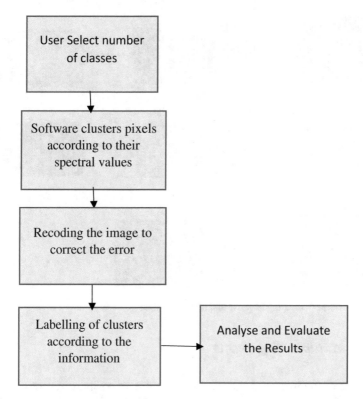

Fig. 15.2   Flow chart explaining the process of hybrid classification

15.3.3   Area Calculation

Area calculation is also a very important process for this particular study of urbanization. Area can be defined as the total space present inside any shape. Area calculation is also performed in ERDAS imagine. By calculating area of classified image of both the years, we are able to get the area of each classified field i.e. water body, settlement, mountainous region, agricultural land and river bed. With the help of this area we can perform future calculations in our study.

15.3.4   Image Reclassification

This is the vital process in this study with which we can reclassify the classified image into various number of classes according to our need. This particular process is performed in Arc GIS. For this study only two classes are reclassified in the classified images of both the years. Those classes are built-up and non-built up, as our study only requires built up area to find out urban sprawl (Fig. 15.3).

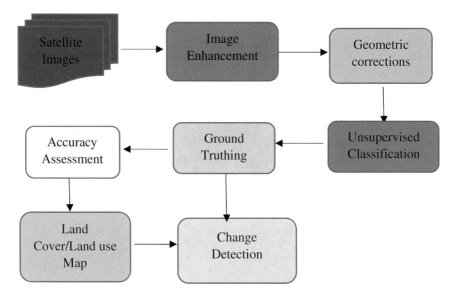

Fig. 15.3 Process of creating land use and land cover map

15.3.5 *Multiple Ring Buffer*

This is one of the important process in estimating urban sprawl of the study area. Creating buffer is also a reclassification process which is based on distance. Buffers are created around a particular input feature, and are created around vector data i.e. around point, line, polygon. Multiple ring buffers are multiple number of buffers created at a time around a particular feature at equal distances. Multiple ring buffer is an analyst tool of ArcGIS. By creating multiple buffers of 2 km around a point feature in study area for both the years ae have divided the built-up and non-built up zone in 85 classes.

15.3.6 *Zonal Statistics Table*

Zonal statistics table is a tool in Arc GIS to obtain mean for the given area of the study area. Mean is calculated for this study as mean is the average that is used to analyse the central tendency of the data. As the tool calculates the mean table of the area of each class, we can convert the table into excel format by using table to excel conversion tool.

15.3.7 Shannon's Entropy

Shannon's Entropy is the approach to measure of randomness i.e. how randomly a certain phenomenon has occurred over a given space (Vani and Prasad 2020; Zachary and Dobson 2021). This process is used to compute the urban growth or urban sprawl of any area. It depicts that urban growth is dispersed or dense and concentrated. Shannon's entropy is symbolised as Hn, which is calculated by the following equation,

$$H_n = -\sum P_i \, \log n(P_i) \qquad (15.1)$$

P_i = It stands for the proportion of built-up area in ith zone
n = total number of zones present
$\log n$ = maximum limit of entropy.

The value obtained from Shannon's entropy lies between the range of 0 to log n.

0—indicates very compact urban areas
log n—signifies built up areas.

15.4 Results

15.4.1 Land Cover, Land Use

Land cover, land use of the Dehradun city is prepared through unsupervised classification and the correctness of the former is checked by accuracy assessment which is a very vital step. Accuracy assessment generates the results by calculating error matrix. For the given area the overall accuracy for the year 2008 is 100% and for 2016 is 100%. The Kappa Coefficients for 2008 is 1.00 and 2016 is 1.00. The minimum accuracy value for a reliable land cover land use map is more than 85%. The accuracy of this particular study area is therefore acceptable. The total built up area is 129.97 sq. miles for the year 2008 and 148.053 sq. miles for the year 2016 (Fig. 15.4).

15.4.2 Area Calculation

The total built up area for year 2008 was 129.97 sq. miles and for the year 2016 the total built up area was 148.053 sq. miles; it shows area under settlements or built up is increased between the span of 8 years which also depicts very rapid increase in urban growth. Following Tables 15.1 and 15.2 depicts the total area covered by each class that are identified in land cover land use (Figs. 15.5, 15.6 and 15.7).

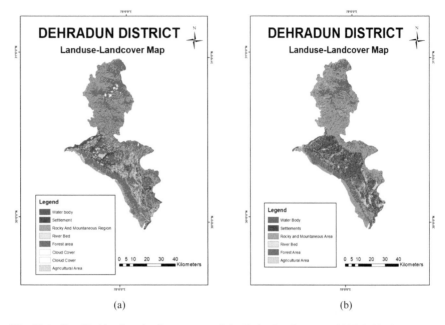

Fig. 15.4 Classified land use land cover map of city Dehradun for year **a** 2008, **b** 2016

Table 15.1 Area for year 2008

Classes	Total area (sq. miles)
Water body	17.31
River bed	35.88
Forest area	511.42
Agricultural land	90.61
Rocky and mountainous region	347.77
Settlement	129.97

Table 15.2 Area for year 2016

Classes	Total area (sq. miles)
Water body	6.119
River bed	35.48
Forest area	361.07
Agricultural land	67.16
Rocky and mountainous region	534.20
Settlement	148.05

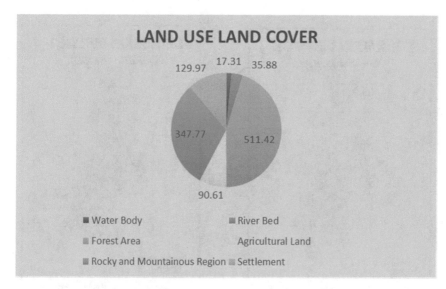

Fig. 15.5 LULC distribution for the year 2008

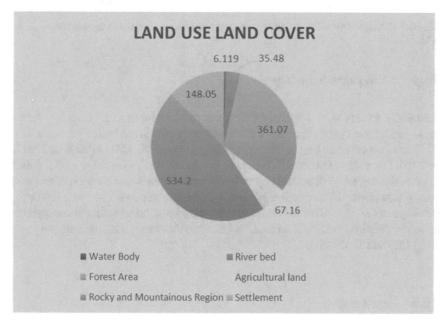

Fig. 15.6 LULC distribution for the year 2016

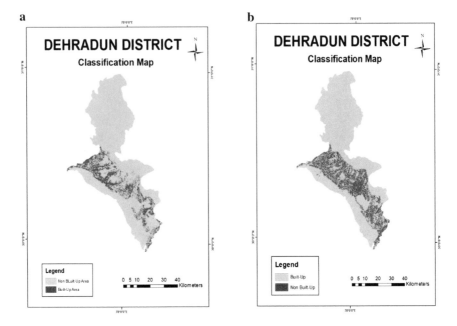

Fig. 15.7 Classification of study area with respect to built up and non-built up area for the year **a** 2008, **b** 2016

15.4.3 Shannon's Entropy

Shannon's entropy was calculated for built area of individual zones. Zones are created through multiple ring buffer. There are 39 zones for the given built up area i.e. $n =$ 39. The value obtained for the entropy of the given area are 1.598 in 2008 and 0.877 in 2016. The value of year 2016 is near to the upper limit of log n i.e. 1.591, which implies the degree of dispersed expansion of built-up area in the region. The result shows that degree of dispersion in year 2008 is less whereas there is a dispersed development in year 2016. This implies the expansion of urban area. Following table shows population, total built up area, and Shannon's entropy value of both the years (Fig. 15.8 and Table 15.3).

15.5 Discussion

The land use land cover map of the city Dehradun is classified into six different classes that are forest area, mountainous and rocky area, water body, river bed, settlement or built up, agricultural land, whole city space is broadly categorized into these six classes. The northern part of the city is completely covered by Himalayas, and there is no built up in that region. The extreme east of the city is also covered by mountains

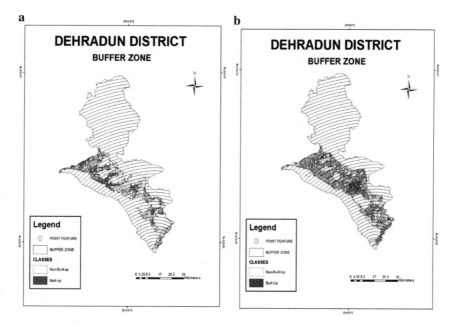

Fig. 15.8 Multiple ring buffer for study area for the year **a** 2008, **b** 2016

Table 15.3 Results of Shannon's entropy

Year	Population (in lakhs)	Total built up area	Shannon's entropy value	Log(n) value
2008	12.82*	129.97	0.877	1.591
2016	16.97**	148.05	1.598	1.591

*Total population Census of India (2001), ** Total population Census if India (2011)

and rocky. The built up is mainly strenuous in the central and southern region of the city. And cultivated area is clustered around and near the settlements and near water bodies. Forested area is spread all over the city. Forest are mainly dense in western and northern part of the city. But there is a big change in forest cover from year 2008 to 2016, the forest cover has been reduced drastically, which can be the result of increase in built up area which is in turn the result of increase in population. The expansion of urban area has increased between two years. Mainly the built-up increase is witnessed by districts of Vikasnagar, Doiwala, Sahaspur, and Chakrata.

As per the result of Shannon's Entropy the value for year 2016 for the study area shows more dispersed growth whereas 2008 value depicts more compact urbanisation. There are various factors which are responsible for this type of dispersed growth. After the creation of Uttarakhand as a state in 2000, Dehradun became it's headquarter. Due to its political and administrative upgradation with its existing suitable geographical and climatic conditions this city became centre of attraction for various inhabitants, which resulted in increase in migration, and on the other hand there is population growth going on in the region. City is also witnessing economic as well as infrastructural growth which are also the reason behind urban expansion. On one hand urban area is expanding rapidly, the environmental aspects of the city are suffering from the drawbacks of urbanization. Urban expansion has been deteriorated the forest area of the city, which is a matter of serious concern.

With all the urbanisation taking place in the city, government has also considered city of Dehradun under its smart city project. As government also consider the urban expansion in the city makes it suitable and potential to implement smart city project.

15.6   Conclusion

Urbanization being a very dynamic phenomena need to be studied time to time with different approaches which are also upgraded with time and space. Geoinformatics and Remote Sensing are emerging sciences and techniques which cab effectively help in the studies of processes like urbanization, urban growth, urban sprawl etc. Remote sensing provides spatial and temporal data which is very appropriate for the study of land development, urban expansion, land use land cover change etc. Such efficient information when combined with approach like Shannon's entropy can aid in studying the degree of urban dispersion and sprawl. This accumulated information can be very helpful in understanding urban arrangements also with respect to available infrastructural facilities and services which provides an efficient information base for further planning at both regional and global level.

The Shannon's entropy value of particular study which are 0.877 for the year 2008 and 1.598 for the year 2016 give clear evidence of highly dispersed urban growth in 2016 in city of Dehradun. And on the other-hand area calculation shows that on one side built up area has increased at faster pace whereas the area under forest has decreased with similar fast pace which is an environmental threat. This proves that Dehradun city requires effective and efficient urban planning to have sustainable development as well as urban growth in right direction. If not planned in right way there can be many social, economic and environmental consequences.

Therefore, it is evident it is crucial to have appropriate data and information for unerring and sustainable planning of urban area. And Geoinformatics and remote sensing are unmatchable approaches in this context.

References

El Garouani A, Mulla DJ, El Garouani S, Knight J (2017) Analysis of urban growth and sprawl from remote sensing data: case of Fez, Morocco. Int J Sustain Built Environ 6(1):160–169. https://doi.org/10.1016/j.ijsbe.2017.02.003
Farooq S, Ahmad S (2008) Urban sprawl development around Aligarh city: a study aided by satellite remote sensing and GIS. J Indian Soc Rem Sens 36(1):77–88. https://doi.org/10.1007/s12524-008-0008-0

Hamad R (2020) A remote sensing and GIS-based analysis of urban sprawl in Soran District, Iraqi Kurdistan. SN Appl Sci 2(1):1–9. https://doi.org/10.1007/s42452-019-1806-4

Ji W, Ma J, Twibell RW, Underhill K (2006) Characterizing urban sprawl using multi-stage remote sensing images and landscape metrics. Comput Environ Urban Syst 30(6):861–879. https://doi.org/10.1016/j.compenvurbsys.2005.09.002

Joorabian Shooshtari S, Silva T, Raheli Namin B, Shayesteh K (2020) Land use and cover change assessment and dynamic spatial modeling in the Ghara-su Basin, Northeastern Iran. J Indian Soc Rem Sens 48(1):81–95. https://doi.org/10.1007/s12524-019-01054-x

Kpienbaareh D, Luginaah I (2020) Modeling the internal structure, dynamics and trends of urban sprawl in Ghanaian cities using remote sensing, spatial metrics and spatial analysis. African Geograph Rev 39(3):189–207. https://doi.org/10.1080/19376812.2019.1677482

Krishnaveni KS, Anilkumar PP (2020) Managing urban sprawl using remote sensing and GIS. In: International archives of the photogrammetry, remote sensing and spatial information sciences—ISPRS Archives, vol 42(3/W11), pp. 59–66. https://doi.org/10.5194/isprs-archives-XLII-3-W11-59-2020

Kumari M, Sarma K (2017) Changing trends of land surface temperature in relation to land use/cover around thermal power plant in Singrauli district, Madhya Pradesh, India. Spatial Inf Res 25(6):769–777

Mahboob MA, Atif I, Iqbal J (2015) Remote sensing and GIS applications for assessment of urban sprawl in Karachi, Pakistan. Sci Technol Dev 34(3):179–188. https://doi.org/10.3923/std.2015.179.188

Meer MS, Mishra AK (2020) Remote sensing application for exploring changes in land-use and land-cover over a district in Northern India. J Indian Soc Rem Sens 48(4):525–534. https://doi.org/10.1007/s12524-019-01095-2

Mishra VN, Rai PK (2016) A remote sensing aided multi-layer perceptron-Markov chain analysis for land use and land cover change prediction in Patna district (Bihar), India. Arab J Geosci 9(4):249

Mishra VN, Prasad R, Kumar P, Gupta DK, Agarwal S, Gangwal A (2019) Assessment of spatio-temporal changes in land use/land cover over a decade (2000–2014) using earth observation datasets: a case study of Varanasi district, India. Iran J Sci Technol Trans Civil Eng 43(1):S383–S401

Mishra VN, Rai PK, Prasad R, Punia M, Nistor MM (2018) Prediction of spatio-temporal land use/land cover dynamics in rapidly developing Varanasi district of Uttar Pradesh, India using geospatial approach: a comparison of hybrid models. Appl Geomatics 10(3):257–276

Rahman A, Aggarwal SP, Netzband M, Fazal S (2011) Monitoring urban sprawl using remote sensing and GIS techniques of a fast growing urban centre, India. IEEE J Sel Topics Appl Earth Observ Rem Sens 4(1):56–64. https://doi.org/10.1109/JSTARS.2010.2084072

Rahman MT (2016) Detection of land use/land cover changes and urban sprawl in Al-Khobar, Saudi Arabia: an analysis of multi-temporal remote sensing data. ISPRS Int J Geo Inf 5(2):15. https://doi.org/10.3390/ijgi5020015

Sonde P, Balamwar S, Ochawar RS (2020) Urban sprawl detection and analysis using unsupervised classification of high resolution image data of Jawaharlal Nehru Port Trust area in India. Rem Sens Appl Soc Environ 17:100282. https://doi.org/10.1016/j.rsase.2019.100282

Sudhira HS, Ramachandra TV, Jagadish KS (2004) Urban sprawl: metrics, dynamics and modelling using GIS. Int J Appl Earth Obs Geoinf 5(1):29–39. https://doi.org/10.1016/j.jag.2003.08.002

Tewolde MG, Cabral P (2011) Urban sprawl analysis and modeling in Asmara, Eritrea. Rem Sens 3(10):2148–2165. https://doi.org/10.3390/rs3102148

Vani M, Prasad PRC (2020) Assessment of spatio-temporal changes in land use and land cover, urban sprawl, and land surface temperature in and around Vijayawada city, India. Environ Dev Sustain 22(4):3079–3095. https://doi.org/10.1007/s10668-019-00335-2

Zachary D, Dobson S (2021) Urban development and complexity: Shannon entropy as a measure of diversity. Plan Pract Res 36(2):157–173. https://doi.org/10.1080/02697459.2020.1852664

Chapter 16
Geospatial Approach for Mapping of Significant Land Use/Land Cover Changes in Andhra Pradesh

Ch. Tata Babu, K. Applanaidu, M. Vanajakshi, and K. V. Ramana

Abstract Ever-increasing population and industrialization are mainly responsible for the conversion of significant amount of change in land use. The change in land use pattern is considered as key parameter to evaluate global change in different spatiotemporal scales. The quantitative analysis of changes in land use pattern is needed to assess the impacts of change in the natural vegetation on the earth's environment for sustainable utilization of the resources. The aim of the study is to map and analyze the dynamics of land use/land cover changes using IRS LISS-III data for the years 2011–2012 and 2015–2016 of Andhra Pradesh State, India. On-screen visual interpretation techniques have been used to delineate the land use/land cover classes in ArcGIS environment and cross-tabulation used for quantifying the changes in land use pattern. The study reveals that built-up area, agriculture land, and water bodies have been increased about 0.21% (343.06 km^2), 0.11% (176.21 km^2), and 0.02% (32.08 km^2), respectively, while area under other land categories such as forest area, wastelands, and wetlands have decreased about 0.02% (107.44 km^2), 0.20% (333.38 km^2), and 0.07% (110.53 km^2), respectively. The results of this study would be helpful for planners, decision-makers, and administrators to plan and implement appropriate decisions in order to sustainable resource utilization.

Keywords Land use/land cover · Land use change · Change detection · Dynamics · Andhra Pradesh

16.1 Introduction

Land management and land planning require knowledge of the current state of the landscape. Understanding the current land use/land cover and how it is being used,

Ch. Tata Babu (✉) · K. Applanaidu · M. Vanajakshi
Andhra Pradesh Space Applications Centre, Vijayawada, Andhra Pradesh, India

K. V. Ramana
National Remote Sensing Centre, ISRO, Hyderabad, Telangana, India

© The Author(s), under exclusive license to Springer Nature Singapore Pte Ltd. 2022   317
P. K. Rai et al. (eds.), *Geospatial Technology for Landscape and Environmental Management*, Advances in Geographical and Environmental Sciences, https://doi.org/10.1007/978-981-16-7373-3_16

along with an accurate means of monitoring change over time, is vital for land administration. Through the land use/land cover mapping, land administrators can be easily measure the current condition and changes of land resources. The land use classification quantifies current land resources into a series of thematic categories such as agriculture, built-up, forest, water, and wastelands. The land use/cover pattern of a region is an outcome of natural and socio-economic factors and their utilization by man in time and space (Mishra and Rai 2016; Halimi et al. 2018; Shastri et al. 2020). Information on land use/cover is required for different kinds of spatial planning at a local level and possibilities for their optimal use is essential for the selection, planning, and implementation of land use schemes to meet the increasing demands for basic human needs and welfare (Vishwakarma et al. 2016; Mishra et al. 2016; Mishra et al. 2018; Miheretu and Yimer 2018). Moreover, the data are used as basic information for sustainable management of natural resources. Hence, they are fundamental for guiding decision making at various geographical levels (Babu et al. 2018). Understanding landscape patterns, changes and interactions between human activities and natural phenomenon are essential for proper land management and decision improvement (Rai and Kumra 2011; Jaiswal and Verma 2013).

Land use refers to the various uses under which land is brought by a multitude of human activities whereas land cover refers to the natural covering or setting of land without interference of any human activity. Land use/land cover (LULC) is dynamic in nature and requires regular monitoring to understand areas of rapid change and to ascertain the reasons/drivers for the change (Yesuph and Dagnew 2019). The change in land use/land cover is the resultant of many interacting processes operating on the natural resources. This will enable planners and administrators to initiate the appropriate measure for preventing/arresting the degradation of natural resources. For ensuring sustainable development, it is necessary to monitor the ongoing changes in land use/land cover pattern over a period of time (Iyyer 2009). These changes result from population growth and migration of poor rural people to urban areas for economic opportunities. Recent technological advances made in the domain of spatial technology caused considerable impact in planning activities. Today, earth resource satellites data are useful for land use/cover change detection studies (Yuan et al. 2005). Remote sensing (RS), geographical information system (GIS), and global positioning system (GPS) are now providing new tools for natural resources mapping, management and planning.

16.2   Objective

The objective of the study is to generate land use/land cover map of 2011–2012 and 2015–2016 for the entire State of Andhra Pradesh using three season's data. The specific objectives include:

- To generate spatial database on land use/land cover for 2015–2016

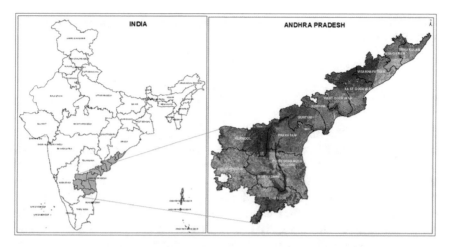

Fig. 16.1 Location map

- To generate land use/land cover change database, along with change matrix with respect to 2011–2012
- To identify areas of major change during the study period.

16.3   Study Area

The study area comprises State of Andhra Pradesh which is situated on the southeastern coast of the country and lies between 12° 33′ and 19° 09′ of North latitude and 76° 38′–84° 42′ of East longitude with a geographical area of 162,970 km². It is located east of Karnataka, south of Telangana, southwest of Odisha, and north of Tamilnadu. The state comprises two regions, i.e., Kosta and Rayalaseema, and it is divided into 13 districts. These 13 districts are further divided into 50 revenue divisions. The 50 revenue divisions are in turn divided into 670 mandals. There are 110 urban local bodies and 12,918 are gram panchayats in the state (Fig. 16.1).

16.4   Materials and Methods

16.4.1   Data Used

The geometrically corrected IRS Resourcesat-2 LISS III data within the framework of NNRMS specified standards form the primary input for updating the land use/land cover. Multitemporal data acquired from September to November, December to March, and April to May corresponding to *kharif*, *rabi,* and *zaid* cropping seasons were used for deriving information on land use/land cover during the year 2011–2012 and 2015–2016.

16.4.2 Methodology

The methodology essentially includes geo-rectification of temporal satellite data of k*harif*, *rabi*, and *zaid* seasons of study area. The visual image interpretation techniques have been used for delineation of land use/land cover categories through on-screen digitization in GIS environment (NRSC 2012). Cross-tabulation has been used for summarizing the category-wise land use/land cover changes. The methodology essentially is based on editing of the previous cycle of land use/land cover vector data layer created using 2011–2012 IRS R2 LISS-III data. Vector polygons for the change areas will only be edited wherever interpretation conflicts are observed. Domain knowledge and site adaptation are used to ascertain land use units. Ground truth verification has been performed by using mobile app, which is provided by NRSC to achieve better accuracy. The methodology adopted for land use/land cover updation and identification of changes during 2011–2012 and 2015–2016 is presented in the form of a flow chart in Fig. 16.2.

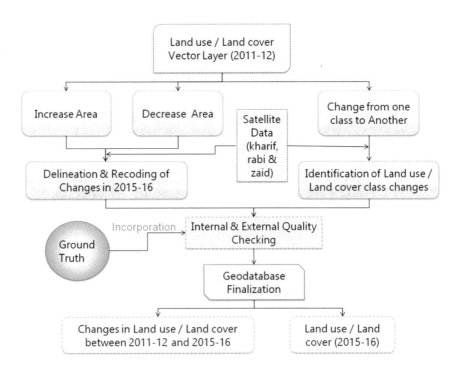

Fig. 16.2 Methodology

16.5 Results and Discussions

16.5.1 Spatial Distributions of Land Use/Land Cover During 2011–2012 and 2015–2016

The major common land use categories such as built-up (3.28%), agriculture (59.26%), forest (21.52%), wastelands (9.59%), wetlands (0.85%), water bodies (5.44%), grass and grazing land (0.06%), and shifting cultivation (0.01%) were identified and mapped using on-screen interpretation techniques during 2015–2016. The study area has been classified into 50 level-III land use classes and grouped into 27 level-II classes and further groped into 8 level-I classes. The predominant category is agriculture land followed by forest. The spatial distribution of land use/land cover map of Andhra Pradesh state during 2015–2016 is presented on Fig. 16.3a, b and the area statistics under level-1 classification during 2011–2012 and 2015–2016 are shown in Table 16.1.

The land use/land cover categories of the study area were mapped using IRS R2 satellite data of 1:50,000 scale. The satellite data were visually interpreted, and after making thorough field check, the map was finalized. Area covered under different categories in level-I for years 2011–2012 and 2015–2016 are shown in Table 16.1 and % changes of each category are shown in Fig. 16.4. The broad categories thus mapped such as built-up, agricultural, forest, wastelands, and water bodies are discussed in detailed in the following section. Table 16.2 shows land use/land cover statistics in level-III classification during 2011–2012 and 2015–2016.

Built-Up: An area of 5342.41 km^2 (3.28%) has been mapped under the built-up land category. This category has been further subdivided into urban compact, sparse, vegetated/open area, rural, industrial, ash/cooling pond, active mining, abandoned

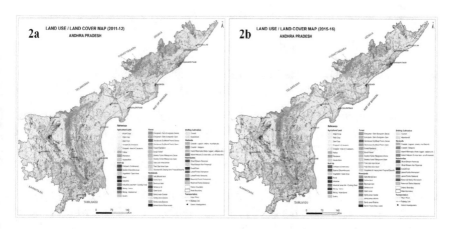

Fig. 16.3   a, b Land use/land cover map—2011–2012 and 2015–2016

Table 16.1 Summary of land use/land cover statistics in level-I classification of Andhra Pradesh during 2011–2012 and 2015–2016 (Area in km^2)

S. No.	Category	Year 2011–2012	(%)	Year 2015–2016	(%)	Change (±)	% diff
1	Built-up	4999.35	3.07	5342.41	3.28	343.06	0.21
2	Agricultural land	96,386.41	59.14	96,562.62	59.25	176.21	0.11
3	Forest	35,170.86	21.58	35,063.42	21.52	−107.44	−0.07
4	Grass and grazing land	101.59	0.06	101.59	0.06	0.00	0.00
5	Wastelands	15,959.55	9.79	15,626.17	9.59	−333.38	−0.20
6	Wetland	1503.68	0.92	1393.14	0.85	−110.53	−0.07
7	Water bodies	8832.53	5.42	8864.61	5.44	32.08	0.02
8	Shifting cultivation	15.66	0.01	15.66	0.01	0.00	0.00
	Total	162,969.6	100.0	162,969.6	100.0	0.0	0.0

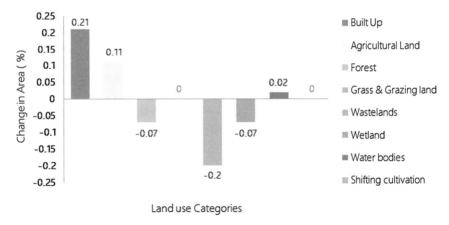

Fig. 16.4 % changes of land use/land cover during 2011–2012 and 2015–2016

mining, and mine quarry area. Amravati, the state capital and all the district headquarters and some mandal headquarters based on size and population have been mapped under urban category. Many vacant lands with layouts and fencing are coming up on the outskirts of the major towns for real estate development. These areas which are clearly visible in the satellite imagery have been identified and mapped under vegetated/mixed built-up category. The predominant built-up category is built up rural category and occupied an area of 2991.18 km^2 during 2015–2016, which is about 1.84% of the state total geographical area. The area under built-up category is estimated as 4999.35 km^2 in 2011–2012, whereas it has been increased to 5342.41 km^2 in

Table 16.2 Land use/land cover statistics in level-III classification of **Andhra Pradesh** during 2011–2012 and 2015–2016 (area in km^2)

S. No.	Category	Year 2011–2012	(%)	Year 2015–2016	(%)	Change (±)	% diff
Built-Up		**4999.35**	**3.07**	**5342.41**	**3.28**	**343.06**	**0.21**
1	Built-up compact (continuous)	736.08	0.45	754.79	0.46	18.71	0.01
2	Built-up sparse (discontinuous)	257.04	0.16	277.51	0.17	20.47	0.01
3	Vegetated/open area	237.08	0.15	311.96	0.19	74.88	0.05
4	Rural	2953.78	1.81	2991.18	1.84	37.39	0.02
5	Industrial	293.45	0.18	401.09	0.25	107.64	0.07
6	Industrial area-ash/cooling pond	18.39	0.01	24.93	0.02	6.54	0.00
7	Mining—active	82.49	0.05	84.35	0.05	1.86	0.00
8	Mining—abandoned	13.02	0.01	15.35	0.01	2.33	0.00
9	Quarry	408.01	0.25	481.25	0.30	73.24	0.04
Agricultural Land		**96,386.41**	**59.14**	**96,562.62**	**59.25**	**176.21**	**0.11**
10	Kharif crop	29,311.60	17.99	26,737.74	16.41	−2573.86	−1.58
11	Rabi crop	13,674.29	8.39	12,364.66	7.59	−1309.63	−0.80
12	Zaid crop	122.42	0.08	209.04	0.13	86.62	0.05
13	Cropped in two seasons	33,369.83	20.48	37,344.21	22.91	3974.37	2.44
14	Cropped more in two seasons	1474.65	0.90	1379.62	0.85	−95.03	−0.06
15	Fallow	9150.62	5.61	9111.56	5.59	−39.06	−0.02
16	Plantation	7477.01	4.59	7308.83	4.48	−168.18	−0.10
17	Aquaculture	1805.99	1.11	2106.97	1.29	300.98	0.18
Forest		**35,170.86**	**21.58**	**35,063.42**	**21.52**	**-107.44**	**-0.07**
18	Evergreen—dense	157.58	0.10	157.58	0.10	0.00	0.00
19	Evergreen—open	97.04	0.06	97.04	0.06	0.00	0.00
20	Deciduous dense	11,483.31	7.05	11,422.77	7.01	−60.53	−0.04
21	Deciduous open	10,349.39	6.35	10,327.97	6.34	−21.42	−0.01
22	Forest plantation	685.45	0.42	714.50	0.44	29.05	0.02
23	Scrub forest	8370.44	5.14	8310.61	5.10	−59.83	−0.04
24	Littoral/swamp forest (mangrove)-dense	279.98	0.17	276.21	0.17	−3.77	0.00
25	Littoral/swamp forest (mangrove)-open	149.52	0.09	175.70	0.11	26.18	0.02
26	Tree clad area-dense	1780.57	1.09	1770.26	1.09	−10.31	−0.01
27	Tree clad area-open	1817.58	1.12	1810.77	1.11	−6.81	0.00

(continued)

Table 16.2 (continued)

S. No.	Category	Year 2011–2012	(%)	Year 2015–2016	(%)	Change (±)	% diff
Grass and Grazing		**101.59**	**0.06**	**101.59**	**0.06**	**0.00**	**0.00**
28	Grass and grazing land-tropical/desertic	101.59	0.06	101.59	0.06	0.00	0.00
Wastelands		**15,959.55**	**9.79**	**15,626.17**	**9.59**	**−333.38**	**−0.20**
29	Salt-affected land	1308.81	0.80	1272.30	0.78	−36.50	−0.02
30	Gullied land	95.73	0.06	85.63	0.05	−10.10	−0.01
31	Ravinous land	150.20	0.09	143.67	0.09	−6.53	0.00
32	Dense scrub	6642.65	4.08	6507.42	3.99	−135.24	−0.08
33	Open scrub	5401.88	3.31	5276.29	3.24	−125.60	−0.08
34	Sandy area-desertic	3.43	0.00	3.43	0.00	0.00	0.00
35	Sandy area-coastal	266.19	0.16	260.12	0.16	−6.07	0.00
36	Sandy area-riverine	21.87	0.01	18.94	0.01	−2.93	0.00
37	Barren rocky/stony waste	2068.78	1.27	2058.37	1.26	−10.42	−0.01
Wetlands		**1503.68**	**0.92**	**1393.14**	**0.85**	**−110.53**	**−0.07**
38	Inland natural (Ox-bow lake, waterlogged, etc.)	176.05	0.11	164.07	0.10	−11.98	−0.01
39	Inland Manmade (water logged, saltpans, etc.)	179.05	0.11	121.13	0.07	−57.92	−0.04
40	Coastal—lagoon, creeks, mud flats, etc	987.17	0.61	960.94	0.59	−26.23	−0.02
41	Coastal—saltpans	161.40	0.10	147.00	0.09	−14.40	−0.01
Water bodies		**8832.53**	**5.42**	**8864.61**	**5.44**	**32.08**	**0.02**
42	River/stream-perennial	1059.01	0.65	1019.11	0.63	−39.90	−0.02
43	River/stream-non perennial	2236.36	1.37	2263.41	1.39	27.05	0.02
44	Canal/drain	641.94	0.39	678.14	0.42	36.21	0.02
45	Lakes/ponds-permanent	0.44	0.00	0.95	0.00	0.50	0.00
46	Lakes/ponds-seasonal	1.42	0.00	3.19	0.00	1.77	0.00
47	Reservoir/tanks-permanent	1188.59	0.73	1185.08	0.73	−3.51	0.00
48	Reservoir/tanks-seasonal	3704.77	2.27	3714.73	2.28	9.96	0.01
Shifting cultivation		**15.66**	**0.01**	**15.66**	**0.01**	**0.00**	**0.00**
49	Shifting cultivation-current	14.53	0.01	14.53	0.01	0.00	0.00
50	Shifting cultivation-abandoned	1.12	0.00	1.12	0.00	0.00	0.00
Grand Total		**162,969.6**	**100.0**	**162,969.6**	**100.0**	**0.0**	**0.0**

2015–2016. It is found that there is an increase of 343.06 km² (0.21%) with reference to the year 2015–2016.

Agriculture: Agriculture is the main source of income of the state's economy and predominant land use/land cover category in the state. Agriculture occupies an area of 96,567.91 km², which is about 59% of the total geographical area of the state during 2015–2016. The agricultural cropland is further subdivided into kharif, rabi, zaid, two crop areas, more than two crop areas, fallow, plantations, and aquaculture. The agricultural activity starts from June and extends up to mid-November and coincides with the southwest monsoon and is known as the Kharif season. The kharif cropland, also known as the rain-fed agriculture, is spread over an area of 26,737.74 km² (16.41%). The main crops grown during kharif are rice, sugarcane, jowar, maize, redgram, cotton, castor, etc. Rabi cropping season starts from November and extends up to March. Mainly irrigated crops like rice, sugarcane, and chillies are grown during rabi season by utilizing canal, tanks, and groundwater resources. In some places, lift irrigation is also practiced in the state. Rabi cropland occupied an area of 12,364.66 (7.59%) during the year 2015–2016. Black soil areas mainly in Anantapuramu, Kurnool, Prakasam, and Guntur have been mapped under rabi cropland. Summer crops grown from April to June are mapped under zaid cropland and occupied an area of 209.04 km² (0.13%) during the study period. Areas under crop during any two seasons are mapped under cropped in two seasons and it occupies an area of 37,344.21 km² (22.92%). These are found in the Godavari and Krishna delta areas with assured irrigation from canals, tanks, and groundwater. More than two crop areas are those lands with a crop like sugarcane present throughout the year is mapped under cropped more in two seasons category. This category occupies an area of 1379.62 km² (0.85%) of the total geographical area of the state. About an area of 9111.56 km² (5.59%) is classified under fallows during 2015–2016. These are areas devoid of crop during both the cropping seasons for various reasons and are found in small extent throughout the state. Plantations like mango, cashew, coconut, casuarinas, eucalyptus, bamboo, teak, etc., are grown throughout the state. This category occupies an area of 7308.83 km² or 4.48% of the total geographical area of the state. Cashew plantations intermixed with mango and coconut are found in the coastal areas of Srikakulam and Visakhapatnam districts. Mango plantations are predominant in Krishna, East Godavari, and Vizianagaram districts. Conversion of paddy fields, mangroves, marshy and swampy areas into aquaculture is found in the districts of West Godavari, Krishna, Guntur, Prakasam, and SPS Nellore. The area under aquaculture had been estimated as 1805.99 km² (1.11%) in 2011–2012 and it increased to 2106.97 km² (1.29%) during 2015–2016. The area around Kolleru lake in West Godavari and Krishna districts has been converted into aquaculture and possesses an environmental threat to the lake which is a noted wetland area and famous as a bird sanctuary.

Forest: The forest area occupied an area of 35,170.86 km² (21.58%) in the year 2011–2012, whereas it has been decreased to 35,063.43 km² (21.52%) during 2015–2016. This is due to the conversation of agricultural lands at fringe areas of forest lands. These are found in north, northeast, and south of the state where several forest species grow. The important species are teak, nalla maddi, rosewood, devadari, sanders, etc. The forest areas are further classified under evergreen/semi-evergreen, deciduous, forest plantations, scrub forest, mangroves, grasslands, etc. The evergreen dense forest and open forest occupied 157.58 km^2 (0.10%) and 97.04 km^2 (0.06%) during 2011–2012, respectively, and found no change in the evergreen category during 2015. The deciduous forests are found in northeastern belt covering agency areas of East Godavari and Visakhapatnam and the southern belt of forests comprising the districts of Chittoor, Kadapa, and Kurnool are covered by five hill ranges namely Seshachalam, Palakonda, Veligondas, Lankamalas, and Nallamalas. In addition to timber, these forests also contain a large variety of fauna like tigers, leopards, wolves, bears, etc. The deciduous forests are further divided into deciduous dense and deciduous open based on the forest vegetation vigor and the area occupied 11,483.31 km^2 (7.05%) and 10,349.39 km^2 (6.35%), respectively, during the year 2011–2012 whereas in 2015–2016, the categories are 11,422.77 km^2 (7.01%) and 10,327.97 km^2 (6.34%), respectively. Forest plantations mainly teak, bamboo, casuarinas, etc., have been delineated with an area of 685.45 km^2 in 2011–2012 and it increased to 714.50 km^2 during 2015–2016. Scrub forest accounted for 8370.44 km^2 (5.14%) in 2011–2012 whereas 8310.61 km^2 in 2015–2016. It is found that most of the scrub forest fringes are converted to forest plantations under Vanam-Manam scheme in the state. The tidal forests are found in the coastal tracks and comprise several littoral species that are popularly known as mangroves. These mangrove forests are found mainly in the districts of East Godavari, West Godavari, Krishna, and Guntur. These are further divided into dense and open mangroves. They occupied an area of 276.21 km^2 and 149.52 km^2 during 2011–2012, respectively. It is found that about 26 km^2 of the area has been increased in open mangrove category and about 4 km^2 has been decreased in dense mangrove category during the year 2015–2016. Areas with tree cover lying outside the notified forest area with the woody perennial plant are delineated under tree clad area. The dense tree clad area occupied 1770.26 km^2 and open area is 1810.77 km^2 during the year 2015–2016, whereas 1780.57 km^2 was under dense tree clad and 1817.58 km^2 under the open tree clad area in the year 2011–2012. The decrease is due to the area utilized for agricultural activity. It is observed that most of the tree clad areas are located adjacent to forest fringes. These areas are being converted into kharif crop land, which constitutes an area of about 14.43 km^2. This is primarily due to the implementation of the Recognition of Forest Rights (RoFR) scheme in the State for economical upliftment of scheduled tribes and other traditional forest dwellers.

Grass and Grazing land: These areas are described as the natural potential (climax) plant cover as being composed of principally native grasses, forbes, and shrubs. The grasslands are found in the hilly region of East and West Godavari districts, located in uplands and hillslopes. The tropical grasslands accounted 101.59 km^2 (0.06%) and no change is found during the study period.

Wastelands: Wastelands refer to degraded land which can be brought under vegetative cover with reasonable efforts and which is currently underutilized or land which is deteriorating due to lack of appropriate water and soil management or on account of natural causes (NRSA 2006). The wastelands occupied an area of 15,626.17 km^2 corresponding to 9.59% of the total geographical area of the state during 2015–2016, compared to 15,959.55 km^2 (9.79%) in 2011–2012. This may be due to the implementation of developmental programs in the state by various line departments. The predominant wasteland category is the scrubland, and it is found all over the state. It occupies an area of 11,783.70 corresponding to 7.23% of the total geographical area of the state. Scrubland is further subdivided into dense scrub (6507.42 km^2) and open scrub (5276.29 km^2) based on the presence of vegetation during 2015–2016. It was estimated as an area of 12,044.54 km^2 (7.39%) during 2011–2012, of which dense scrub is 6642.65 km^2 (4.08%) and open scrub is 5401.88 km^2 (3.31%). Barren rocky areas have been observed as rocky outcrops in the forest and scrubland. This class occupied an area of 2058.37 km^2 (1.26%) during the year 2015–2016, whereas 2068.78 km^2 (1.27%) in 2011–2012. It is found that most of the barren rocky areas are being quarried for various construction activities. The salt-affected lands were mapped with an area of 1308.81 km^2 (0.80%) in 2011–2012, whereas it has been decreased to 1272.31 km^2 (0.78%) during the year 2015–2016. These are significant in south coastal districts of SPS Nellore, Prakasam, Guntur and Rayalaseema districts of Anantapuramu, YSR Kadapa, and Kurnool. The gullied/ravinous lands totaling 229.31 km^2 (0.14%) of which gullied lands are about 85 km^2 (0.05%) and ravinous lands are 143.67 km^2 (0.09%) during the year 2015–2016, whereas 245.93 km^2 (0.15%) in the base year 2011–2012. These areas are found in the districts of SPS Nellore, Chittoor, Prakasam, Visakhapatnam, Vizianagaram, and West Godavari. Sandy area comprising coastal, desertic, and riverine sands occupied an area of 282.50 km^2 (0.17%) during 2015–2016, whereas 291.49 km^2 (0.18%) in 2011–2012. The coastal sands category is predominant with an area of 260.12 km^2 (0.16%) and 266.19 km^2 (0.16%) during 2015–2016 and 2011–2012, respectively. These are found along the coastal stretch of the state.

Wetlands: These are the lands covered by water or all submerged or water-saturated lands, natural or man-made, inland or coastal, permanent or temporary, static or dynamic which necessarily have a land–water interface are delineated as wetlands. These include ox-bow lakes, swamp, marshy, cut-off meanders, lagoons, creek, backwaters, bay, tidal flat, etc. These areas occur mainly in the coastal region. Pulicat lagoon in Nellore district and Kolleru lake in West Godavari and Krishna districts are the noted wetlands and bird sanctuaries in the state. The total wetland category has been estimated as 1503.68 km^2 (0.92%) in 2011–2012, and it has been decreased to 1393.14 km^2 (0.85%) during 2015–2016. It is found that coastal wetlands are the predominant category during the study period.

Water Bodies: This category comprises areas with surface water, either impounded in the form of ponds, lakes, and reservoirs or flowing as streams, rivers, canals, etc., are delineated (NRSA 2006). These are seen clearly on the satellite image in blue to dark blue or cyan color depending on the depth of water. This category comprises river, stream, canal, lakes, ponds, reservoir, and tanks. The major rivers are The Godavari, Krishna, Pennar, Nagavali, and Vamsadhara. Around 40 major, medium, and minor rivers which flow across the state are delineated. The water bodies which include all the subcategories are about 8832.53 km^2 (5.42%) in 2011–2012, and it has increased to 8864.61 km^2 (5.44%). The increase may be due to construction of new irrigation projects and canals. The water spreads in the tanks/reservoirs during kharif, rabi, and zaid seasons have been mapped. Tanks are the major source of irrigation which is about 3% followed by river/stream (2%). **Shifting Cultivation:** It is a form of agriculture used especially in hilly regions. In this process, an area of ground is cleared and cultivated for a few years and abandoned for a new area until its fertility has been naturally restored. Shifting cultivation occupied an area of 15.66 km^2 (0.01%) during the study period. These areas have been observed mainly in the hilly regions of East Godavari district. No change was observed in this category during the study period.

16.5.2 Land Use/Land Cover Change During 2011–2012 and 2015–2016

From the analysis, it is observed that there is significant change in built-up, wasteland, and agriculture categories. Built-up increased from 4999.35 km^2 in the year 2011–2012 to 5342.41 km^2 in 2015–2016 with a tune of 343.06 km^2. This is due to the growth of settlements, layouts at urban fringe and industrial areas. Obviously such growth in built-up land has been at the cost of agricultural land. The temporal changes in land use/land cover were analyzed and presented in Table 16.3. In Table 16.3, the diagonal yellow color denotes no change area during the study period and table represented in very small due to more number of categories. The wasteland category with an area of 333.38 km^2 has been decreased from the year 2011–2012 to 2015–2016, of which 61 km^2 has been converted into built-up and 273.8 km^2 converted into agriculture land. Under agricultural category, the kharif crop accounted for 29,311.60 km^2 in the year 2011–2012, and it has been decreased to 26,737.74 km^2 during 2015–2016. The reduction in kharif crop area has been converted to cropped in two seasons (5641 km^2), rabi crop (406 km^2), plantation (147 km^2), current fallow (1397 km^2), built-up (100 km^2), aquaculture (90 km^2), and water bodies (16 km^2). The transformation of rabi crop into khairf is 452.5 km^2, cropped in two seasons is 1964.6 km^2, and fallow is about 173 km^2 from 2011–2012 to 2015–2016. The cropped in two seasons area has been converted to kharif crop (2785 km^2), rabi crop (590.6 km^2), zaid crop (111.3 km^2), fallow (364.6 km^2), aquaculture (139.7 km^2), built-up (76 km^2), respectively, during the study period. The area of fallow land is converted into built-up (47.6 km^2), kharif crop (1548.1 km^2), rabi crop (192 km^2), cropped in two seasons (276.2 km^2) and plantations (127.5 km^2). There is a significant change in cropland due to suitable irrigation systems. The plantation area has been decreased with an area of 168.18 km^2 from 2011–2012 to 2015–2016, and it converted to built-up (26.7 km^2), cropland (604 km^2), and aquaculture (4.98 km^2) during 2011–2012 and 2015–2016. This is due to development of special economic zones (SEZs), National Investment and Manufacturing Zones (NIMZs), industries and industrial corridors in the state. Aquaculture has been increased to about 301 km^2 during the study period with the cost of agriculture land. This is found in southern and central coastal districts of the state. The dense, open,

Table 16.3 Land use/land cover change matrix of Andhra Pradesh during 2011–2012 and 2015–2016 (Area in km^2)

Code legend:

Built Up
1 Compact/Continuous
2 Sparse/Discontinuous
3 Vegetated / Open Area
4 Industrial
5 Mining - Active
6 Mining - Abandoned
7 Quarry

Agricultural Land
10 Kharif Crop
11 Rabi Crop
12 Zaid Crop
13 Cropped in 2 seasons
14 Cropped more in 2 seasons
15 Fallow
16 Plantation
17 Aquaculture

Forest
18 Evergreen / Semi Evergreen-Dense
19 Evergreen / Semi Evergreen-Open
20 Deciduous (Dry/Moist) Dense
21 Deciduous (Dry/Moist) Open
22 Forest Plantation
23 Scrub Forest
24 Littoral/Swamp Forest (Mangrove)-Dense
25 Littoral/Swamp Forest (Mangrove)-Open
27 Tree Clad Area-Open

Wastelands
31 Salt affected land
32 Gullied land
33 Barren rocky
34 Open scrub
35 Sandy area (Desrtic)
36 Coastal
37 Riverine
38 Barren Rocky/Stony waste

Wetland
40 Inland-Natural (Ox-bow lake, waterlogged etc.)
41 Inland-Manmade (Water logged, tanks etc.)
42 Coastal-Lagoon, creek, mud flats etc.
43 Coastal-Saltpans

Grass- Grazing
28 Grass & Grazing land-Alpine/Sub Alpine
29 Grass & Grazing land-Temperate / Sub Tropical
30 Grass & Grazing land-Tropical/Dryzone

Water bodies
44 River/Stream-Perennial
45 River/Stream-Non Perennial
46 Reservoir/Tanks-Perennial
47 Lakes/Ponds-Permanent
48 Lakes/Ponds-Seasonal
49 Reservoir/Tanks-Permanent
50 Reservoir/Tanks-Seasonal

Snow: Shifting cultivation & Snow
51 Snow
52 Shifting cultivation-Current
53 Shifting cultivation-Abandoned
54 Ice

scrub forest and forest plantation together are slightly decreased from 35,171 km^2 in the year 2011–2012 to 35,063 km^2 in 2015–2016. About 113 km^2 of forest area has been converted to crop land during the study period. It is observed that the fringe area of the forest cover has been converted to agriculture lands due to the implementation of Recognition of Forest Rights (ROFR) scheme in the state. The decrease in forest area indicates unhealthy trends of land use pattern. The reduction in forest cover may be due to the constant felling of trees for fuel, fodder, and other agriculture practices. The wetlands together are slightly decreased from 1503.68 km^2 in the year 2011–2012 to 1393.14 km^2 in 2015–2016 with a decrease of 110.53 km^2. The area under water bodies has been increased about 32 km^2 from 2011–2012 to 2015–2016. This is because of the normal rainfall and implementation of irrigation projects during the period.

Acknowledgements The authors are grateful to the NRSC, ISRO for their financial and technical support of land use cycle projects. The authors are greatly acknowledged to Dr. B.Sundar, IFS, Vice Chairman, APSAC for his constant encouragement and support. The authors wish to extend sincere appreciation and encouragement given by the management and the staff concerned of APSAC, ITE&C Dept., Govt. of Andhra Pradesh during the course of the study. We would like to thank anonymous reviewers for their valuable suggestions which have improved the manuscript enormously.

References

Babu CT, Vanajakshi M, Ramana KV (2018) Quantitative analysis of land use change using remote sensing and GIS techniques in Guntur Urban Mandal Andhra Pradesh State. Ind Geogr J 93(1):66–71

Halimi M, Sedighifar Z, Mohammadi C (2018) Analyzing spatiotemporal land use/cover dynamic using remote sensing imagery and GIS techniques case: Kan basin of Iran. GeoJournal 83:1067–1077. https://doi.org/10.1007/s10708-017-9819-2

Iyyer C (2009) Land management: challenges and strategies. Global India Publications Pvt Ltd., New Delhi, p 267

Jaiswal JK, Verma N (2013) The study of the Land use/land cover in Varanasi district using remote sensing and GIS. Trans Inst Ind Geogr 35:201–212

Miheretu BA, Yimer AA (2018) Land use/land cover changes and their environmental implications in the Gelana sub-watershed of Northern highlands of Ethiopia. Environ Syst Res 6:7. https://doi.org/10.1186/s40068-017-0084-7

Mishra VN, Rai PK, Kumar P, Prashad R (2016) Evaluation of land use/land covers classification accuracy using multi-temporal remote sensing images. Forum Geogr J 15(1):45–53

Mishra VN, Rai PK, Rajendra P, Puniya M, Nistor MM (2018) Prediction of spatio-temporal land use/land cover dynamics in rapidly developing Varanasi district of Uttar Pradesh India, using geospatial approach: a comparison of hybrid models. Appl Geomatics. https://doi.org/10.1007/s12518-018-0223-5

Mishra V, Rai PK (2016) A remote sensing aided multi-layer perceptron-Marcove chain analysis for land use and land cover change prediction in Patna district (Bihar), India. Arab J Geosci 9(1):1–18. https://doi.org/10.1007/s12517-015-2138-3

NRSA (National Remote Sensing Agency) (2006) User manual of national land use land cover mapping using multi-temporal satellite data, Hyderabad

NRSC (2012) User manual of national land use/land cover mapping using multi temporal satellite data (2nd cycle). National Remote Sensing Agency, Hyderabad, Govt. of India

Rai PK, Kumra VK (2011) Role of geoinformatics in urban planning. J Sci Res 55:11–24

Shastri S, Singh P, Verma P, Rai PK, Singh AP (2020) Assessment of spatial changes of land use/land cover dynamics, using multi-temporal Landsat data in Dadri Block, Gautam Buddh Nagar, India. Forum Geogr XIX(1):72–79. https://doi.org/10.5775/fg.2020.063.i

Vishwakarma CS, Thakur S, Rai PK, Kamal V, Mukharjee S (2016) Changing land trajectories: a case study from India using remote sensing. Eur J Geogr 7(2):63–73

Yesuph AY, Dagnew AB (2019) Land use/cover spatio-temporal dynamics, driving forces and implications at the Beshillo catchment of the Blue Nile Basin, North Eastern Highlands of Ethiopia. Environ Syst Res 8:21. https://doi.org/10.1186/s40068-019-0148-y

Yuan F, Sawaya KE, Loeffelholz BC, Bauer ME (2005) Land cover classification and change analysis of the Twin Cities (Minnesota) Metropolitan Area by multitemporal Landsat remote sensing. Rem Sens Envi 98:317–328. https://doi.org/10.1016/j.rse.2005.08.006

Chapter 17
Assessing the Impact of Delhi Metro Network Towards Urbanisation of Delhi-NCR

Diksha Rana, Deepak Kumar, Maya Kumari, and Rina Kumari

Abstract The unprecedented urbanisation of Megacities like Delhi at an irreversible rate has played a significant role expansion of Delhi Metro projects in the public urban transit system in the Delhi-NCR region. The current work tried to envisage the study on urbanisation transitions due to the launch of the metro rail network in the NCR region. Apart from urban, it has influenced the periphery and rural areas of Delhi. These regions are now hastening towards development at a higher pace. Apart from these, the metro network is contributing to the linear development or parallel development of infrastructures along with the network. Adjoining cities of Uttar Pradesh and Haryana states containing Gurugram, Noida, Bahadurgarh and Faridabad have become major intersections of the metro network. These are also contributing to urbanisation. Thus the current work tries to visualize urbanisation with the social scenario, as it has a massive network of nearly 288 km length comprising of six different lines. The people from different parts are commuting daily via the metro network. In conclusion, it proposes to identify the push and pull to further analyse the urbanisation pattern due to the expansion of the network in the region, as the growing population in the region requires further expansion. Hence, metro projects are expanding continuously and providing a new stimulus towards the increasing urbanisation.

Keywords Corridors · Delhi metro · Delhi-NCR · Expressway · Green cover · Change detection

D. Rana · M. Kumari
Amity School of Natural Resources and Sustainable Development (ASNRSD), Amity University, Sector 125, Noida, Uttar Pradesh 201313, India

D. Rana · D. Kumar (✉)
Amity Institute of Geoinformatics and Remote Sensing (AIGIRS), Amity University, Sector 125, Noida, Uttar Pradesh 201313, India

R. Kumari
School of Environment and Sustainable Development (SESD), Central University of Gujarat, Sector-30, Gandhinagar, Gujarat 382030, India

Abbreviations/Acronyms

BU	Built Up
CLC	Classified land cover
GST	Geo-Spatial Technology
ESRI	Environmental Systems Research Institute
NIR	Near infrared
NDVI	Normalized difference vegetation index
NDBI	Normalized difference built-up index
UHI	Urban heat island
UBS	Urban Blue Space
DMR	Delhi Metro Rail Network
RVI	Ratio Vegetation Index

17.1 Introduction

Urbanisation in megacities has led to an escalation of the development rate in the surrounding region (Khairkar 2011; Kumar and Shekhar 2015; Thomas et al. 2007). Similarly, this pattern of urbanisation around Delhi-NCR has been also influenced by the extension of the Delhi Metro network. Just after the inauguration of the Delhi Metro network in the early twenty-first century, the traffic and transportation pattern of the city were full influenced (Follmann et al. 2018; Powles and Yu 2010). The extension of the network across the neighbouring areas comprising Gurugram, Noida, Ghaziabad, Faridabad has been observed, and these have continuously exhibited similar infrastructural development (Errampalli et al. 2020; Follmann et al. 2018). Currently, the extension of the metro network across the Delhi and adjoining cities have completely changed the life of transportation network with advanced and convenient transportation service to the commuters. Apart from these, the physical landscape of transportation around the city is also apparently transforming with its expansion (Wang et al. 2018). The utilisation of total urban areas is also being influenced due to the growth of population and push factor of migration due to the influence of the development of planned and unplanned residence in and around the city (Follmann et al. 2018; Mohammadi et al. 2012; Rajput and Arora 2017; Tamis et al. 2009). The metro network prominently in Delhi has eight prominent metro lines, namely Redline, Yellow line, Greenline, Blueline, Violet line, Pink line, Magenta line and Orange line *(Airport express line)* with a gross length of around 288 kms. The Delhi Metro network has observed speedy growth of population around the adjoining cities like Gurugram, Noida, Ghaziabad and Faridabad and is capable to provide the easiest convenient public transit system. In this context, the current work tries to address the research objectives like (a) To assess the trends of urbanisation along the metro corridors, (b) To detect the buffer area influenced by the expansion of metro corridors, and (c) To evaluate the expected amount of urbanisation along upcoming metro corridors.

Developments due to the urbanisation pattern are altering their scopes at a very high pace. The basic indicators of expansion comprise transformation of modern infrastructure, developing economy, congested roads, sleepless nights and many more (Guan 2011; Loukanov et al. 2020; Yeh and Li 2001). These indicators of urbanisation can have different meanings as per the region. This urbanisation also creates several problems related to the development of slums, environmental problems, congestion and declining land-man ratio (Pratomo et al. 2016; Lee et al. 2012; Shekhar 2014). Hence, there is a dearth need for researches on urbanisation, with a socio-environmental perspective to evaluate the factors which significantly contribute to the deterioration of the environment (Das et al. 2002; Feizizadeh et a l. 2017; Larsen and Hertwich 2010). Air and water quality become the key factor in climate change and emerging health issues (Caragliu et al. 2011; Chen et al. 2006; Elbar and Hassan 2019; Deepak Kumar 2016, 2020). Also, other researches highlight on evaluation of capability in terms of capacity, travel time and accessibility of the system to evaluate their indices to reflect commuter's perspective. Some other research work on CO_2 emission reduction due to metro rail transport system focuses emission saving due to metro network (Bose and Srinivasachary 1997; Khandelwal et al. 2017; Zhang et al. 2019). The impact of the metro rail construction work zone shows systematic work zone scheduling and traffic management.

Researches on economic and equity evaluation of Delhi Metro express that the metro network has a positive impact on equity (of mobility and accessibility) and many articles in the newspaper reports about the influence of rampant urbanisation on house shortages to become a convenient and dignified mode of transport for Delhiites (Bassuk et al. 2015; Kumar and Aggarwal 2019; Meisen et al. 2006). But there is an absence of studies on the valuation of urbanisation patterns due to metro rail network development. Thus, the current work focuses on the analysis of satellite datasets to understand the competitiveness of this research. The data and analysis of all the existing metro try to identify the influence of the metro on infrastructure development leading to urbanisation.

17.2 Materials and Methods

(a) **Study Area**

The current work is attempted at the capital of India *(i.e. New Delhi)*. It is also the second-largest metropolis in India (in terms of population), which is positioned at 28.7041° N latitude and 77.1025° E longitudes. It encompasses another region of NCR comprising Noida, Ghaziabad, Gurugram and Farid-abad. The metro network at Delhi contains eight lines named with different colours and covers not only Delhi but also shares a conjoint border with satel-lite towns. Figure 17.1 shows the location of the study area and extends of the metro rail network across the region.

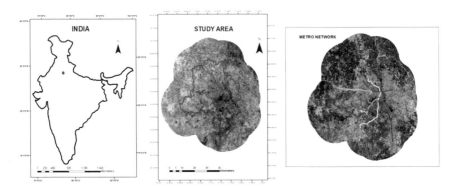

Fig. 17.1 Location of the study area

Geography and physical features: The city of Delhi can be broadly divided into three segments (based on the physical and geographical properties), *namely the Yamuna flood plain, the Ridge and the Plain*. The general properties of Yamuna flood plains lie somewhat in a low-lying area with sandy soils, and these are subjected to recurrent floods. These areas are sometimes also known as Khadar. The ridge portion of the area consists of the most leading physiographic features. It originates from the Aravalli hills of Rajasthan and entering the union territory from the south extends in a north-eastern direction. It encloses the city on the northwest and west. The point near Bhatti has a height of approximately 1045 ft. above sea level has Tughlakabad fort. The entire area except for the Yamuna flood plain (khadar) and the ridge of the national capital territory of Delhi can be characterised as Bangar or the plain. The main part is a plain area, and the city of Delhi is located on this. It contains the NCR region including Noida, Ghaziabad, Gurugram and Faridabad. The metro network of Delhi comprises of eight main lines namely with different colours, and it covers adjoining border areas with satellite towns of the Delhi city.

Climate: The city of Delhi experiences an extreme climate with very cold winters and hot summers. The cold season starts from November and is at its peak nearby the New Year and the first half of January. Subsequently, the weather turns warmer around the middle of March, and sooner it converts hot from April to June. Later, it experiences extreme heat with a temperature around 45 degrees Celsius. The month of monsoon reaches around the end of June. The city experiences a rainy season in winter, which is helpful for the farmers of the village around the union territory because the rabbi crops are benefitted from it. The general weather of the region is generally dry except for 2–3 months of humidity.

(b) *Datasets Used*

The current work utilised the freely available open archive datasets namely Resourcesat and Sentinel datasets. The Resourcesat-1 satellite of AWIFS sensor is available at a spatial resolution of 56 m and Sentinel-2 data is available at 10 m spatial resolution. Table 17.1 summarises the brief details of datasets being used for

Table 17.1 Summary of data sources

Satellite	Download Link	Spatial resolution (m)	Acquisition date
Sentinel-2 A/B	https://scihub.copernicus.eu/ (ESA)	10	19 June 2018
Sentinel-2 A/B	https://scihub.copernicus.eu/ (ESA)	10	21 March 2017
Resourcesat	https://bhuvan.nrsc.gov.in/home/ index.php	56	06 February 2010
Resourcesat	https://bhuvan.nrsc.gov.in/home/ index.php	56	12 November 2007

the current work. It also accounts for the different resolutions of datasets used for accomplishing the current work. Table 17.2 captures the summary of Delhi Metro network lines.

(c)   *The extent of Delhi Metro Rail Network*

Year 2007

Figure 17.2 illustrates the extent of the metro network in the year 2007, and it comprises three major metro lines consisting of Red metro line covering a stretch of 21.3 km expanding from Shahdara to Inderlok, Yellow metro line with 10.84 km of the length extending Vishwa Vidyalaya to Central Secretariat and Blue metro with a length of 32.2 km extending Dwarka to Indraprastha. All metro network lines are used to perform buffer operations of 0.5, 1, 2, 3, 4 and 5 km and were overlaid on the image to visualise the influence of metro lines on urbanisation due to the extension of metro network covering the central region of Delhi.

Year 2010

Figure 17.3 displays the extent of the metro network in the year 2010. It comprises five metro lines with a length of approximately 93.06 km and comprises of metro lines, namely Red metro line, Blue metro line, Yellow metro line, Green metro line

Table 17.2 Colour code schematics of the metro network

METRO LINES	LENGTH (in km)	EXTENT OF METRO LINE-ROUTE	COLOURS
RED	25.09	DILSHAN GARDEN TO RITHALA	
YELLOW	49.31	SAMAYPUR BADLI TO HUDA CITY CENTER	
GREEN	29.64	INDERLOK/ KIRTI NAGAR TO CITY PARK	
VIOLET	43.4	KASHMERE GATE TO ESCORTS MUJESAR	
ORANGE	22.7	NEW DELHI TO DWARKA SECTOR 21	
MAGENTA	37.72	BOTANICAL GARDEN TO JANAKPURI WEST	
PINK	29.66	MAJLIS PARK TO DURGABAG DESHMUKH SOUTH CAMPUS	

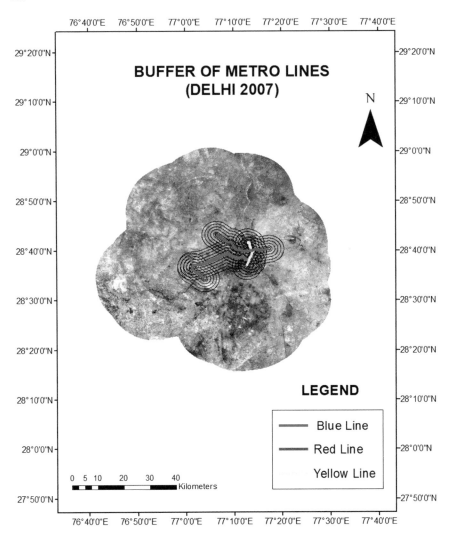

Fig. 17.2 The extent of the metro network in the year 2007 with multiple ring buffers

and Violet metro line. It covered most inner parts of Delhi. To visualise the impact of the metro network on urbanisation, the buffer zones with 0.5, 1, 2, 3, 4 and 5 km distance were created along the metro line. These showed that the metro has influenced urbanisation in NCR regions like Noida.

Year 2017

Figure 17.4 indicates the metro network in the year 2017 and comprises seven metro lines covering a length of approximately 214.22 km. It comprises major metro lines including the Red metro line, Blue metro line, Yellow metro line, Green metro line,

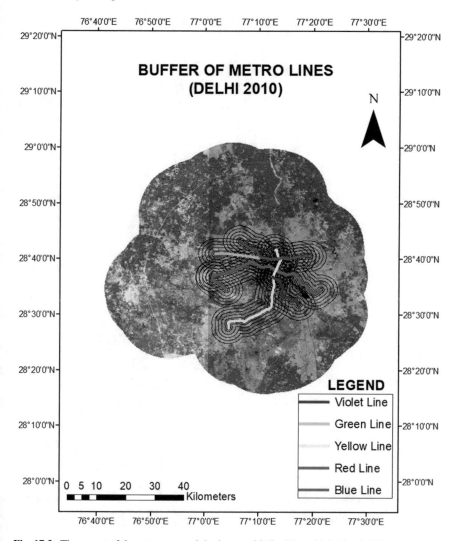

Fig. 17.3 The extent of the metro network in the year 2010 with multiple ring buffers

Violet metro line, Magenta metro line and Orange metro line. It covers the major part of Delhi comprising the NCR region like Gurugram, Noida, Faridabad, Ghaziabad etc. Buffer zones are created about metro lines to visualise the influence in nearby areas within the proximity of 0.5 to 5 km due to metro and cost of influencing urbanisation in the capital region.

Year 2018

The extent of the metro network in the year 2018 can be seen in Fig. 17.5, which comprises eight metro lines covering a length of approximately 214.22 km. It encloses

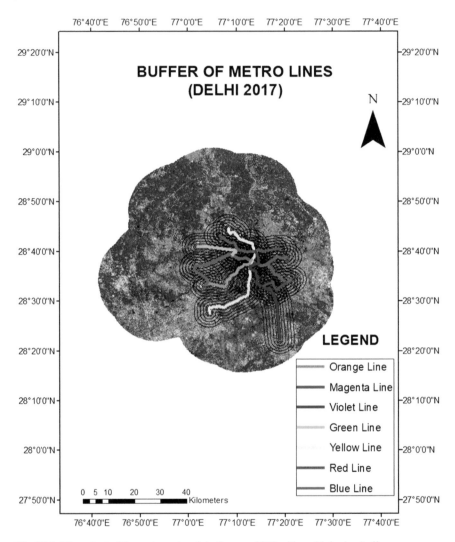

Fig. 17.4 The extent of the metro network in the year 2017 with multiple ring buffers

the Red metro line, Blue metro line, Yellow metro line, Green metro line, Violet metro line, Magenta metro line and Pink metro line cover most of the part of Delhi and also sharing its network in the NCR region. The overlaid buffer zones in the map try to visualise the influence of the metro network in nearby areas.

(d) *Software Used*

The current work utilises commercial remote sensing and GIS softwares, namely ERDAS IMAGINE 201 × for various operations. Pre-processing of the raw satellite images were performed with the help of ERDAS Imagine software and mapping,

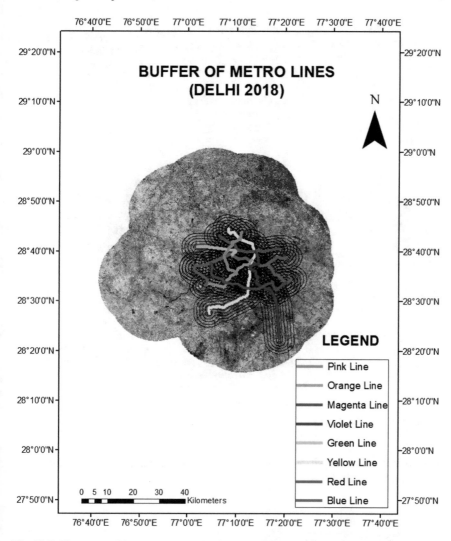

Fig. 17.5 The extent of the metro network in the year 2018 with multiple ring buffers

spatial calculations and analysis were performed on Arc Desktop to create various maps. The same software was also used for geostatistical computations to generate create several layers. Google Earth Pro was used for vectoring the metro network of Delhi from the corresponding satellite image.

Methodology of the work

In order to assess/estimate the influence of Delhi Metro network expansion on the urbanization pattern of the NCR region, it uses state-of-the-art geospatial/ technology

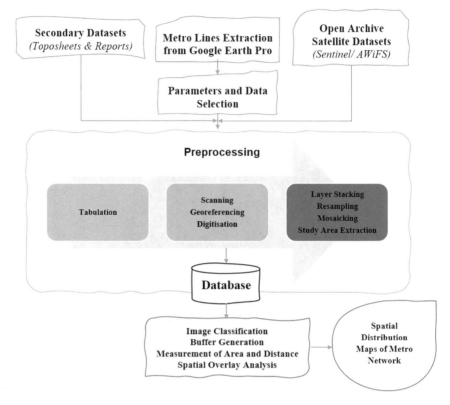

Fig. 17.6 Brief methodology

to accomplish the analysis of the work. Figure 17.6 elucidates the brief methodology for the thorough work.

17.3 Results and Analysis

(a) *Yearly land use variations with respect to distance from Yellow line*

Figures 17.7, 17.8, 17.9 and 17.10 signify the influence of the metro network on urbanisation. It can be observed that the built-up area is insignificant near the metro line in the year 2007. Figure 17.7 also exhibits that if we move away from the metro network, the number of urbanisations increases. Figure 17.8 shows that built-up is increasing from the buffer of 1 to 5 km away from the metro line. In the year 2017, it can be observed in Fig. 17.9 that the built-up area increases if we move away from the metro line. Figure 17.10 also shows a similar trend that built-up areas increase if we move away from the metro line due to the continuous extensions of the metro network throughout the NCR region.

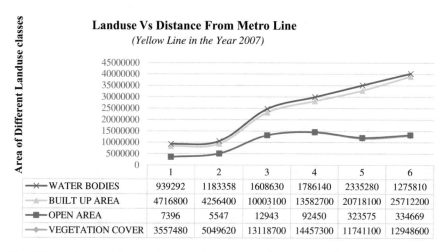

Fig. 17.7 Land use at different distances from the network in the year 2007

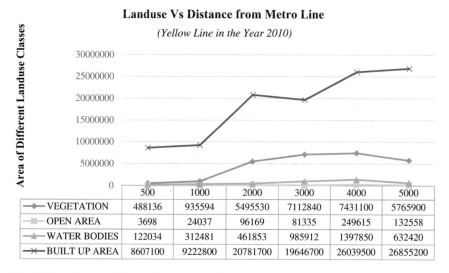

Fig. 17.8 Land use at different distances from the network in the year 2010

(b) *Yearly land use variations with respect to distance from Redline*

Figure 17.11 illustrates that in the year 2018, after the introduction of the red line, the built-up area started increasing. The curve shows that if we are moving away from the metro network, the amount of urbanisation is increasing, which clearly shows the influence of the metro network over the urbanisation pattern.

(c) *Yearly land use variations with respect to distance from Greenline*

Figure 17.12 exhibits that that built-up area is negligible near the metro line in the year 2018, but if we move away, it shows an increasing trend in terms of an urban

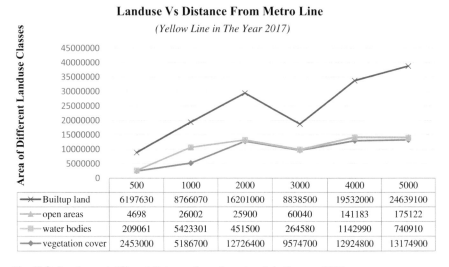

Fig. 17.9 Land use at different distances from the network in the year 2017

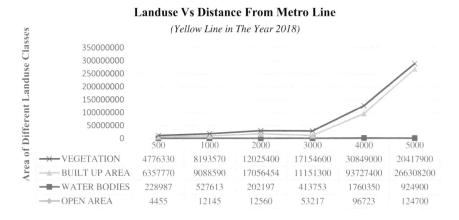

Fig. 17.10 Land use at different distances from the network in the year 2018

area, which infers that land use area magnificently increased due to instruction of green metro line network.

(d)   *Yearly land use variations with respect to distance from Pink line*

Figure 17.13 exhibits that the built-up area is negligible near the metro line in the year 2018, but if we move away, it shows an increasing trend in terms of the urban area, which infers that land use area admirably increased due to instruction of pink line metro network. It also infers that every buffer zone namely 500, 1000, 2000, 3000, 4000 and 5000 m open area shows negligible increase, but the build-up area exhibits comparatively very high growth throughout in NCR region.

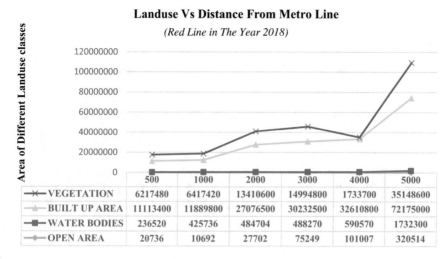

Fig. 17.11   Land use at different distances from the network in the year 2018

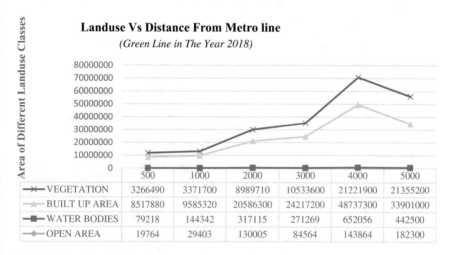

Fig. 17.12   Land use at different distances from the network in the year 2018

(e)   *Yearly land use variations with respect to distance from Blue line*

After the introduction of the blue line metro network, the whole scenario of the metro transit system has changed. In this context, Fig. 17.14 illustrates that in the year 2018, the built-up area shows a sharp slog in the pattern of development. The curve indicates that the metro network has contributed a lot towards the urbanisation of the region.

Land use area curves of various metro lines exhibit that there is very high growth in the built-up area near the Red metro network in the year 2018, and it also included

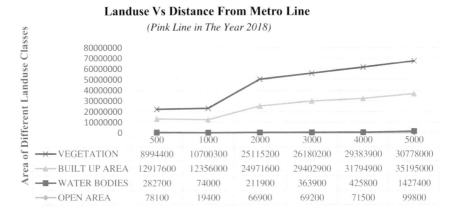

Fig. 17.13 Land use at different distances from the network in the year 2018

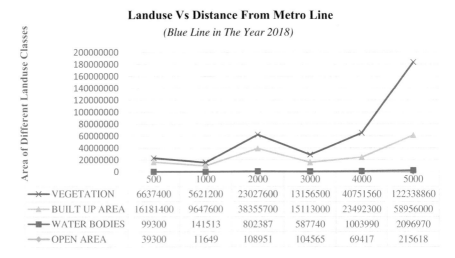

Fig. 17.14 Land use at different distances from the network in the year 2018

areas of old Delhi and its dense settlement. The built-up area near the Green line metro network in 2018 is less than the Blue and Red line metro network, but more than the pink line metro network because it includes dense settlement regions like Inderlok, Nangloi etc. The open area in all four metro lines is negligible due to its influence on urbanisation. The built-up area near the Blue metro line is more than the pink and Green metro line as it covers dense built-up areas like Tilak Nagar, Laxmi Nagar etc., and these places are highly populated. The built-up area near the pink line is less than other metro lines due to being a new metro line, and it conjuncts with various old metro lines. Due to less urbanisation near the pink line metro network and the inclusion of a few DDA parks, including Yamuna river and nearby green cover, Jayantpark, Delhi cantonment green cover, it is rich in vegetation cover in compared to other features.

17.4 Conclusion

Delhi Metro Rail is transiting towards rapid expansion to cater to a huge area to offer a faster transportation system to the daily commuters. In this manner, the urbanisation landscape is also increasing due to the faster development of the metro network. The subsequent regions located within the proximity of the metro lines are also exhibiting rapid infrastructural advancement. The push factors and emerging frugality of this region are also showing the cumulative growth due to urbanisation. The commuters are using the metro network due to the reliability, punctuality and speedy transit and many of the commuters are from the peripheral regions of the NCR. Therefore, these detrimental growths are analysed to understand and ascertain the real influence of the metro network on urban expansion. The metro network needs to expand the network in a way that it should not impact the progress in certain, and it should hold the force of migration. Therefore, the current study attempts to augment a new perspective to the urban studies in context to the development caused by the Delhi Metro rail network.

References

Bassuk NL, Universite AB, Jean M, Universite C, Theoretical L, Politics U, Bibliography AA (2015) On using landscape metrics for landscape similarity search. Landsc Urban Plan 117(1):1–12. https://doi.org/10.1038/srep11160

Bose RK, Srinivasachary V (1997) Policies to reduce energy use and environmental emissions in the transport sector: a case of Delhi city. Energy Policy 25(14–15):1137–1150. https://doi.org/10.1016/s0301-4215(97)00106-7

Caragliu A, Bo CD, Kourtit K, Nijkamp P (2011) Comparative performance assessment of smart cities around the North Sea basin. Network Industries Quarterly 13(3):15–17. https://doi.org/10.1017/CBO9781107415324.004

Chen X, Zhao H, Li P, Yin Z (2006) Remote sensing image-based analysis of the relationship between urban heat island and land use / cover changes. 104:133–146.https://doi.org/10.1016/j.rse.2005.11.016

Das G, Mandal KK, Mandal S (2002) Energy system., 34. Retrieved from http://environ.andrew.cmu.edu/m3/s3/energy_sys.pdf

Elbar ARA, Hassan H (2019) Experimental investigation on the impact of thermal energy storage on the solar still performance coupled with PV module via new integration. Sol Energy 184(April):584–593. https://doi.org/10.1016/j.solener.2019.04.042

Errampalli M, Patil KS, Prasad CSRK (2020) Evaluation of integration between public transportation modes by developing sustainability index for Indian cities. Case Stud Transp Policy 8(1):180–187. https://doi.org/10.1016/j.cstp.2018.09.005

Feizizadeh B, Blaschke T, Tiede D, Moghaddam MHR (2017) Evaluating fuzzy operators of an object-based image analysis for detecting landslides and their changes. Geomorphology 293(August 2016):240–254. https://doi.org/10.1016/j.geomorph.2017.06.002

Follmann A, Hartmann G, Dannenberg P (2018) Multi-temporal transect analysis of peri-urban developments in Faridabad India. J Maps 14(1):17–25. https://doi.org/10.1080/17445647.2018.1424656

Guan K (2011) Surface and ambient air temperatures associated with different ground material: a case study at the University of California, Berkeley, 14. Retrieved from http://nature.berkeley.edu/classes/es196/projects/2011final/GuanK_2011.pdf

Pratomo J, Kuffer H, Martinez J, Kohli D (2016) Uncertainties in analysing the transferability of the generic slum ontology. Geoforum 1

Khairkar VP (2011) Quantifying land use and land cover change using geographic information system : a case study of Srinagar city. Jammu and Kashmir India 2(1):110–120

Khandelwal S, Goyal R, Kaul N, Mathew A, Li ZL, Tang BH, Kaloop MR (2017) Detecting urban growth using remote sensing and GIS techniques in Al Gharbiya governorate Egypt. Egyptian J Remote Sens Space Sci 20(1):571–575. https://doi.org/10.1016/j.ijsbe.2015.02.005

Kumar D, Aggarwal S (2019) Analysis of women safety in indian cities using machine learning on tweets. In: Proceedings—2019 amity international conference on artificial intelligence, AICAI 2019. https://doi.org/10.1109/AICAI.2019.8701247

Kumar D, Shekhar S (2015) Statistical analysis of land surface temperature-vegetation indexes relationship through thermal remote sensing. Ecotoxicol Environ Saf. https://doi.org/10.1016/j.ecoenv.2015.07.004

Kumar D (2016) Adaptive hierarchical cell sub-division (AHCS) method for enhanced surface radiance temperature variability analysis. Model Earth Syst Environ 2(3). https://doi.org/10.1007/s40808-016-0194-7

Kumar D (2020) Statistical image processing for enhanced scientific analysis. In: Smart innovation, systems and technologies, vol 141. Springer, pp 1–11. https://doi.org/10.1007/978-981-13-8406-6_1

Larsen HN, Hertwich EG (2010) Identifying important characteristics of municipal carbon footprints. Ecol Econ 70(1):60–66. https://doi.org/10.1016/j.ecolecon.2010.05.001

Lee T-W, Lee JY, Wang Z-H (2012) Scaling of the urban heat island intensity using time-dependent energy balance. Urban Climate 2:16–24. https://doi.org/10.1016/j.uclim.2012.10.005

Loukanov A, El Allaoui N, Omor A, Elmadani FZ, Bouayad K, Seiichiro N, He HS (2020) Effects of neighborhood building density, height, greenspace, and cleanliness on indoor environment and health of building occupants. Environ Res 106(February):213–222. https://doi.org/10.1016/j.buildenv.2018.06.028

Meisen P, Quéneudec E, Yuan M, Nara A, Bothwell J, Ramirez L, Ramkrishna B (2006) Overview of renewable energy potential of India. Modeling Earth Syst. Environ. 2(October):1–20. https://doi.org/10.1016/j.compenvurbsys.2015.03.002

Mohammadi J, Zarabi A, Mobaraki O (2012) Urban sprawl pattern and effective factors on them : the case of Urmia City, Iran. J Urban Regional Anal IV:77–89

Powles SB, Yu Q (2010) Evolution in action : plants resistant t herbicides. https://doi.org/10.1146/annurev-arplant-042809-112119

Rajput S, Arora K (2017) Sustainable smart cities in India. In: Sustainable smart cities in india: challenges and future perspectives, (March), pp 369–382. https://doi.org/10.1007/978-3-319-47145-7

Shekhar S (2014) Improving the slum planning through geospatial decision support system. ISPRS—Int Archives Photogramm Remote Sens Spatial Info Sci XL–2(October):99–105. https://doi.org/10.5194/isprsarchives-XL-2-99-2014

Tamis WLM, van Dommelen A, de Snoo GR (2009) Lack of transparency on environmental risks of genetically modified micro-organisms in industrial biotechnology. J Cleaner Prod. https://doi.org/10.1016/j.jclepro.2008.11.004

Thomas I, Frankhauser P, De Keersmaecker ML (2007) Fractal dimension versus density of built-up surfaces in the periphery of Brussels. Pap Reg Sci 86(2):287–308. https://doi.org/10.1111/j.1435-5957.2007.00122.x

Wang K, Liu G, Zhai M, Wang Z, Zhou C (2018) Building an efficient storage model of spatial-temporal information based on HBase. J Spat Sci 8596:1–17. https://doi.org/10.1080/14498596.2018.1440648

Yeh AG, Li X (2001) Measurement and monitoring of urban sprawl in a rapidly growing region using entropy. Photogramm Eng Remote Sens 67(1):83–90

Zhang T, Shen WB, Wu W, Zhang B, Pan Y (2019) Recent surface deformation in the Tianjin area revealed by Sentinel-1A data. Remote Sensing 11(2):1–25. https://doi.org/10.3390/rs11020130

Chapter 18
Analysis of Urban Heat Island Effect in Rajkot City Using Geospatial Techniques

Mit J. Kotecha, Shruti Kanga, Suraj Kumar Singh, Ritwik Nigam, Karthik Nagarajan, and Achala Shakya

Abstract Rapid urbanization and unsustainable industrialization have shown negative impact on climate, which has led to climate change and ultimately leading towards global warming. It is estimated that presently global warming is increasing at the rate of 0.2 °C per decade, eventually making urban areas warmer. Haphazard development in the Indian cities have made them prone to urban heat island (UHI) phenomenon. Higher temperature in core urban areas due to concretization and excessive energy usage in comparison to its rural surroundings is known as the UHI effect. Higher UHI intensity in certain urban core areas put the population at a great risk of morbidity, and mortality makes the UHI assessment prerequisite during current times. This study has assessed the spatiotemporal effect of UHI in Rajkot city using LANDSAT5TMand LANDSAT 8 OLI remote sensing data. The study distinguished the Land use/ Land cover (LULC) using Landsat images for the year 2009 and 2017 in order to perform maximum livelihood classification. Utilizing classified

M. J. Kotecha
Department of Housing, Rajkot Municipal Corporation, Rajkot, Gujarat 360001, India

S. Kanga (✉)
Centre for Climate Change and Water Research, Suresh Gyan Vihar University, Jaipur 302025, India
e-mail: shruti.kanga@mygyanvihar.com

S. K. Singh
Centre for Sustainable Development, Suresh Gyan Vihar University, Jaipur 302025, India
e-mail: suraj.kumar@mygyanvihar.com

R. Nigam
Earth Science, School of Earth, Ocean and Atmospheric sciences (SEOAS), Goa University, Taleigao Plateau, Goa 403004, India
e-mail: earthscience.ritwik@unigoa.ac.in

K. Nagarajan
Department of Civil Engineering, Pillai HOC College of Engineering and Technology, Raigad, Rasayani, Maharashtra 410207, India
e-mail: knagarajan@mes.ac.in

A. Shakya
Department of Computer Engineering, National Institute of Technology, Kurukshetra, India

P. K. Rai et al. (eds.), *Geospatial Technology for Landscape and Environmental Management*, Advances in Geographical and Environmental Sciences,
https://doi.org/10.1007/978-981-16-7373-3_18

results and Normalized Difference Vegetation Index (NDVI), land surface temperature (LST) was derived using mono-window algorithm. Subsequently, ambient air temperature was scrutinized and isotherm was derived for three locations in Rajkot city such as Madhapar chowk, Trikon Baugh and Atika industrial area of different typology. Later on, discrepancy between LST and Ambient air temperature was figured out. Some environmental factors such as the concentrations of carbon dioxide and carbon monoxide, which contribute in UHI effect, were also analysed for the above-mentioned locations. On the basis of various results derived and analysis of temperature trend of past 60 years, it was determined that UHI effect was more prominent in the central business district (CBD) area of the selected regions. The results also revealed that the study region has experienced an increase of 0.3 °C in ambient air temperature in past 60 years. The built-up area and LST for LULC classes have also increased by 8.42% between 2009 and 2017 in Rajkot. The reasons behind increment in temperature can be: Rajkot, being the largest city of Saurashtra region has experienced rapid urbanization, higher energy consumption, rural to urban migration, which has modified the LU/LC of the city and eventually resulted into haphazard development that subsequently increase land surface temperature (LST).

Keywords Ambient air temperature · Climate change · Land surface temperature · Land use land cover · Rajkot · Urban heat island

18.1   Introduction

Twenty-first century is the era of urbanization. According to (UNFPA 2011), an actual report from the United Nations Population Fund stated that roughly, 3.1 billion people live already in cities and by 2017, that number will have risen to 4.1 billion. The growth of cities against time is abided by increase in population, and urban development has raised the already existing gap in demand and supply of essential infrastructure services. This flip side of urban system disrupts their ability and affect the cities' resilience to climate.

The IPCC Working committee (Report 2001) witnessed that recently the climate is changing in regions which have already overwhelmed physical system of region, biological system of region and in Homo sapien systems. Water supply, energy consumption and material consumptions has raised owing to raise in population and rapid urbanization (Da Silva and Moench 2014). Climate variability and change risks depend on many sensitive factors such as the baseline infrastructure of cities and quality of services, resource linkages, specifically water, energy and economic growth.

Urban resilience is 'The capacity of urban systems, communities, individuals, organizations and businesses to reclaim their function and thrive in the aftermath of a shock or a stress, regardless its impact, frequency or magnitude' (Parikh and Magotra 2014). Urban resilience is not a new concept. In 1970, the concept of urban resilience was introduced in scientific literature. The ecological research on urban area consists of the concept of thinking the ecological complexity with social-ecological system and its vulnerabilities. (Frantzeskaki 2016; Tripathi et al 2020). When a resilient system loosed by humans, it turned into highly vulnerable to disturbances that hitherto could be absorbed (Frantzeskaki 2016; Kanga et al. 2017).

There are scientific evidences that climate change occurs. According to the IPCC, increase in temperature has been observed since last 20–30 years due to human intervention, generation of anthropogenic heat, maximum use of fossil fuels, deforestation, etc., which increase greenhouse gas (GHG). GHG emission is highly observed in city area that leads to rise in temperature in the urban area; thus, urban heat island (UHI) effect can be a discern in the urban area.

There are clear evidences that the UHI effect affects the people living the in urban area. UHI within city boundary can lead to raise due to high concentration of ground-level ozone which increases air pollution problems (EPA 2008). Other parameters such as hasty industrialization, more numbers of automobiles and burning of fossil fuels for heating or electricity generation in city which also contribute in UHI. High-density inhabitants caused lower wind speeds in the city, UHI intensity increases by warm air stagnates in urban canyons and pollution remains due to evaporation cooling. However, reduction in concentration of GHG emission or CO_2 level within city limit can subsequently reduce effect of UHI. Other resilient strategies like cool roof, cool pavement, etc. would also help in reduction of UHI effect.

Nowadays, Rajkot is facing hotter days and less winter which has led to the occurrence of heat island effect. High temperature decreased the precipitation, that can reduce water supply, and drought can occur. In summer, there is shortage of water and prices of food increases. Owing to more population and migration, numbers of vehicles increased which is responsible for greenhouse gas emission and directly affects the climate. Rapid urbanization increases the narrow urban canyon in city area and reduces the green cover. Higher albedo effect due to impermeable soil and excessive use of material like concrete, asphalt, etc. which trap the heat. High rate of non-reflective and water-resistant surfaces are more in Rajkot. The intensity of UHI can also upraised by anthropogenic heat into the urban atmosphere.

The objectives of this study were; (i) To understand the concept of urban heat island (UHI) in context with climate change. (ii) To appraise the urban heat island through satellite image of Rajkot across a decade. Moreover, evaluate urban heat island through ambient air temperature and (iii) to ascertain the relationship of land surface temperature (LST), ambient air temperature, and albedo and air quality in UHI. Although, the UHI effect is the result of the interplay of various climatic components like temperature, precipitation, humidity, wind speed etc. However, the fluctuation in the temperature plays the vital role among all the components; thus, the study has considered 'temperature' and the summer season (which is highly uncomfortable) in Rajkot city which have been covered under the scope of the study.

18.1.1 Study Area

As per population and urban area, Rajkot is the fourth biggest city in Gujarat. Before independence, Saurashtra state was there and Rajkot was its capital. It was added into Gujarat state during May 1960, which was earlier in Bombay state. The distance from state capital Gandhinagar is 245 km from Rajkot. Rajkot is placed at the centre of Saurashtra peninsula in the central plains of Gujarat State. The elevation of Rajkot is 138 m above mean sea level. Further demographic and climate details are mentioned in Table 18.1 (Fig. 18.1).

The climate of Rajkot is hot and dry. The average maximum temperature observed over the last 40 years is 43.5 °C, and the average minimum temperature observed over the last 40 years is 24.2 °C. Rajkot has the average annual rainfall of 500 mm. Rainfall in Rajkot has been below normal during last 20 years.

104.86 km^2 area of Rajkot city consists the population density of 12,735 people/km^2 with a total literacy rate of 82.20%; in the year 2011, the city has comparatively higher literacy rate than 69.96% of State literacy rate as well as 54.16% of National literacy rate. Rajkot has Sex ratio of 908 females per 1000 males according to 2011 census, which is low compared to the state figure of 921 (Figs. 18.2 and 18.3; Table 18.2).

Rajkot city is having multiple land use. 77% of total area is developed for urban activities, half of this area is occupied by residential use, while industry occupies 5% and commercial zone occupies less than 20%.

18.2 Material and Methodology

Below mentioned methodology was adopted for the present study, which involved 'satellite data collection', 'classification of imagery', 'development of Land use Land

Table 18.1 Demographic and climatic details

Demographics	
Area	104.86 km^2
Population	1,288,599
Total household	286,838
Population density	12735/ km^2
Sex ratio	908
Literacy rate	82.20%
Climate: semi-arid	
Average rainfall	500 mm
Average humidity	20–30%
Temperature range	24–47 °C

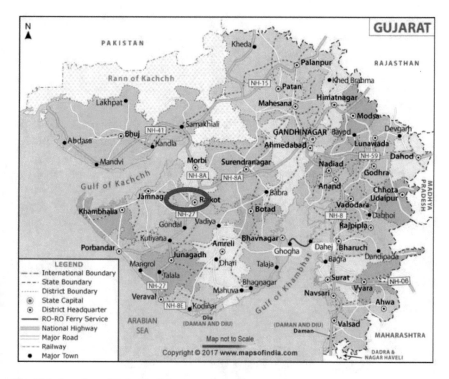

Fig. 18.1 Location of Rajkot in Gujarat

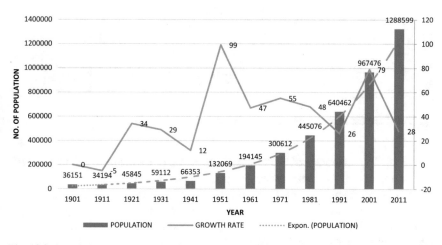

Fig. 18.2 Population trend of Rajkot. *Source* Census of India 2011

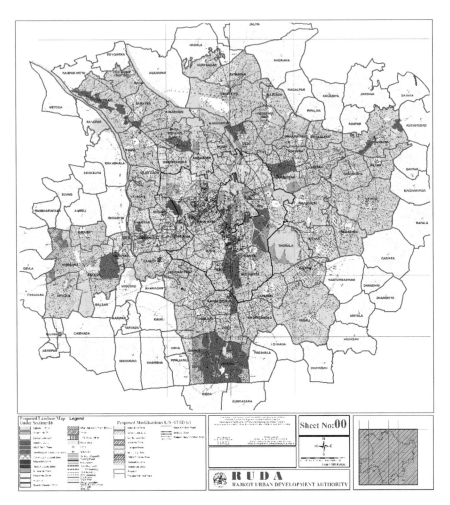

Fig. 18.3 Land use map of Rajkot as per Development Plan 2011. *Source* RUDA

cover maps', 'preparation of NDVI maps', 'retrieval of Land Surface Temperature', and 'correlation of Air temperature' in specific spots within study region.

USGS Earth Explorer website was used to download the cloud-free Landsat satellite data of 2009 and 2017 for the study area. The pre-processed data is projected to the Universal Transverse Mercator (UTM) projection system. The detail of satellite data collected are shown in Table 18.3.

Air temperature data was also used in this study which acquired from Indian Meteorological Department (IMD). Past 20-year temperature and humidity data of Rajkot city was used for this study. Census of India is considered as a reference and data collection of demographic details. For correlation of urban heat island in specific spots within city area was selected based on different characteristics of particular spot

Table 18.2 Land use of the Rajkot City

Land use	Land use as per 2001		Land use as per DP 2011	
	Area in hectare	Percent	Area in hectare	Percentage
Residential	4247	40.50	5502	52.47
Commercial	209	2.0	279	2.66
Industrial	628	5.99	738	7.04
Transportation	1400	13.35	1650	15.74
Public purpose	149	1.42	249	2.38
Recreational space	123	1.17	523	4.99
Agriculture	995	9.49	800	7.63
Water body	236	2.25	236	2.25
Vacant land	1510	14.40	–	–
Other	988	9.42	508	4.84
Total	10484	100%	10486	100%

Source RUDA

Table 18.3 Characteristics of landsat satellite data used in this study

Satellite	Acquisition date	Path and raw	Spatial resolution (m)	Cloud cover (%)
Landsat 5 TM	14 May 2009	150/045	30	1
Landsat 8 OLI	15 May 2017	150/044	30	1.78

and temperature was measured by normal thermometer at an interval of 24 h from January 2018 to April 2018.

18.2.1 Calculation of Spatial and Temporal Land Use Land Cover (LULC)

ERDAS Imagine is used for the supervised classification of LULC for 2009 and 2017 and change detection. This land use classification helped in delineating various temperature zones within study area. Respective classes were assigned to the feature signatures of similar characteristics. Each rectified images were independently classified to fit a common land typology. The classified resulting land cover maps were overlaid and compared on a pixel-by-pixel basis.

18.2.2   Calculation of Land Surface Temperature (LST) and Various Related Parameters

The land surface temperature (LST) was obtained from Landsat TM and Landsat OLI to study UHI. There are many algorithms such as mono window and single channel. Here, the mono-window algorithm proposed by (Qin and Karnieli 2001) was used for LST calculation. Here, two different sensors TM and OLI are used, so two different methods were used to compute the LST.

18.2.2.1   Conversion of Digital Numbers to Radiance

With the objective of conversion to the DN data of band 6 of Landsat 5 TM and band 10 of Landsat 8 into spectral radiance, the equations (one and 2) can be written in raster calculator of ArcGIS 10.3 as (Fig. 18.4).

For Landsat 5 TM

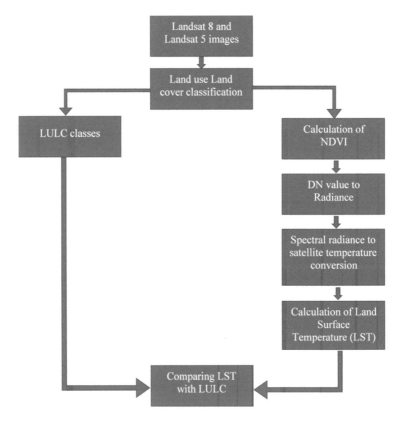

Fig. 18.4. Methodology

$$L_\lambda = \frac{(L_{MAX\lambda} - L_{MIN\lambda}) \times (DN - QCAL_{MIN})}{QCAL_{MAX} - QCAL_{MIN}} + L_{MIN\lambda}$$

where, L_λ is the spectral radiance received by sensor.
L_{MAX} is spectral radiance scaled to $QCAL_{MAX}$.
L_{MIN} is spectral radiance scaled to $QCAL_{MIN}$.
$QCAL_{MAX}$ is maximum quantized calibrated pixel value in DN = 255.
$QCAL_{MIN}$ is maximum quantized calibrated pixel value in DN = 1.

18.2.2.2 For Landsat 8 OLI

$$L_\lambda = M_L Q_{CAL} + A_L$$

where, L_λ is Top of Atmospheric spectral radiance,
M_L is band specific multiplicative rescaling factor,
A_L Band specific additivity rescaling factor.

18.2.2.3 Conversion of Radiance to Degree Kelvin

$$T_K = \frac{K_2}{\ln\left(\frac{K_1}{L_\lambda} + 1\right)}$$

where, T_K is temperature brightness, K_1 and K_2 are constants obtained from metadata, L_λ is spectral radiance at sensor.

18.2.2.4 Calculation of Land Surface Emissivity

Land surface emissivity calculation was done by using NDVI. Following Table 18.4 shows the values of emissivity with respect to NDVI range.

Table 18.4 Land surface emissivity

NDVI range	Emissivity value
NDVI < −0.185	0.995
−0.185 ≤ NDVI < 0.157	0.970
0.157 ≤ NDVI < 0.727	1.0094 + 0.047ln (NDVI)
NDVI > 0.727	0.990

18.2.3 Retrieval of Land Surface Parameters

18.2.3.1 Derivation of Normalized Difference Vegetation Index (NDVI)

NDVI is used to measure vegetation cover. NDVI ranges between -1 to $+1$. The classification of NDVI range such as a dense vegetation canopy (0.3 to 0.8), soil (0.1 to 0.2) reflects near infrared spectral somewhat larger than the red spectral, clear water low reflectance in the both spectral band. NDVI makes use of visible light and near infrared radiation to identify vegetation abundance (Kikon et al 2016). The NDVI expressed as in equation below:

$$NDVI = \frac{(NIR - R)}{(NIR + R)}$$

where, NIR is Band 5 (For Landsat 8) and Band 4 (For Landsat 5), R is Band 4 (For Landsat 8) and Band 3 (For Landsat 5). Below figures shows the NDVI for Rajkot.

18.2.4 Derivation of Land Surface Temperature (LST)

$$LST = \frac{T_K}{1 + \left(\lambda + \frac{T_K}{\rho}\right) \times \ln \varepsilon}$$

where, T_K is temperature brightness,
λ is the wavelength of the emitted radiance that is equal to 11.5 μm,
$\rho = h * c/\sigma$, σ is Stefan Boltzmann's constant which is equal to 5.67×10^{-8} Wm^{-2} K^{-4}, h is Plank's constant (6.626×10^{-34} J Sec), c is velocity of light (2.998×10^8 m/sec) and ε is the spectral emissivity.

18.3 Results and Discussion

The rapid urbanization in Rajkot is observed. A set of Landsat satellite data of 2009 and 2017 (Cloud free) have been prepared and used to retrieve LULC, NDVI and LST of the study area. The generated LULC, NDVI and LST has been shown below.

18.3.1 Retrieval of Land Use/Land Cover

Settlements include residential, commercial, industrial and mixed-use type of built form. Vegetation includes grassland, cultivated land, farms, parks and gardens. There

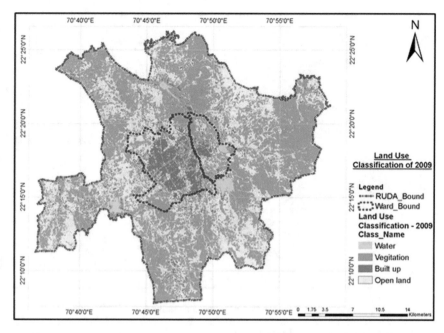

Fig. 18.5 LU/LC map of year 2009 through supervised classification of Rajkot, which comprises of RUDA area of 696 km². In addition, RMC area of 104 km². Which has higher land cover of vegetation (44.80%) and lower land cover of water body (1.50%)

is no significant settlements or any type of extensive natural or cultivated green cover in open land area (Figs. 18.5 and 18.6).

The urban area of Rajkot was categorized as urban built-up area, water body, vegetation and open land (Table 18.5).

18.3.2   Retrieval of NDVI

NDVI in year 2009 shows vegetation cover 44.80% with range between 0.634 and −0.238. Higher vegetation cover has been observed near northern part of Aji River (Figs. 18.7 and 18.8).

NDVI in year 2017 shows vegetation cover 51.15% with range between 0.604 and −0.077. Higher vegetation cover has been observed near northern part of Aji River.

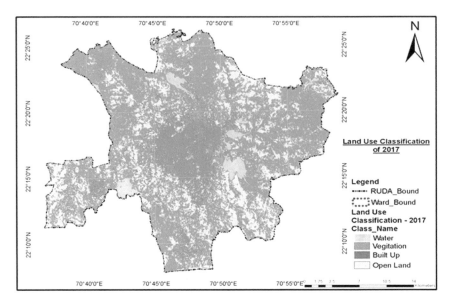

Fig. 18.6 LU/LC map of year 2017 through supervised classification of Rajkot, which comprises of RUDA area of 696 km². And RMC area of 104 km². Which has higher land cover of vegetation (51.15%) and lower land cover of water body (3.18%)

Table 18.5 Land use classification details

Class distribution	LC area in years (%)		LC change detection (%)
	2009	2017	2009–2017
Built-up	12.10	20.52	8.42
Open land	37.72	25.13	−12.59
Vegetation	44.80	51.15	6.35
Water bodies	1.50	3.18	1.68

18.3.3 Retrieval of LST

As per methodology, final LST was calculated by math algebra in ArcGIS 10.3. Final LST image is as shown in Figs. 18.9 and 18.10.

Below images of LST show the highest temperature of about 44.2° is existing at urban built-up and lowest temperature of about 27.6° is existing at vegetative area in 2009. Similarly, highest temperature of about 44.5° exists at urban built-up, and lowest temperature of about 28.3° is existing at vegetative area in 2017.

Built-up area consist of 12.10% majorly in municipal boundary which shows higher land surface temperature difference in comparison to other types of land cover. Thereby, heat island has been observed to be dense in the municipal boundary of study area.

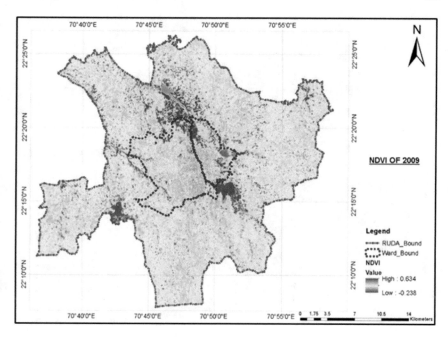

Fig. 18.7 NDVI of year 2009

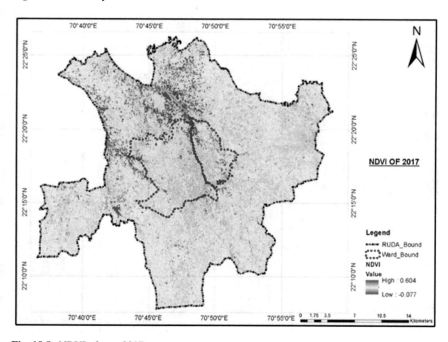

Fig. 18.8 NDVI of year 2017

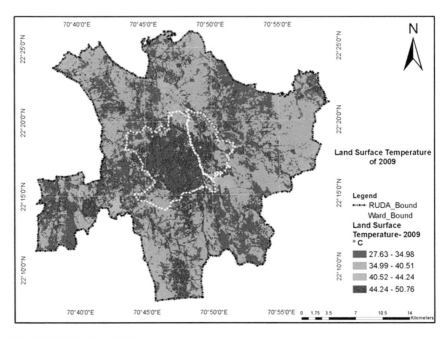

Fig. 18.9 LST of year 2009

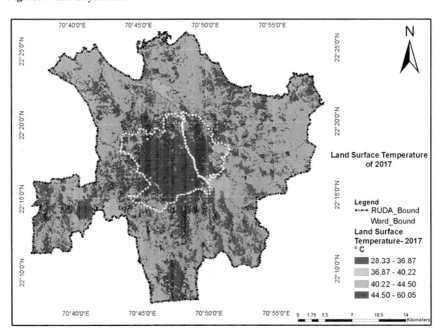

Fig. 18.10 LST of year 2017

Built-up area consist of 20.52% majorly in municipal boundary which shows higher land surface temperature difference in comparison to other types of land cover. Thereby, heat island has been observed to be concentrated in the municipal boundary and some region of RUDA boundary of study area.

18.3.4   Ambient Air Temperature of Rajkot

Rajkot is located on drought prone Saurashtra peninsula. The summers are very hot with sever day temperature ranging from 39 to 43 °C. The winters are idyllic having night temperatures 10 °C in January.

18.3.4.1   Process of Measurement

Indian Meteorological Department (IDM) measures the ambient air temperature at an interval of 3 hours. IDM uses the Stevenson Screen to measure temperature. Stevenson screen is used as a temperature-measuring instrument, which is called as 'A standard shelter for meteorological instruments'. It has wet and dry bulb thermometers that are used to record humidity and air temperature. For avoiding extreme temperature gradients at ground level, it should kept 1.2 m above the ground by legs. The louvered sides are mounted to encourage the free passage of air. The colour of this instrument is always white so that it can reflect heat radiation; it measures the temperature of the air in the shade, not of the sunshine (Figs. 18.11 and 18.12).

Fig. 18.11   Stevenson screen

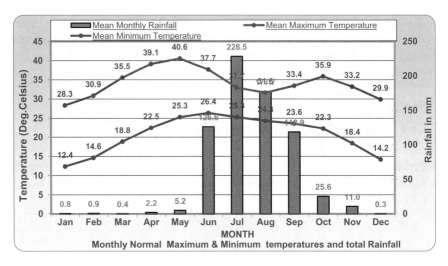

Fig. 18.12 Temperature and rainfall monthly prognosis

18.3.4.2 Temperature Prognosis

IDM measures the temperature using Stevenson screen. Precipitation is also measured through rain gauge. As per Fig. 18.13, average temperature of Rajkot city is increasing at a rate of 0.3 °C. Last 20 years data said that normal temperature of Rajkot was 34.3 °C in 1995 that is risen by 34.6 °C in 2017. Which means normal temperature of Rajkot may reached at 35 °C by 2030.

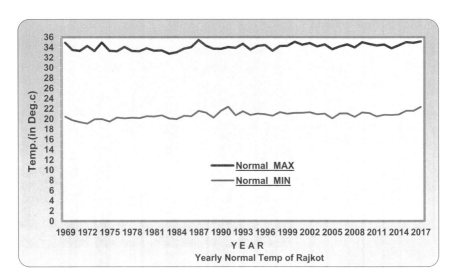

Fig. 18.13 Yearly normal temperature

Following temperature and rainfall data were recorded from the year 1969 to 2017, showing temperature and rainfall variations month wise. The highest temperature and rainfall were observed in the month of May and July, respectively, and lowest temperature in August. The maximum monthly temperature variation of 9 °C and monthly rainfall variation of 200mm have been observed.

Yearly normal temperature from 1969 to 2017 is shown in above chart, which shows that there is a slight variation in temperature. Maximum highest temperature is observed in the year of 1988 and 2014 to 2017. Maximum normal temperature varies in the range from 32 to 36 °C. Normal maximum temperature trend shows that temperature is increased at a rate of 0.3 °C in last two decades. Minimum lowest temperature has been observed in the year of 1972 and 2005 having temperature of 19 °C. Minimum temperature is increasing in the last five years and reached up to 23 °C.

18.3.5 Isotherm of Rajkot

Figure 18.14 shows the area-specific heat map of ambient air temperature for RMC area. Temperature is measured by air monitoring system installed by RMC at particular location. This map shows the average temperature from January to April 2018.

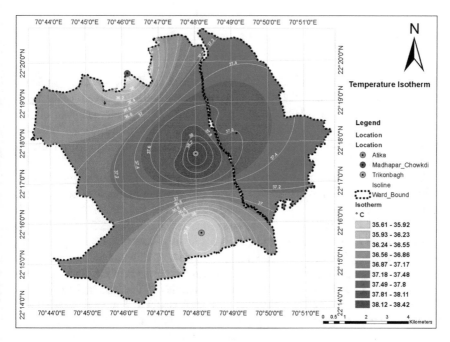

Fig. 18.14 Temperature isotherm map

Table 18.6 Temperature difference

Location	Ambient temperature (°C)	Surface temperature (°C)
Madhapar chowk	35.7	44.35
Trikon Baugh	38.2	45.85
Atika industrial area	35.6	42.85

It has been observed that there is minor temperature variation between above-mentioned locations. Temperature difference between Madhapar chowk to Trikon Baugh is 2.68 °C, 0.13 °C has been observed between Madhapar chowk and Atika industrial area, 2.81 °C has been observed between Trikon Baugh and Atika industrial area. Maximum temperature has been observed at Trikon Baugh because it is highly congested area and centralized bus stand is situated within CBD. In addition, some other factors considered such as air quality index, carbon emission, etc. isotherms are indicated in above heat map (Table 18.6).

Within the stretch of P2 to P1 of 5 km, there is a significant change in ambient air temperature, which is approximately 2.5 °C, and land surface temperature difference is approximately 1.5 °C. On the stretch of P2 to P3 of 3.6 km, there is a significant change in ambient air temperature which is approximately 2.68 °C and land surface temperature difference of approximately 3 °C has been observed.

18.4 Conclusion

The analysis has found that the UHI effect is next to global warming, and it is need to consider seriously. Climate change poses the serious problem to the urban development. Urban heat island is the more documented phenomenon of climate change causing energy and environmental problems of the city. The main intension of this research work is to make Rajkot urban heat resilient through spatiotemporal analysis and ambient air temperature isotherms.

Spatiotemporal analysis was conducted for detecting UHI. Mono-window algorithm was used for retrieving LST of Rajkot and surrounding area using Landsat 5 and Landsat 8. LULC change is evaluated over the last decade. Also, the trend in LST with four different LULC classes was examined. From the LST estimation, it is concluded that distribution of UHI in Rajkot is mainly located in central part of city due to compact urban activities like high population, buildings, reduced vegetation, pavement roads and anthropogenic heat. There is inverse relation found between LST and NDVI that if there is less vegetation then the LST is high and vice versa.

Ambient air temperature prognosis was carried out for winter, summer and monsoon season to observe temperature trend for study area. Ambient air temperature has risen by rate of 0.3 °C. Comparative analysis was conducted for surface temperature and ambient air temperature. There is a difference of 2.5 to 3 °C between surface and air temperature. Air pollutant can also be considered in the analysis to see the effect of UHI and air quality.

Subsequently, temperature isotherm was carried out from ambient air temperature measured by the IMD. Isotherm indicates the location where ambient air temperature is high; maximum 38.5 °C average temperature has been discerned at Trikon Baugh. Trikon Baugh is the CBD area and central bus station is nearer to Trikon Baugh and is a highly congested area in Rajkot so reason for high temperature at Trikon Baugh. Consequently, this study has shown the variations in UHI components, which may affect human health, flora–fauna and environment in total.

References

Ahmed S (2018) The Egyptian Journal of Remote Sensing and Space Sciences Assessment of urban heat islands and impact of climate change on socioeconomic over Suez Governorate using remote sensing and GIS techniques. Egypt J Remote Sens Space Sci 21(1):15–25. https://doi.org/10.1016/j.ejrs.2017.08.001

Da Silva J, Moench M (2014) City resilience framework. Arup, 2014 (November), http://www.sea changecop.org/files/documents/URF_Bo. Retrieved from http://www.seachangecop.org/files/documents/URF_Booklet_Final_for_Bellagio.pdf%5Cn http://www.rockefellerfoundation.org/uploads/files/0bb537c0-d872-467f-9470-b20f57c32488.pdf%5Cnhttp://resilient-cities.iclei.org/fileadmin/sites/resilient-cities/files/Image

EPA (2008) Compendium of strategies urban heat island basics. Reducing Urban Heat Islands, 1–22. Retrieved from http://www.epa.gov/heatisland/resources/pdf/BasicsCompendium.pdf%5Cn http://www.epa.gov/heatisland/about/index.htm%5Cnpapers2://publication/uuid/E82 A9E0C-E51A-400D-A7EE-877DF661C830

Frantzeskaki N (2016) Urban Resilience a concept for co-creating cities of the future 1, 1–19. Retrieved from http://urbact.eu/sites/default/files/resilient_europe_baseline_study.pdf

Joshi R, Raval H, Pathak M, Prajapati S, Patel A, Singh V, Kalubarme MH (2015) Urban heat island characterization and isotherm mapping using geo-informatics technology in Ahmedabad City, Gujarat State, India. Int J Geosci 6(3):274–285. https://doi.org/10.4236/ijg.2015.63021

Kanga S, Tripathi G, Singh SK (2017) Forest fire hazards vulnerability and risk assessment in Bhajji forest range of Himachal Pradesh (India): a geospatial approach. J Remote Sens GIS 8(1):1–16

Kikon N, Singh P, Singh SK, Vyas A (2016) Assessment of urban heat islands (UHI) of Noida City, India using multi-temporal satellite data. Sustain Cities Soc 22:19–28. https://doi.org/10.1016/j.scs.2016.01.005

Kim Y-H, Baik J-J (2005) Spatial and temporal structure of the urban heat island in Seoul. J Appl Meteorol 44(5):591–605. https://doi.org/10.1175/JAM2226.1

Kotharkar R, Surawar M (2016) Land Use, land cover, and population density impact on the formation of canopy urban heat islands through traverse survey in the Nagpur Urban Area, India. J Urban Plan Dev 142(1):4015003. https://doi.org/10.1061/(ASCE)UP.1943-5444.0000277

Kumar K (2007) Minimizing urban heat island effect and imperviousness factor in Bangalore. In: Minimizing urban heat island effect and imperviousness factor in Bangalore, 38. Retrieved from http://www.teriin.org/projects/apn_capable/pdf/bang_present/Kiran_Kumar.pdf

Kumar KS, Bhaskar PU, Padmakumari K (2012) Estimation of land surface temperature to study urban heat island effect using landsat Etm+ image. Int J Eng Sci Technol 4(2):771–778

Kumar S, Panwar M (2017) Urban heat island footprint mapping of Delhi using remote sensing. Int J Emerg Technol 8(1):80–83

Kumari M, Joshi N (2015) Spatiotemporal analysis of urban growth, sprawl and structure of Rajkot, Vadodara and Surat (Gujarat-India) based on geographic information systems, in relation to the sustainability pentagon analysis. Indian J Sci Technol 8(October):1–6

Parikh J, Magotra R (2014) Towards resilient cities

Qin Z, Karnieli A (2001) A mono-window algorithm for retrieving land surface temperature from landsat TM data and its application to the Israel-Egypt border 22(18):3719–3746

Report S (2001) Climate change 2001: Synthesis report synthesis report. Synthesis 24–29

Shahmohamadi P, Che-Ani AI, Abdullah NAG, Maulud KNA, Tahir MM, Mohd-Nor MFI (2010) The conceptual framework on formation of urban heat island in Tehran metropolitan, Iran: A focus on urbanization factor. In: International conference on electric power systems, high voltages, electric machines, international conference on remote sensing—proceedings, (Emmanuel), pp 251–259. Retrieved from http://www.scopus.com/inward/record.url?eid=2-s2.0-79958741448&partnerID=40&md5=5e9f3a78131c0b181e508c05f1777158

Surawar M, Kotharkar R (2017) Assessment of urban heat island through remote sensing in Nagpur urban area using landsat 7 ETM + Satellite Images. Int J Urban Civ Eng 11(7):868–874

Taylor P, Kim HH (2007) International journal of remote sensing urban heat island. Int J Remote Sens (November):37–41

Tripathi G, Pandey AC, Parida BR, Shakya A (2020) Comparative flood inundation mapping utilizing multi-temporal optical and sar satellite data over North Bihar region: a case study of 2019 flooding event over North Bihar. In: Spatial Information science for natural resource management. IGI Global, pp 149–168

UNFPA (2011) The state of world population 2011. In: United Nations Population Fund, pp 1–132. ISBN 978–0–89714–990–7

Chapter 19
Multispectral Remote Sensing for Urban Planning and Development

Anubhav Bhartiya, Deepak Kumar, and Praveen Kumar Rai

Abstract As in the time of economic development, various rural towns are developing into urban towns, and hence, for a balanced and a proper development, a planning is required as it is a process through which proper development can be executed accordingly to the requirements. There are various criterions which are to be followed to create a proper development plan. Remote sensing is the acquiring of the data about the item without contacting it or without physically being present there. Due to advanced technology and new innovations, satellite imaging has enabled to collect and interpret various data which earlier was done physically and consumed a lot of time. A surface analysis is conducted with the help of remote sensing which gives a lot information regarding various aspects, whereas it also interprets the physical data with other socioeconomic data. This interpretation helps in getting a link to the planning process. The information collected through satellites helps planning in various formats such as time, efficiency and other ways. Therefore, it enables to a lot of things for a better planning.

Keywords Optical/Multispectral Remote Sensing · Images · Planning · Data

19.1 Introduction

Nowadays, more settlements are developing into a bigger and better state due to a development in the economy, A plan to execute a proper development is much required (Jain et al. 2016; Leiva-Murillo et al. 2013). To create a plan, critical information is required to be assessed which may include the components of the settlements or the environment in the surrounding (Hazarika et al. 2015; Khandelwal et al. 2017). The availability of the land data is very important as it may act as the base

A. Bhartiya · D. Kumar (✉)
Amity Institute of Geo-Informatics and Remote Sensing (AIGIRS), Amity University Uttar Pradesh, Sector 125, Gautam Buddha Nagar, Noida, Uttar Pradesh 201313, India

P. K. Rai
Department of Geography, Khwaja Moinuddin Chishti (K.M.C.) Language University, Lucknow, Uttar Pradesh 226013, India

© The Author(s), under exclusive license to Springer Nature Singapore Pte Ltd. 2022   371
P. K. Rai et al. (eds.), *Geospatial Technology for Landscape and Environmental Management*, Advances in Geographical and Environmental Sciences,
https://doi.org/10.1007/978-981-16-7373-3_19

for planning (Rai and Kumra 2011; Mishra and Rai 2014; Ghorbanian et al. 2020; Loukanov et al. 2020); it would act as a main guideline for the plan maker. As this data is very critical, it is to be updated in the database frequently so that the data bears the updates to show the latest changes (Galetsi et al. 2019; Mohammed 2013; Weng 2011). When this data is collected using conventional methods, it costs a ton of time, effort and money to fulfil the needs of the increasing demand. For example, underground structure, surface layout, size and shape of the settlements, population trend and predictions, things such as sewerage line, traffic layout, road network, water and electricity supply and various other things. Remote sensing can ease out the work and can provide a lot of critical data which is significant for planning (Shrivastava and Rai 2015; Liu et al. 2016; Loukanov et al. 2020; Shahmohamadi et al. 2011). As remote sensing is a part of modern technology, which provides images/data of the area in a much better perspective with the help of using various equipments (Bokaie et al. 2016; Hu et al. 2011). Hence, the demand of remote sensing for planning is increasing as the efficiency of collecting the data is increased, whereas the accuracy of the data is also increased and is much better. Remote sensing is a very valuable research tool for application purpose; when the survey encompass a very large area, output for the same takes a lot of time when done by the use of conventional methods, the results are available much faster using remote sensing method (Jeansoulin 2019; Kong and Nakagoshi 2006; Panda et al. 2016). Despite the tough terrains or unreachable locations, with the use of remote sensing, data of the designated area may be collected easily and efficiently. The benchmarks of collecting data can be very use-specific and can be collected in different types such as land use-specific or settlement-specific which makes it easy to analyse the data for further usage (Babí Almenar et al. 2021; Mishra et al. 2006). With the assistance of remote sensing, the data could be collected at short intervals easily, and hence, the database would always be updated and the time to create a plan would greatly decrease and be more efficient. For instance, if Border Road Organisation[1] (BRO) wants to build a road in a hilly terrain which was till now inaccessible, they would have two methods to collect the data for further analysis and to create a road plan, conventional method or the use of remote sensing (satellite imaging). Conventional method would be a tedious task: consumes lot of time and workforce to collect the data. It would also be risky to collect data in an inaccessible terrain, whereas with the use of remote sensing, data can be collected in a very less amount of time in comparison to conventional methods, whereas the accuracy and the extent of data captured would differ and would be better in comparison (Jain et al. 2016; Liu et al. 2015; Mishra et al. 2006). It is to know how usage of remote sensing in planning can be beneficial in various aspects and how it can change or increase the efficiency of the work done.

---

[1] An organisation which builds roads In India's strategic important places, as well as few other countries such as Bhutan.

19.1.1 How a Planning is to Be Done

What is a Plan? A plan is a detailed proposal, It is made by looking up to some kind of data/information available to the plan maker; a plan is utmost one of the most important procedure for various activities. It acts as an blueprint/guideline for the work. Planning is done to set a work timeline and to set a work frame procedure how the work would be initiated and what ways are to be followed. A plan basically answers all the W&H questions which may occur at the time of work. A plan is made by analysing all sort of data which is available, or which can be made available, data is the key for most things to create a plan. While executing a task, there may be chances of facing difficulties or barriers to completion; hence, to overcome those barriers, a plan is made using various methods and technology. To create a plan, there are a few basic requirements which one may ask for that includes the latest data available to know the latest trends and then create a plan according to those trends, keeping in mind all the factors that may be effected.

19.1.2 Types of Ways to Collect Data

There are mainly two methods of collecting land use data, conventional method or remote sensing method. Both the methods have differences amongst them (Bauer et al. 2008; Narayanan 2013). The conventional method is the basic method which does not require much of the technology and works in the olden ways surveyors physically go and collect the data required. This method is very time-consuming and is not much efficient to survey huge areas (Babí Almenar et al. 2021; Lu et al. 2014; Mishra et al. 2006). There are various limitations to this method of collecting data as humans physically go and conduct the survey of the area; therefore, conducting the survey of few areas is very difficult such as inaccessible terrains or small cluster (Li et al. 2019; Turkar et al. 2012; Zavadskas and Antucheviciene 2007). The accuracy while collecting the data is low, and it may have an overall impact while creating a plan. Remote sensing is the use of technology to capture the data without visiting the place physically (Jain et al. 2016; Zhang et al. 2012). There are various ways of collecting the data, varying from ground-based remote sensing—aerial remote sensing to spaceborne remote sensing. All these methods are selected depending upon the data (Jain et al. 2016; Wang 2015). For small area such as a farm ground-based remote sensing fulfils the requirement, whereas spaceborne remote sensing is capturing the earth image through different satellites which covers a huge area at a time and have repeat the path at fixed intervals. Remote sensing is the outcome of technical innovations to capture image through various methods such as using electro-magnetic radiation, laser sensors or through infrared rays.[2] All these methods collect data of various kinds and are further processed depending upon the required usage (Bassuk et al. 2015; Shahabfar et al. 2012; Völker and Kistemann 2015). With

---

[2] Bhatta, B. *Remote Sensing and GIS*. Oxford University Press, 2011.

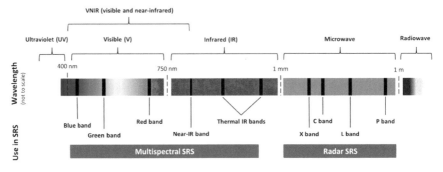

Fig. 19.1 Wavelength Bands used in Remote Sensing[3]

the assistance of remote sensing, a large area may also be covered in a shorter time, and the data is capable of interpreting various information such as land use or what type of land is there, soil type, extent of settlement. Collecting data with the help of remote sensing takes less time due to which the data can be collected frequently by which the changes can be seen, and the data would always be updated. A latest set of data is beneficial for all sort of planning as with the help of recent data as well as repetitive data enables various aspects of analysis, such as the differences between times can be traced different patterns of expansion and development can be traced according to which a plan can be made with suitable measures.

19.1.3 Optical and Multi-Spectral Remote Sensing

"Remote Sensing is the science of acquiring information about earth's surface without actually being in contact with it. This is done by sensing and recording reflected or emitted energy and processing, analysing, and Applying that information."- Definition by Canada Centre for Remote Sensing (CCRS).

Remote sensing is a general seven-step process starting from the emittance of the energy and ending at the application use data. There are primarily two types of remote sensing, active and passive remote sensing. Both have a varied outcome and have different benefits; after capturing the reflected energy, the data is stored and stored in a database management system (DBMS) which is further processed as per the use requirement after which the analysis is done, and hence, the data is ready for application use (Fig. 19.1).

The sensors mostly work by capturing the electromagnetic rays mostly with the R.G.B. band to microwave band ranging from 0.4 μm to 1.0 m. There are several factors which may affect the image quality of the data captured such as scattering, absorption, cloud cover and more. Even through which mostly the data accuracy is better, and the data can be used for analysis. The main disadvantage of remote

---

[3] https://dai-global-digital.com/visualizing-remotely-sensed-data-true-color-and-false-color.html.

sensing is the cost of launching the image capturing sensor, The cost may vary upon the platform it is sent on, but the higher the cost, it would be more beneficial in the future as it would be a onetime expense for a long period and would give much higher returns.

19.1.4   Advantages of Remote Sensing

There are various benefits of using remote sensing for collecting data. As a spatial image would have a large spatial coverage area in comparison to the conventional method as the satellite would cover a larger area in a single piece of data. The repetivity factor of a satellite is very useful as the data would be updated at a given interval through which a general change can be figured out, and a plan can be shaped according to the changes. The data which is captured through the remote sensing method can be analysed for various purposes, such as a single piece of data may interpret the terrain the settlement expanse, land use feature, soil types and other useful information. The level of accuracy is very high in spatial imaging as Maxar is the first company to provide a spatial resolution of 30cms.[4] The biggest help what remote sensing provides is collection of data without being physically present at the location that makes the collection of data of a harsh/inaccessible terrain much more easier as the risk and time consumption of reaching the place by a surveyor is erased out. In terms of cost launching, a sensor is a onetime investment following which the data can be further collected for a much longer period and the collected data can be used for various purposes.

19.1.5   Optical and Multi-Spectral Remote Sensing Data
in Planning

A plan can be required for various purposes like creating a policy framework, mapping out the transportation network, sanitary management, water supply management, industrial development, revenue generation, policing and surveillance and various other things. (Kulkarni, 2020). Remote sensing offers tools and data to ease out the planning to create plans for such things. For instance, to map out a transport network using the spatial image and mapping out a route is much more convenient, easy and interactive. The network can be mapped out on the data available, and, hence the demand can be further marked with all this the route creation, and modification becomes much easier in comparison to the data available by conventional method. To layout or to see a water pipeline a underground pipeline map is to be drawn, taking the measurements physically and seeing the soil type of the land would cost a lot of time and human capital as long maps are to be marked and taking physical measurements

---

[4] https://www.digitalglobe.com/company/about-us.

may even affect the accuracy, whereas with the assistance of remote sensing, the data of the particular area can be traced out; with the help of the repetivity, latest updates can be tracked. In case of an emergency or for a plan of an upgrade of the pipeline, the status of the pipeline can be traced about the construction nearby the line. With the assistance of remote sensing data, a road planning can be done by planning a proper route by seeing the traffic patterns at intervals and then constructions or alternate routes can be planned. Planning of an emergency is laid out with the help of remote sensing in the form of Global Positioning System (G.P.S.); it uses the satellite images in forms of maps and gives real-time traffic updates.

19.1.6   Significance of Remote Sensing in Planning

As we see nowadays, the demand for creating a plan to execute several works is increasing, as the development for various urban and rural settlements is increasing. A demand for a plan is very high as it is used in various things; therefore, to create a plan, a set of data is required by which the planning can be done. Therefore, to collect the data usage of remote sensing is at a Boom, as there are several advancements in the technology of remote sensing which makes it much better in acquiring the data. There are several projects such as mining which acquire their critical data from the use of remote sensing whether it would be through a use of a drone or a satellite. The image/data received from the sensing is of various spectral resolutions and each data can be customised to identify the features of a data, such as land use or water body, it may even be used to figure out the soil quality. With the help of remote sensing, the government bodies even identify the areas prone to forest fire. They find the before and after effects of a disaster, through which they plan out upon what can be done to prevent or to be prepared. Such as, in a case of a forest fire, the authority can collect the data, and with the help of the data, they will find the hotspot and try to extinguish the fire at that place; they would even see the widespread of the fire and plan out the evacuation procedure. The most advanced thing about remote sensing is that it provides an in-depth knowledge of the area we want to know about without physically being present there, and with the help of various software, a much detailed information could be interpreted from the data. Hence, in the forthcoming times, the usage of remote sensing is increasing and will be crucial; even nowadays, the usage of remote sensing is very important.

19.1.7   Optical and Multi-Spectral Remote Sensing

Figure 19.2 visualises a remote sensing data showing a town area in Varanasi and its surroundings. With the help of this data, we can understand and interpret a lot of information for various knowledge aspects. For instance, we can see a main road which can be an interstate highway, or a main road near which there are few settlements

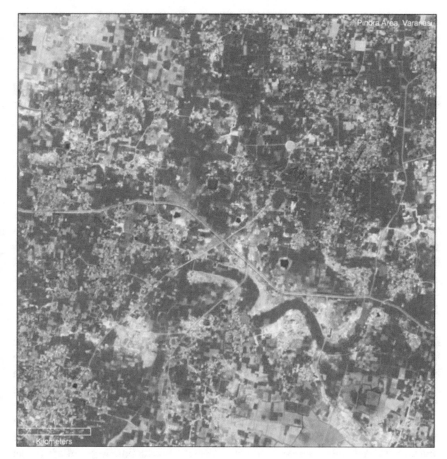

Fig. 19.2 Spatial Data of Pindra Area, Varanasi[5]

and agricultural lands. This particular data is a false colour data which collected data using infra-red rays and decoded it in RGB spectrum; hence, we can understand a lot about the ground. In this data, we can even see a paleo channel, several farmlands and division of areas for various uses. We even get to see about the road network, how well the farms and houses are accessible to a road. With the help of remote sensing data we have, we can delineate a lot of information and a plan can be created for any activity, such as if government wants to create a water supply channel or a reservoir, so they can analyse the data and look through it for the availability of water bodies in and around, by which they can carry out the planning for location and area to be covered, The farms or the land which have access to water or which do not have access can be segregated. The terrain of the area can also be assessed with this data as we can see this is a plain area, and there was a flow of river in the olden times

---

[5] Downloaded Data of Landsat from Earth Explorer.

which can be interpreted through the Paleo channel, as the path and the identification of paleo channel would be difficult through the ground level or through conventional methods of data collection. Therefore, use of remote sensing in planning is very helpful, and it eases out the work and reduces the time, the collection of data using remote sensing can be use-specific to carry out a particular task, whereas a general survey can be done to collect data which can be further processed for a particular task.

19.1.8 Case Studies

The purpose of using remote sensing is to provide an added dimension to data analysis as it acts as a step up to visualising the compound patterns and relationships that simulates real-world problem of policy and planning. There are various cities which used remote sensing for planning and developed the cities to smarter cities.

19.1.8.1 Bhopal

Bhopal city of Madhya Pradesh, India, (A) is located between 77°19′E and 77°31′E longitude and 23°09′N and 23°21′N latitude. The boundary of Bhopal is considered by the area which comes under the Bhopal Municipal Corporation (BMC) (B). To study the area and collect the data a small test site was created for ward 30 of BMC (C). Hence, the data was collected of the test site by Cartosat-1 (Fig. 19.3).

As that Satellite carried best of the time Panchromatic (PAN) cameras. GeoTIFF format files were acquired and referenced to WGS84 datum and elipsoid. The total roof and available area was captured and collected for planning. This collected data would further help in estimating the all available smart solutions for infrastructural problems and location plannings of a city. The data which was collected for BMC went under an accuracy assessment. This assessment was conducted by comparing the total roof area by acquired by automatic extracted data and manually digitized data. The automated data had positive result and identified 89.57% of the manually mapped roof area.[6]

19.1.8.2 Indore

Indore a city in the western part of Malwa Plateau in the state of Madhya Pradesh located at 76′ 42′ E longitude, 22′ 43′ N latitude used GIS and remote sensing for creating a base map to develop the city to make it a smart city. For making the base

---

[6] Saha, Kakoli. "A Remote Sensing Approach to Smart City Development in India." *Proceedings of the Special Collection on EGovernment Innovations in India - ICEGOV '17*, 2017, https://doi.org/10.1145/3055219.3055232.

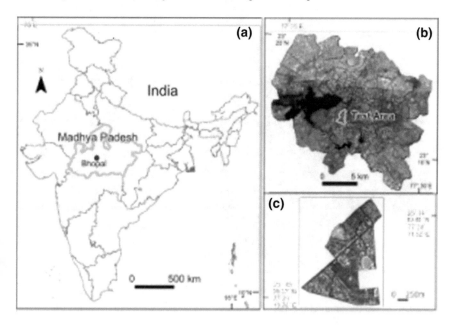

Fig. 19.3 Location of the Bhopal area

map of ABD Area Worldview-2 Satellite data was used with a spatial resolution of 0.5 m. This was used mainly to draw out baseline data collected and other information related to natural resources. Base Map was created for ABD Area (742 Acres), where there were existing features such as roads and buffers spaces in between roads and boundaries. The data collected was kept with Indore Smart City Development Limited (ISCDL) and for the geo-rectification of this data, 21 corresponding ground control points were marked; this consisted of street intersections. This was done to digitally align the satellite data with the real features for better accuracy. A digital elevation model (DEM) was created to look for terrain surface using ArcGIS. With the data, land use categories were divided and further facilities, utilities waterbodies and transportation networks were mapped out. This led to better utilisation of resources and the occurrence of problems were known. By analysing this data, the development authority came up with parking spaces, designated parking lots and structures and located different buildings which area has what height of buildings. This all led to infrastructure assessment gap where sewer lines are located and where all the water supply pipelines are there. Therefore, the ISCDL had used remote sensing to find out all the problems and a solution for better development and turning their city into smart cities they took help of GIS and Remote Sensing data (Fig. 19.4).[7]

---

[7] Indore Smart City: ABD master Plan.

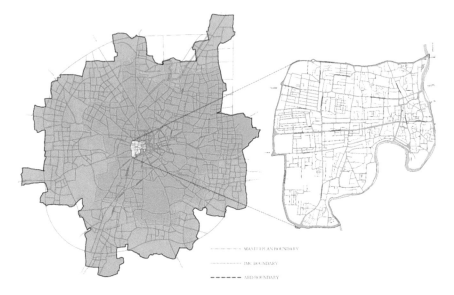

Fig. 19.4 Indore City and ABD Area

19.2   Conclusion

Remote sensing is the outcome of the advancement in technology and new innovations as well as it enables the collection of land data much easily and in a much
better manner with much better outcome. It is a very effective tool for mapping as
it is multispectral gives vivid knowledge about the land use surface as well as the
repetivity which enables a broader knowledge of the area and the work happening
there. Remote sensing gives a much better information, has better accuracy and is
more efficient in terms of time and capital. The reachability to capture the data is
high as the satellite-based sensor can capture the image of each place and would
capture the data. The process of mapping has become much easier, whereas the data
captures can be moulded as per the use requirement to show the things required.
The multispectral band enables to interpret a lot of knowledge about the land and
enables to plan by keeping various factors in mind. Applications of remote sensing
have gained popularity in the recent past and have an increasing demand in the near
future as it is a very important source of data which provides very critical data at
various levels and different perspectives. Remote sensing provides data in various
types such as the spatial resolution may differ according to the needs different scales
of data for different levels of planning which may help in the details of the planning
and would have a greater impact. Remote sensing is much better in comparison to
the conventional methods and is much better and helpful in terms of planning, the
captured data/information by the sensors are very critical for planning; hence, the
use of remote sensing is increasing and is giving a better outcome for better planning
and increasing the efficiency and accuracy of the planning.

References

Babí Almenar J, Elliot T, Rugani B, Philippe B, Navarrete Gutierrez T, Sonnemann G, Geneletti D (2021) Nexus between nature-based solutions, ecosystem services and urban challenges. Land Use Policy 100:104898 (April 2019). https://doi.org/10.1016/j.landusepol.2020.104898

Bassuk NL, Universite AB, Jean M, Universite C, Theoretical L, Politics U, Bibliography AA (2015) On using landscape metrics for landscape similarity search. Landsc Urban Plan 117(1):1–12. https://doi.org/10.1038/srep11160

Bauer ME, Loffelholz B, Wilson B (2008) Estimating and mapping impervious surface area by regression analysis of Landsat imagery. Remote Sens Impervious Surf 612–625. https://doi.org/10.1201/9781420043754.pt1

Bokaie M, Zarkesh MK, Arasteh PD, Hosseini A (2016) Assessment of Urban Heat Island based on the relationship between land surface temperature and land use/land cover in Tehran. Sustain Cities Soc 23:94–104. https://doi.org/10.1016/j.scs.2016.03.009

Galetsi P, Katsaliaki K, Kumar S (2019) Values, challenges and future directions of big data analytics in healthcare: a systematic review. Soc Sci Med, 241:112533. https://doi.org/10.1016/j.socscimed.2019.112533

Ghorbanian A, Kakooei M, Amani M, Mahdavi S, Mohammadzadeh A, Hasanlou M (2020) Improved land cover map of Iran using Sentinel imagery within Google Earth Engine and a novel automatic workflow for land cover classification using migrated training samples. ISPRS J Photogramm Remote Sens 167(July):276–288. https://doi.org/10.1016/j.isprsjprs.2020.07.013

Hazarika N, Das AK, Borah SB (2015) Assessing land-use changes driven by river dynamics in chronically flood affected Upper Brahmaputra plains, India, using RS-GIS techniques. Egypt J Remote Sens Space Sci 18(1):107–118. https://doi.org/10.1016/j.ejrs.2015.02.001

Hu W, Li M, Liu Y, Huang Q, Mao K (2011) A new method of restoring ETM + SLC-off images based on multi-temporal images. IEEE (41001046), pp 0–3

Jain M, Dawa D, Mehta R, Pandit APDMK (2016) Monitoring land use change and its drivers in Delhi, India using multi-temporal satellite data. Model Earth Syst Environ 2(1):1–14. https://doi.org/10.1007/s40808-016-0075-0

Jeansoulin R (2019) Multi-source geo-information fusion in transition: a summer 2019 snapshot. ISPRS Int J Geo Inf 8(8):330. https://doi.org/10.3390/ijgi8080330

Khandelwal S, Goyal R, Kaul N, Mathew A, Li ZL, Tang BH, Kaloop MR (2017) Detecting urban growth using remote sensing and GIS techniques in Al Gharbiya governorate Egypt. Egypt J Remote Sens Space Sci 20(1):571–575. https://doi.org/10.1016/j.ijsbe.2015.02.005

Kong F, Nakagoshi N (2006) Spatial-temporal gradient analysis of urban green spaces in Jinan China. Landscape and Urban Planning 78(3):147–164. https://doi.org/10.1016/j.landurbplan.2005.07.006

Leiva-Murillo JM, Gomez-Chova L, Camps-Valls G (2013) Multitask remote sensing data classification. IEEE Trans Geosci Remote Sens 51(1):151–161. https://doi.org/10.1109/TGRS.2012.2200043

Li L, Yang J, Wu J (2019) A method of watershed delineation for flat terrain using sentinel-2A imagery and DEM: a case study of the Taihu basin. ISPRS Int J Geo-Inform 8(12). https://doi.org/10.3390/ijgi8120528

Liu J, Liu Q, Yang H (2016) Assessing water scarcity by simultaneously considering environmental flow requirements, water quantity, and water quality. Ecol Ind 60:434–441. https://doi.org/10.1016/j.ecolind.2015.07.019

Liu X, Bo Y, Zhang J, He Y (2015) Classification of C3 and C4 vegetation types using MODIS and ETM+ blended high spatio-temporal resolution data. Remote Sens 15244–15268. https://doi.org/10.3390/rs71115244

Loukanov A, El Allaoui N, Omor A, Elmadani FZ, Bouayad K, Seiichiro N, He HS (2020) Effects of neighborhood building density, height, greenspace, and cleanliness on indoor environment and health of building occupants. Environ Res 106(February):213–222. https://doi.org/10.1016/j.buildenv.2018.06.028

Lu D, Li G, Kuang W, Moran E (2014) Methods to extract impervious surface areas from satellite images. Int J Digit Earth 7(2):93–112. https://doi.org/10.1080/17538947.2013.866173

Mishra M, Mishra KK, Subudhi AP, Ravenshaw MP (2006) Urban sprawl mapping and land use change analysis using remote sensing and GIS. 13

Mohammed FG (2013) Satellite image gap filling technique. Int J Advancements Res Technol 2:348–351

Mishra VN, Rai PK, Mohan K (2014) Prediction of land use changes based on land change modeler (LCM) using remote sensing: a case study of Muzaffarpur (Bihar), India. J Geogr Inst, Jovan Cvijić" SASA (Serbia) 64(1):111–127. https://doi.org/10.2298/IJGI1401111M

Narayanan P (2013) Analysing the Urban sprawl through entropy of Gulbarga city and its spatial promoters of growth through Geoinformatics. Cartosat Imagery of Gulbarga IRS 1D Pan Imagery of Gulbarga 1998 Extract Built up through ENVI Extract Prepare sector grids for covering. XXXIII

Panda S, Chakraborty M, Misra SK (2016) Assessment of social sustainable development in urban India by a composite index. Int J Sustain Built Environ 5(2):435–450. https://doi.org/10.1016/j.ijsbe.2016.08.001

Rai PK, Kumra VK (2011) Role of geoinformatics in urban planning. J Sci Res 55:11–24

Shahabfar A, Ghulam A, Eitzinger J (2012) Drought monitoring in Iran using the perpendicular drought indices. Int J Appl Earth Obs Geoinf 18(1):119–127. https://doi.org/10.1016/j.jag.2012.01.011

Shahmohamadi P, Che-Ani AI, Etessam I, Maulud KNA, Tawil NM (2011) Healthy environment: the need to mitigate urban heat island effects on human health. Procedia Engi 20:61–70. https://doi.org/10.1016/j.proeng.2011.11.139

Shrivastava N, Rai PK (2015) An object based building extraction method and classification using high resolution remote sensing data. Forum Geogr J 14(1):14–21. https://doi.org/10.5775/fg.2067-4635.2015.045.i

Turkar V, Deo R, Rao YS, Mohan S, Das A (2012) Classification accuracy of multi-frequency and multi-polarization SAR images for various land covers. IEEE J Sel Top Appl Earth Obs Remote Sens 5(3):936–941. https://doi.org/10.1109/JSTARS.2012.2192915

Völker S, Kistemann T (2015) Health & place developing the urban blue : comparative health responses to blue and green urban open spaces in Germany. Health Place 35:196–205. https://doi.org/10.1016/j.healthplace.2014.10.015

Wang Y (2015) Impervious surface mapping using satellite data and runoff modelling in Amersfoort. Utrecht University, Utrecht, Netherland, NL

Weng Q (2011) Remote sensing of impervious surfaces in the urban areas : requirements, methods, and trends. https://doi.org/10.1016/j.rse.2011.02.030

Zavadskas EK, Antucheviciene J (2007) Multiple criteria evaluation of rural building's regeneration alternatives. Build Environ 42(1):436–451. https://doi.org/10.1016/j.buildenv.2005.08.001

Zhang Y, Yiyun C, Qing D, Jiang P (2012) Study on Urban heat island effect based on normalized difference vegetated index: a case study of Wuhan City. Procedia Environ Sci 13:574–581. https://doi.org/10.1016/j.proenv.2012.01.048

Chapter 20
Analysis of Urban Green Spaces Using Geospatial Techniques—A Case Study of Vijayawada Urban Local Body Andhra Pradesh, India

Vani Timmapuram and Priyal Bhatia

Abstract In the past, urban green space was regarded as one of the most important aspects of a healthy city. However, with rapid urbanisation, urban sprawl, and population growth, there has been a sharp decline in open green spaces in cities, particularly in metropolitan areas. Also, over the last few decades, there has been a shift in land use and land cover, with a decrease in the area of green spaces, agricultural lands, and urban greenery. This study was carried out to assess green spaces in Vijayawada's Urban Local Body. Because of better economic opportunities, the city has seen a surge in population inflow. Furthermore, the city's outskirts are more vulnerable to transition due to the presence of the Krishna River. As a result of these factors, there has been a decrease in urban green spaces from 2012 to 2020. This has skewed per capita greenness of the city. It is only 16 m^2 at present. The results show that there exists a negative correlation of -0.46 between per capita green and population; therefore, with every one unit increase of population, the demand for built up and urban amenities will increase, thereby, impacting the per capita green and overall greenness index of the city negatively. The study also compares the normalised difference vegetation index (NDVI) to the transformed difference vegetation index (TDVI). It is a new index that is not widely used. Because it does not saturate, TDVI has proven to be superior to NDVI for urban green analysis. NDVI shows vegetation of 21.25 km^2 whereas TDVI shows vegetation of 16 km^2. Not only this, there has been an increase of merely 2% of vegetation in past 8 years span in the Vijayawada city.

Keywords Urban landscape · NDVI · TDVI · Urban green space · Per capita greenery · Greenness Index

V. Timmapuram (✉)
Andhra Pradesh Space Applications Centre, Vijayawada, Andhra Pradesh 520008, India

P. Bhatia
Department of Natural Resources, TERI SAS, New Delhi 110070, India

© The Author(s), under exclusive license to Springer Nature Singapore Pte Ltd. 2022   383
P. K. Rai et al. (eds.), *Geospatial Technology for Landscape and Environmental Management*, Advances in Geographical and Environmental Sciences,
https://doi.org/10.1007/978-981-16-7373-3_20

20.1 Introduction

In the past few decades, developing countries like India have experienced a rapid rate of urbanisation. According to the 2011 census of India, 31.16% of India's population lives in urban areas, and this figure is expected to rise by 40% by 2026. According to NRSC (2008), India's urban population is expected to reach 575 million by 2030. As a result, existing geographical areas face significant challenges. Urbanisation fosters economic development, a higher standard of living, and more opportunities. As a result, it causes in-migration from rural to urban areas, which contributes to urban sprawl, land pressure, and urban heat island, encroachment of open and green spaces, ground water depletion, air pollution, thereby, damaging the environment (Rai and Kumra 2011). With ongoing developments, metro cities now provide higher living standards but poor health conditions due to a reduction in green lungs, or vegetative cover. As a result, people in higher economic strata now prefer to live in suburbs because they provide better environmental conditions, cleaner air, more vegetative cover, and greenness.

Urban green spaces are a very essential part of sustainable cities and their importance can't be neglected by policymakers and planners. These green spaces are capable of having positive impacts on key areas of human lives and ecosystem services. Urban green spaces (UGS) include private and government owned or community parks, gardens, stadiums and grounds, recreational venues, industrial and institutional sites, water and river fronts, railway and road corridors, open spaces surrounding the monuments, vacant plots, schoolyards (Venn and Niemela 2004). With respect to land use land cover, urban green space can be defined as any government or private land like university's playground, private garden, or community which includes plant life or vegetation of any kind. It also includes water fronts thereby, also referred to as blue spaces. However, any national and state parks are not included under urban green space (WHO 2017). Andhra Pradesh has large amount of urban greenness due to backyard trees of the old constructed houses and buildings (Timmapuram et al. 2017). An urban environment is a combination of green spaces (natural and man-made) and built up (man-made). Therefore, it is right to say that it includes both 'green spaces' and 'grey spaces'. UGS includes everything within a city either of private or public nature, which has vegetation of any form (tall trees, shrubs, parks) (Singh 2018 cited Urban green space availability in Bathinda City).

Exposure to these green spaces in urban areas are very vital for overall growth and development of young children, serves as recreational venue, important for elderly citizens and holds environmental, ecological, economic and physical benefits, thereby, enhances one's quality of life in urban environments.

Lately, due to the global pandemic of Covid-19, many reports show that people accessed these green spaces more than usual for walking around the neighbourhood in order to get some fresh air, leisure and community life, after the restrictions relaxed relatively. The improvement of quantity as well as quality of greenness is another important aspect of sustainable cities, and it can only be achieved if planners and the community work together to coordinate their efforts. However, the expansion, monitoring and management of green spaces have to be in proportion. Studies conducted in similar fields suggest that there is a transparent trade-off between built-up region and green spaces. The areas which report high density of settlements and impervious surfaces registers a low green space ratio (Singh 2018; Dallimer et al. 2011). Recently, it has been witnessed that urban greenness is being managed and restored, but at the cost of poor, e.g. in Mumbai, India, the protection of the mangroves forest has led to displacement of slums and people living there; however, the high-end developmental projects still continue in the region leading to encroachment of mangrove forest lands (Grinspan et al. 2020).

Various standards are issued by different organisations for open green spaces in urban environment (Singh et al. 2010 quoted Sukopp et al. 1995 and Wang 2009) which suggests than in the twentieth century, countries like Germany and Japan had an international norm for urban green spaces as 40 m² with acceptable high-quality standards in order to curb down and maintain a balance between oxygen and other gases of the atmosphere. In the developing counties, the standard was 20 m² of park or garden per capita. Table 20.1 shows green spaces' standards globally cited by various authors.

With advancements in the fields of GIS and remote sensing over the last two decades, processes such as monitoring, assessing and mapping of various land use categories have become more common, easy, accurate, reliable and cost-effective (Singh and Rai 2017; Mishra et al. 2014, 2016, 2018; Mishra and Rai 2016). It aids in studying change phenomenon and helps understand which regions are more vulnerable to environmental degradation with respect to reduction in green spaces. This study is unique way as in India, a very few studies have been conducted on assessing current green spaces and potential site selection of urban green spaces.

Table 20.1 Standards for green spaces in urban regions globally

S. No	Region	Green Spaces Standards (m² or %)	Author
1	China	32.4%	(Singh et al. 2010 quoted Sukopp et al. 1995 and Wang 2009)
2	Reggio di Calabria, Italy	1.9%	(Singh et al. 2010)
3	Ferrol, Spain	46%	
4	Wellington, New Zealand	200 m² per person of green space	
5	Turkey	Urban parks: 20 m², community parks: 10 m², playgrounds: 6 m², neighbourhood parks and sports complexes: 8 m² each	(Martin 2012)
6	Japan	40 m²	(Timmapuram et al. 2017)
7	World	9 m²	WHO 2017

There is great scope and potential of intense research on urban green spaces and its functions as well as impacts, especially for Indian cities. Moreover, in India, most of the studies focus on metropolitan cities. In this work, the first section discusses the studies and findings of other authors done in similar fields.

20.2 Study Area

The study area includes the Urban Local Body of Vijayawada (ULB). ULB is a small body that governs a town with specified population. The main role of these bodies is to look into functions such as public welfare and safety, developmental activities, health facilities and taxation work, etc. (www.appublichealth.gov.in, n.d.). Vijayawada is well connected roadways, airways and railways. In fact, it is one of the most important railway station junctions of India.

Vijayawada city is lying in Krishna district; it is a business and educational hub of Andhra Pradesh. Vijayawada city is the second largest and populous city in the state after Visakhapatnam city with an area of 61.88 km^2. The latitudinal and longitudinal coordinates of the city is 16.5193°N 80.6305°E, respectively. The city lies beside one of the most important and perennial river bank of Krishna River and is covered by low-lying hills. It is also well known for its cultural history. Vijayawada city has three major canals running through out for delta area from the river of Krishna the city.

The mean temperature ranges from 23 to 34 °C; month of May is being the hottest in the years, and the climate is tropical (hot and humid). The month of January registers the lowest temperature in the years. Average annual rainfall precipitation receives 977.9 mm in both southwest and northeast rain bearing winds. There are four major types of soils in the city, namely black soil, clay loamy soil, red loamy soil and sandy soil. Major crops and fruits grown here includes paddy, sugarcane, mango and tomato. The occupation of people is agriculture, transportation, artefacts, educational institutes, hospitality, construction companies, IT Hub, petroleum products and major industries.

The city has experienced a rapid rate of urbanisation over past few years. As per the census of India, 2011, the total no. of population was 1,021,806 (DTCP, AP, n.d.). The Vijayawada Urban Agglomeration has registered a growth rate which is higher than that of the state. It is expected that the city along with urban agglomerations will reach a two-million mark in population. As per a report by Oxford Economics, 10 fastest growing cities and economies of the world by 2035 are in India. Vijayawada is one of them and holds the 10th position with an expected growth rate of 8.16% between 2019 and 2035 (Wood 2018).

Therefore, preferred this region as our area of interest to understand the population and land use of present scenario, existence of green cover, policy and planning and change dynamics. Figure 20.1 shows the maps of study area and the administrative map of the region, respectively.

Fig. 20.1 Study area map
and administrative map,
Vijayawada

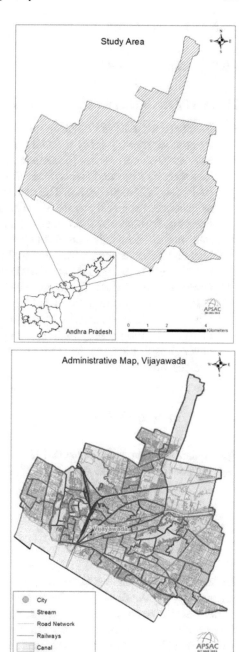

20.3 Materials and Methods

20.3.1 Satellite Data Used

The data set that is used for the study are as follows:

1. **Resource sat-LISS-IV**: Linear Imaging Self-Scanning Sensor 4 is a multi-spectral sensor with a high spatial resolution of 5.8 m, revisit period of five days and swath of 70 km (pan mode). It has 4 bands; 1 band is mono, and 3 bands are multispectral bands (Band 2: 0.52–0.59, (green), Band 3: 0.62–0.68, (red), B4: 0.77–0.86 (NIR)). The data generated from the sensor can be used to generate maps at the scale of 1:12,500 which further has utility for town planning, developmental projects, resource managements.

2. **Sentinel 2**: The satellite provides spatial resolution of 10–60 m over coastal, land and water resources. The temporal resolution is of five days with a swath of 290 km. There are 13 spectral bands with 4 bands at 10 m spatial resolution (Band 2: 0.490 (Blue), Band 3: 0.560(Green), Band 4: 0.665 (Red), Band 8: 0.842 (NIR)), 6 bands at 20 m and 3 bands at 60 m, and is the very first of its kind to include three bands in vegetation red edge. It has great utility in the field of land monitoring, early plant and crop health detection, water quality, land cover classification, etc. due to its high temporal as well as spatial resolution and cost effectiveness.

20.3.2 Software's Used

All the data set is projected to WGS 1984 UTM 44 N coordinate system. The study is carried out using ERDAS Imagine 2015, ArcGIS, and Terrset software.

20.3.3 Image Pre-processing

The Resourcesat LISS-IV (2012) satellite data were obtained from the Andhra Pradesh Space Applications Centre's archive data (APSAC) which is procured under various projects. Similarly, data for Sentinel 2A (2020) were gathered from ESA archives which is made available in the Copernicus Data and Information Access Service, cloud environment for analysis. Both sets of data were layer-stacked. It is a process in which multiple images or bands are combined into a single image. After layer stack, the rasters were rectified with respect to CartoDEM data in order to assure positional accuracy. Datasets are atmospherically corrected. In this process, the effect of atmosphere distractions and noise are removed before stating the analysis. This process gives the spectral enhancement for identification of the objects. In Erdas Imagine Suite, haze reduction tool is used in order to perform this process.

20.3.4   NDVI and Greenness Index

NDVI stands for normalised difference vegetation index. NDVI is an index that works on the red and near infrared band. It is well-known index and widely used to identify vegetation. The value ranges from negative $1(-1)$ to positive 1 (+1). Extremely low values of NDVI (0.1 and below) indicate towards the presence of sand, water body or barren land. Moderate values ranging from 0.2 to 0.3 suggest presence of grassland or a shrub, whereas higher values indicate presence of vegetation. The formula for estimating NDVI is given in Eq. (20.1)

$$NDVI = (NIR - Red\ Band)/(NIR + Red\ Band) \qquad (20.1)$$

NDVI was estimated using Erdas Imagine software, where per pixel of vegetation was identified in pseudo-colour mode and assigned suitable colour code. The values were reclassified using the model maker tool and recode tool. Using the model maker, conditional function was used in order to classify values of NDVI. A code1 was given to values which indicates presence of vegetation. All the other values were assigned a code-0. At last, the results were clipped to Vijayawada Municipal Boundary, and area lying under vegetation was calculated.

Greenness Index is used to evaluate the quality and quantity of urban greenness. It is a proportion of area covered by green vegetation to total administrative area. In order to calculate the percentage of green or total vegetative cover in percentage, the Greenness Index value can be multiplied by 100. This method was proposed by Abutaleb et al. (2020). To estimate green and non-green patches based on classified NDVI, Greenness Index can be estimated by the formula given in Eqs. (20.2) and (20.3)

$$GI = Area\ covered\ by\ green\ vegetation/Total\ study\ area \qquad (20.2)$$

$$\%Greenness = GI*100 \qquad (20.3)$$

20.3.5   Transformed Difference Vegetation Index (TDVI)

Transformed difference vegetation index, also called TDVI, was developed by Bannari et al. (2002). NDVI is derived empirically, therefore it gets saturated with variation is type of soil, moisture content and density of vegetation. However, to improve NDVI, transformed difference vegetation index was developed and introduced. It is useful for monitoring vegetation cover in urban environments. This index does not saturate as NDVI and soil adjusted vegetation index (SAVI). TDVI shows same sensitivity to bare soil as SAVI It can be calculated using the following general formula:

Bannari et al. (2002) in their research suggested that TDVI is nothing but sum of NDVI and SAVI. In the SAVI index, we can adjust the L values as it indicates amount of green cover or we can say canopy background adjustment factor. The value for L ranges from 0 to 1 representing very high vegetative cover and very low vegetative cover, respectively. Therefore, by default, it is taken as 0.5 which is a measure of area of moderate vegetation. It minimizes soil brightness variations and also eliminates the need for calibrations as per different types of soils (Huete 1988; gsp.humboldt.edu, n.d.) Therefore, by referring to literature, we have used an L factor of 2, in order to get optimum results. This factor was decided based on trial-and-error method for the area of interest.

20.4 Results and Discussion

20.4.1 Image Pre-processing

Before any further analysis, the satellite data for both study years of LISS-IV 2012 and sentinel data of 2020 were pre-processed. This aided in ensuring that there are no noises in the satellite data that could lead to skewed results. Following processing, the data were viewed in false colour composite to compare spatial resolution. Figure 20.2a, b depict the false colour composite map of Vijayawada Urban Local Body in both the years, as well as revenue ward divisions.

20.4.2 Normalised Difference Vegetation Index (NDVI) and Greenness Index

The NDVI and Greenness Index are used in the study to assess the current state of Vijayawada urban green spaces. NDVI index shows the value ranges between -1 to $+1$. Table 20.2 contains NDVI threshold values as well as area statistics for the study years. However, the NDVI was obscured by hills and agricultural lands. It was done to promote true urban vegetation. For hills and agricultural lands, layers were added and vector files were created. The recode tool was also used here, and a code of 0 was assigned to the masked feature. After saving the edits and layers, an area column was added to tabulate the new area with masked features. It is a general nomenclature for urban regions that we mask hill area and agricultural lands. After this process was completed, the map composition for the three indices was investigated. Similar procedure was followed for TDVI and NDBI as well.

In the year, 2012, the vegetation covered 24.74% of the total land area. In contrast, it is only 22.67% in 2020. The change of greenery has stalled over the course of eight years. It has dropped by 2.07%. One of the assumptions made before beginning this study was that the amount of greenery must have increased significantly. The

(a)   **(b)**

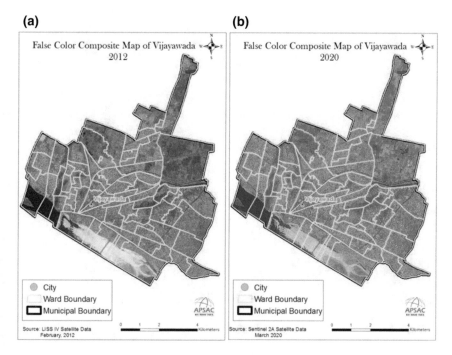

Fig. 20.2 **a** FCC map of Vijayawada, 2012. **b** FCC map of Vijayawada, 2020

Table 20.2 NDVI statistics in Vijayawada for the year 2012 and 2020

S. No	Index year	Threshold values of NDVI	Vegetative area with hills and agriculture (in km^2)	Vegetative area without hills and agriculture (in km^2)	Total vegetative cover (%)	Greenness Index
1	NDVI (2012)	0.19–0.86	28.73	15.31	24.74	0.24
2	NDVI (2020)	0.21–0.99	21.25	14.03	22.67	0.22

reason for this is that many new zones and areas have been designated for urban greenery expansion in Vijayawada's Masterplan of 2021. However, there has been no vegetation increase over the years.

The NDVI index was used to extract the urban green vegetative area. For the year 2012, it is estimated that the area of greenery with hills and agricultural land is 28.73 km^2, and the area of greenery without hill sand agricultural land is 15.31 km^2. In the year 2020, the total estimated greenery is 21.25 km^2 with hills and agricultural lands. There is only 14.03 km^2 of greenay area in the city without hills and agricultural lands in the city. The agricultural lands have been converted for construction and

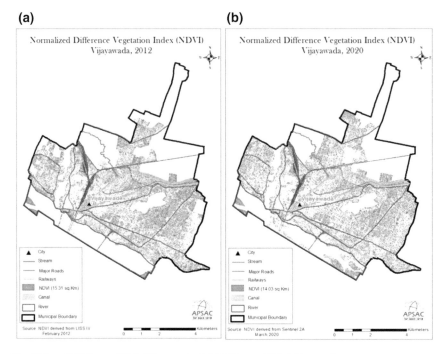

Fig. 20.3 **a** NDVI map 2012, Vijayawada. **b** NDVI map 2020, Vijayawada

other uses. As a result, only 1.28 km^2 of vegetative cover change in the real scenario has been stagnant for 8 years. Figure 20.3a, b depict the NDVI maps for the years 2012 and 2020, respectively.

20.4.3 Transformed Difference Vegetation Index (TDVI)

The TDVI index is calculated for this study area. It is a one-of-a-kind index to find out the vegetative cover in urban environments. This index is expected to provide better and more accurate results, especially in urban environments. The TDVI indeed for both years only shows vegetation beginning with positive values. The total vegetative cover in the year 2012 was 17.80 and 18.37% in the year 2020.

According to the TDVI index for the year 2012, the area under vegetation is 22.22 km^2 (including the hills and agricultural tracts), but it is only 11.02 km^2 area without hills and agricultural tracts. Similarly, the area with hills and agricultural tracts in the year 2020 is 16.92 km^2, whilst the area without hills and agricultural tracts is only 11.37 km^2. As a result, we can see that the growth of urban green in Vijayawada's ULB has stalled. Over the eight years, there is only a net increase of 0.35 km^2 in Vijayawada's ULB. Because TDVI takes into account soil and barren land reflection,

Table 20.3 TDVI statistics in Vijayawada for the year 2012 and 2020

S. No	Name of the indices	Vegetative area with hills and agriculture (in km^2)	Vegetative area without hills and agriculture (in km^2)	Total vegetative cover (%)	Greenness index
1	TDVI (2012)	22.22	11.02	17.80	0.17
2	TDVI (2020)	16.92	11.37	18.37	0.18

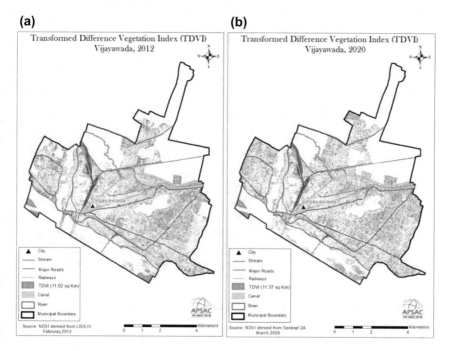

Fig. 20.4 a TDVI map 2012, Vijayawada. b TDVI map 2020, Vijayawada

it is expected to give only the true amount of green vegetation present. Table 20.3 and Fig. 20.4 show the statistics and map for the same.

20.4.4   Per Capita Greenness

Per capita greenness is calculated as total population/total vegetative or green cover. It essentially indicates the city standards as well as overall health of the city and its residents. A city with a higher per capita greenness indicates a healthy functional city where residents can enjoy the benefits of green space and open spaces, whereas a city with a low per capita greenness indicates poor and lowered living standards.

Table 20.4 Per capita greenness, Vijayawada, 2021

S. No	Population (2021)	Total vegetative cover (m^2)	Per capita greenness
1	1,201,005	TDVI (2020)—18,370,000	15.29
2	1,201,005	NDVI (2020)—22,670,000	18.87

Table 20.5 Per capita greenness, Indian cities, 2011

S. No	Indian cities/regions	Per capita greenness (m^2)	Population (2011)
1	Gandhi Nagar	29.77	6,330,000
2	Chennai	18.05	9,880,000
3	Hyderabad	0.93	11,570,000
4	Delhi	10.41	28,500,000
5	Mumbai	9.36	23,500,000
6	Chandigarh	38.00	1,050,000
7	Jaipur	6.67	3,710,000
8	Bangalore	3.31	13,900,000
9	Machlipatnam	30.00	169,892
10	Vizag	10.00	2,035,922
11	Vijayawada (2017)	17.00	1,021,806
12	Vijayawada (2021)	15.00	1,201,005

Source Singh et al. (2010), Timmapuram et al. (2017)

Given below in Table 20.4 is the per capita greenness for the year 2021 for the city of Vijayawada.

In Table 20.5, per capita greenness and population as per 2011 are given for few cities in India. In the year 2011, the greenness in the city of Vijayawada was about 17 m^2. However, for the year 2021, it has reduced to 15 m^2. We can see from the table below that as the population grows, so does the demand for urban amenities such as housing. As a result, there is a clear trade-off between urban and green space expansion. As a result, cities with a higher population have a lower per capita greenness. The figure depicts a graph of cities based on their per capita greenness (Fig. 20.5).

In addition, a correlation matrix was created in Excel to determine whether population and per capita greenness are correlated in any way. As illustrated in Fig. 20.6, there is a −0.46 correlation between population and per capita greenness. It means that there is an inverse relationship between the two; as population grows, per capita green tends to decrease.

According to the World Health Organisation (WHO), a healthy city is one in which community resources are constantly created and expanded to ensure proper life functions in physical and social environments. However, in this regard, open

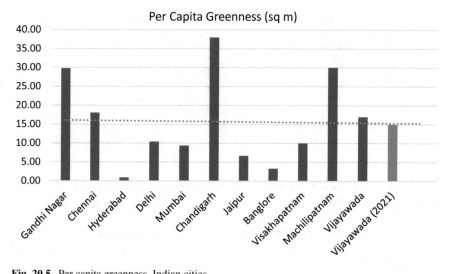

Fig. 20.5 Per capita greenness, Indian cities

	Per Capita Greenness (m2)	Pop (2011)
Per Capita Greenness (m2)	1	
Population (2011)	-0.462388993	1

Fig. 20.6 Correlation between population and per capita greenness

spaces and green spaces such as recreational areas, common areas such as forests, flood plains, and so on are frequently overlooked. For a functional city, it should have 25–35% of area under open and green spaces. The pandemic of 2019 made us realise that isolation with absence of access to adequate open green space causes mental discomfort as these common areas are vital for workout, physical movement and ensure well-being of humans. Historically, we also have evidence that green spaces and the natural environment were important factors in post-pandemic planning and design. The years 2019 and 2020 have focussed our attention on the scarcity of green and blue spaces in urban environments, particularly in India's densely populated cities.

A report by Andhra Pradesh government stated number of tree plantations as per its area for residential and commercial or institutional area (Reddy 2015). Table 20.6 depicts the plot area and no. of trees to be planted.

Bhaskar (2012), in his study found that the city of Pune has witnessed an overall expansion or increase of 43.01 km^2 from year 1999 to 2009 of built up and a decrease of 5.58 km^2 of sparse vegetation and a decrease of 1.66 km^2 dense vegetation in the due course of 10 years. The temporal changes of green spaces in Mumbai from the

Table 20.6 Tree plantation standards, A.P for residential and commercial area (Reddy 2015)

Residential area		Commercial/Institutional Area	
Plot area (m^2)	Number of trees	Plot area (m^2)	Number of trees
Below 100 m^2	3	Below 200 m^2	2
101–200 m^2	5	201–500 m^2	4
201–300	10	501–1000 m^2	6
301 m^2 and above	10 + 5 for per 100 m^2 increase of area	1001 m^2 and above	6 + 2 for per 100 m^2 increase of area

year 1988 to 2018 have been assessed by Rahaman et al. (2020). In the year 1988–1998, a 14% transition of urban green space to different developmental expansions was registered.

The state of Andhra Pradesh has recorded an increase of about 990.40 km^2 of forest cover in 2019 as per Indian State of Forest Report. It stood second in terms of expansion of forest cover after Karnataka (1025 km^2) (Vaishali 2019).

According to the study's findings, over the last eight years, the city's greenery has declined whilst its built-up area has increased dramatically. The masterplan for the city of Vijayawada has designated zones for the expansion of urban green, but the goal has yet to be met. The results of the fieldwork point in the same direction. Also, we can notice, that a very small area is available for expansion of green in the city. Furthermore, with decrease in green, per capita which has disrupted balance of the city.

20.4.5 Field Data Collection and Field Verification

Even when researchers have prior knowledge of the thematic classes and other aspects of the region, there may be some areas of uncertainty in the study area. Ground verification is essential for improving the accuracy of classified maps. All questionable areas must be verified using field verification techniques. Under this procedure, various samples pertaining to different sample points were collected for verification. Proper field planning considering all the available resources and resource constraints, careful selection of the points, etc. has to be well accounted for before going on the field. Not only this, we also must ensure that GIS and remote sensing data collected are checked for the quality.

The same can be done by inspecting the projection and datum for all input layers, as well as the sliver polygons and data quality in terms of noise and cloud cover (NRSC 2011). The following instruments were used to collect precise ground data:

a. GPS-enabled camera for taking photos and later generating coordinate data
b. Hardcopies of the generated sample points for accuracy assessment and verification, as well as satellite images for data collection and verification (Shastri et al. 2020).

Fig. 20.7 Plantation in residential area

c.   Satellite image overlay points on toposheet
d.   A paper copy of the road network map (pertaining information of major and
     minor roads)
e.   A ground verification form or an Excel spreadsheet

As per ISRO, the coordinates collected on field should be recorded in
Degree/Minutes/Seconds format However, they can be recorded in latitude and
longitude extent as well based on SOI toposheets (NRSC 2011).

As a result, for our fieldwork, the study area was randomly divided into four
quadrants, with each quadrant covering all of the thematic LULC classes. For each
of the classes, doubtful points and few random points were generated for verification.
For four days, the field work was carried extensively, covering one quadrant per day
and including personal observations on the field. Field photographs from Figs. 20.7,
20.8, 20.9, 20.10, 20.11 and 20.12 depict the actual environment on the ground.

20.5   Conclusion

With ongoing global climatic changes, it is critical to expand a city's green index
in order to make it more functional and healthier. The rapid pace of urbanisation is

Fig. 20.8 Open plot in residential area

destroying a city's green belt. Even with masterplans, implementation is not always complete. Modern cities are transforming into urban forests, which will have an impact on the overall functionality and health of the residents. As green spaces are important aspect of an urban landscape, it provides numerous benefits. They contribute significantly towards maintaining climatic balance, improves hydrological cycles, etc. But, unplanned and mismanaged expansion as well as development have led to haphazard. It makes the city more vulnerable to both natural and man-made calamities. Over the period of time population has increased and the per capita has decreased. Even with masterplan, the city is still deprived if proper urban green spaces. It is now at an alarming stage. If the same is ignored and neglected over longer time spans, the city will register even more decline in green space, thereby impacting the overall functionality. According to the study, TDVI has proven to produce better results than NDVI in urban settings due to its characteristic of not becoming saturated. In 2020, the NDVI showed a greenery of 22.67 km^2, whilst the TDVI showed a greenery of 18.37 km^2.

When it comes to managing green spaces, some challenges may arise, such as balancing built up and green spaces, each community will have different expectations from the common space, making it cumbersome, once transitioned, the city offers very little area for further expansion, especially with increasing urban population and migration, mismanagement of resources, and poor planning.

Fig. 20.9   Garden area infront of home

It is critical that more emphasis and space be set aside for the expansion of greenery. It requires supportive and integrated policies as well as plans. When developing the city's master plan, a special emphasis should be placed on roof top urban green, urban gardens and parks, playgrounds, avenue plantations and so on, in order to achieve balance and equilibrium. Education on importance of urban green spaces, benefits of watering USG, roadside vegetation, etc. should also be given emphasis on. Also, since Vijayawada is a part of Smart Cities Mission, programmes like such should aim at preparing guidelines, implementation of all the regulations needed much to expand urban green space. More researches should be encouraged in this regard in order to enhance urban green spaces. Certain measures can be implemented at the local and state levels to expand as well as maintain healthy green spaces. Some of them includes:

1   The city has experienced high rate of urbanisation which led to reflectance of its biodiversity and recreational open spaces. Therefore, it is important that the city observes an increase in the same.
2   Proactive approach for integrating land use with other sustainable methods.
3   Vijayawada is surrounded by agricultural dominant region; therefore, land use and planning is extremely crucial.
4   Regulated urban growth,
5   Energy and resource efficient buildings,

Fig. 20.10 Wasteland in residential area

6   Track unauthorised buildings encroached areas,
7   Buildings expansion should be done vertically with more emphasis on sustainable aspect.

Another approach that could be taken for the expansion of green spaces is one similar to that taken by the government of Telengana. Giving various incentives to various wards, districts and Urban Local Bodies can increase urban green space and cover. In Telangana, the government has launched a green space index as part of the Haritha Haaram plantation programme in order to promote increased green space. Geospatial techniques are recommended for determining the extent of green cover. The index aims to increase the number of parks, planation and designs, roadside greenery and so on. In terms of incentives, awards will be given to ULBs that succeed in increasing total green cover, as well as innovative methods used to increase cover and different plantation designs (ET Government 2020).

20.6   Limitations and Future Scope

Although entire internship period and work went smooth with minimal challenges, however, some of the limitations of the study includes:

Fig. 20.11 Government park area

1   Unavailability of full and open-source government data.
2   Analysis is based on remotely sensed data which often varies from the true ground data.
3   Due to unavailability of full cadastral data, ground truthing was carried on only on some government land parcels.

The analysis performed in the study is critical for policymakers and planners. Due to time constraints and the global pandemic of Covid-19, detailed and extensive field work was not carried out. As a result, in the future, the project can be expanded with a comparative study of different cities and ULBs in the state of Andhra Pradesh. Also, Vijayawada being a part of India's Smart City Plan, offers much more scope for analysis and in-depth assessment of the city. Not only urban green space, but city also faces issues such as water shortage, solid waste management, urban population growth and infrastructure mismanagement, etc. Therefore, there is scope for such analysis and study as well.

Fig. 20.12 Wasteland in residential area

Acknowledgements The authors express their sincere gratitude to the Vice Chairman, APSAC for his continuous support for the research study and also the authors are grateful to the editor and anonymous reviewers for suggestions and comments which improved the manuscript significantly.

References

Abutaleb K, Mudede MF, Nkongolo N, Newete SW (2020) Estimating urban greenness index using remote sensing data: A case study of an affluent vs poor suburbs in the city of Johannesburg. Egypt J Remote Sens Space Sci 1–9 (online). Available at: https://www.sciencedirect.com/science/article/pii/S1110982319304211#b0165

Bannari A, Asalhi H, Teillet PM (2002) Transformed difference vegetation index (TDVI) for vegetation cover mapping. IEEE Int Geosci Remote Sens Symp 5(0)

Bhaskar P (2012) Urbanization and changing green spaces in Indian cities (case study—city of Pune). Int J Geol Earth Environ Sci 2(2):1 (online). Available at: t http://www.cibtech.org/jgee.html. Accessed 4 Feb 2021

Censusindia.gov.in (2011) Census of India Website: Office of the Registrar General and Census Commissioner, India (online). Available at: https://censusindia.gov.in/2011-common/censusdata2011.html

Dallimer M, Tang Z, Bibby P, Brindley P, Gaston KJ, Davies ZG (2011) Temporal changes in greenspace in a highly urbanised region. Biol Let 7(5):763–766

DTCP, AP (n.d.) DTCP (online). www.dtcp.ap.gov.in. Available at: http://www.dtcp.ap.gov.in/dtc pweb/DtcpHome.html. Accessed 15 Jun 2021

ETGovernment (2020) Telangana: Govt launches green space index to promote green cover in ULBs—ET Government (online). ETGovernment.com. Available at: https://government.eco nomictimes.indiatimes.com/news/smart-infra/telangana-govt-launches-green-space-index-to-promote-green-cover-in-ulbs/77844841. Accessed 10 Feb 2021

Grinspan D, Pool J-R, Trivedi A, Anderson J, Bouye M (2020) Green space: an underestimated tool to create more equal cities. World Resources Institute (WRI). Available at: https://www.wri. org/blog/2020/09/green-space-social-equity-cities. Accessed 10 Feb 2021

gsp.humboldt.edu (n.d.) Introduction to remote sensing: indices and analysis (online). Available at: http://gsp.humboldt.edu/OLM/Courses/GSP_216_Online/lesson5-1/indices.html#:~:text= Soil%20Adjusted%20Vegetation%20Index%20(SAVI)&text=%22L%22%20is%20a%20corr ection%20factor. Accessed 2 Mar 2021

Huete AR (1988) A soil-adjusted vegetation index (SAVI). Remote Sens Environ 25(3):295–309

Martin J (2012) Methods, tools and best practices to increase the capacity of urban systems to adapt to natural and man-made changes. J Land Mob Environ 10(1):42, 43 (online). Available at: https://www.semanticscholar.org/paper/METHODS%2C-TOOLS-AND-BEST-PRACTICES-TO-INCREASE-THE-Martin/765fc3ae6a62f3035af1e793e4d4d20547a8f337. Accessed 4 Feb 2021

Mishra VN, Rai PK, Mohan K (2014) Prediction of land use changes based on land change modeler (LCM) using remote sensing: a case study of Muzaffarpur (Bihar), India. J Geograph Inst Jovan Cvijić SASA (Serbia) 64(1):111–127. https://doi.org/10.2298/IJGI1401111M

Mishra VN, Rai PK, Rajendra P, Puniya M, Nistor MM (2018) Prediction of spatio-temporal land use/land cover dynamics in rapidly developing Varanasi district of Uttar Pradesh, India, using geospatial approach: a comparison of hybrid models. Appl Geomatics. https://doi.org/10.1007/ s12518-018-0223-5

Mishra VN, Rai PK, Kumar P, Prashad R (2016) Evaluation of land use/land covers classification accuracy using multi-temporal remote sensing images. Forum Geograph (Romania) 15(1):45–53

Mishra V, Rai PK (2016) A remote sensing aided multi-layer perceptron-marcove chain analysis for land use and land cover change prediction in Patna district (Bihar), India. Arab J Geosci 9(1):1–18. https://doi.org/10.1007/s12517-015-2138-3

NRSC (2008) National urban information system (NUIS), manual for thematic mapping using high resolution satellite data and geospatial techniques. NRSC, Urban Studies and Geoinformatics Group, Hyderabad, pp 1–110

NRSC (2011) Space based information support for decentralized planning (SIS-DP)- manual for creating SIS-DP resource layers. NRSC, ISRO, Hyderabad, pp1–226

Oxford (2019) Oxford Economics. (online) Oxford Economics. Available at: https://www.oxford economics.com/

Reddy NCM (2015) Importance of urban greening in the contet of climate change. (online) Vijayawada: A.P. Urban Greening and Beautification Corporation Ltd., pp 1, 32. Available at: https://aphrdi.ap.gov.in/documents/Trainings@APHRDI/2016/09_Sep/Engine ering/Sri%20Prabhakaran.pdf. Accessed 15 Feb 2021

Rai PK, Kumra VK (2011) Role of geoinformatics in urban planning. J Sci Res 55:11–24

Shastri S, Singh P, Rai PK (2020) Å Land covers change dynamics and their impacts on thermal environment of Dadri Block, Gautam Budh Nagar, India. J Landscape Ecol 13(2):1–13

Singh KK (2018) Urban green space availability in Bathinda City, India. Environ Monit Assess 190(11):1–17 (online). Available at: https://doi.org/10.1007/s10661-018-7053-0. Accessed 23 Aug 2019

Singh S, Rai PK (2017) Application of earth observation data for estimation of changes in land trajectories in Varanasi District, India. J Landscape Ecol. ISSN: 1805-4196. https://doi.org/10. 1515/jlecol-2017-0017

Singh VS, Pandey DN, Chaudhry P (2010) Urban forests and open green spaces: lessons for Jaipur, Rajasthan, India. (online) Digital library of the commons, Rajasthan State Pollution Control

Board (RSPCB), p 7. Available at: http://dlc.dlib.indiana.edu/dlc/handle/10535/5458?show=full. Accessed 3 Feb 2021

Timmapuram V, Rao GP, Ramana KV (2017) A technical report on urban green cover in Andhra Pradesh. Vijayawada: Andhra Pradesh Space Applications Centre (APSAC), p 1

Vaishali RA (2019) Andhra Pradesh records 990 sq km increase in forest cover compared to 2017. (online). The New Indian Express. Available at: https://www.newindianexpress.com/cities/vijayawada/2019/dec/31/andhra-pradesh-records-990-sq-km-increase-in-forest-cover-compared-to-2017-2083028.html. Accessed 10 Mar 2021

Venn S, Niemela J (2004) Ecology in a multidisciplinary study of urban green space: the URGE project. Boreal Environ Res 9(6):479, 489

Vijayawada Municipal Corporation (2014) VMC Green Vijayawada Project. (online). www.vmcdm.org Available at: http://www.vmcdm.org/VMC_Green_Vijayawada_Project.html. Accessed 11 Feb 2021

WHO, R.O. for E (2017) Urban green spaces: a brief for action (online). WHO. Available at: https://www.euro.who.int/__data/assets/pdf_file/0010/342289/Urban-Green-Spaces_EN_WHO_web3.pdf%3Fua=1#:~:text=2%20In%20this%20brief%20urban,(%E2%80%9Cblue%20spaces%E2%80%9D). Accessed 3 Feb 2021

Wood J (2018) The 10 fastest-growing cities in the world are all in India (online). World Economic Forum. Available at: https://www.weforum.org/agenda/2018/12/all-of-the-world-s-top-10-cities-with-the-fastest-growing-economies-will-be-in-india

Chapter 21
Magnetic Susceptibility and Heavy Metals Contamination in Agricultural Soil of Kopargaon Area, Ahmadnagar District, Maharashtra, India

S. N. Patil, A. V. Deshpande, A. M. Varade, Pranaya Diwate, A. A. Kokoreva, R. B. Golekar, and P. B. Gawali

Abstract The objective of this work is to investigate the suitability of such measurements for indicating heavy metal contamination. Magnetic susceptibility measurements were carried out of agricultural soil which was collected from 23 locations from Kopargaon area of Ahmadnagar district, Maharashtra State of India, using AGICO-MFK1-FA multifunction frequency Kappabridge KLY4S with low frequency susceptibility (F1) 976 Hz and high frequency susceptibility (F2) 15,616 Hz. The magnetic susceptibility values at low frequency were observed ranging from 16.83×10^{-7} m^3/kg^{-1} to 59.38×10^{-7} m^3/kg^{-1}, whereas at high frequency, magnetic susceptibility found ranged from 16.17×10^{-7} m^3/kg^{-1} to 56.38×10^{-7} m^3/kg^{-1}. This significant

S. N. Patil
School of Environmental and Earth Sciences, Kavayitri Bahinabai Chaudhari North Maharashtra University, Jalgaon (M.S.) 425001, India

A. V. Deshpande
Department of Civil Engineering, Sanjivani College of Engineering, Kopargaon District, Ahmadnagar (M. S.) 423603, India

A. M. Varade
Department of Geology, Rashtrasant Tukadoji Maharaj Nagpur University, Nagpur (M.S.) 440001, India

P. Diwate
Centre for Climate Change and Water Research, Suresh Gyan Vihar University, Jaipur (Raj) 302017, India

A. A. Kokoreva
Faculty of Soil Science, Lomonosov Moscow State University, GSP-1, Leninskie Gory, Moscow 119991, Russian Federation

R. B. Golekar (✉)
Department of Geology, G. B. Tatha Tatyasaheb Khare Commerce, Parvatibai Gurupad Dhere Arts and Shri. Mahesh Janardan Bhosale Science College, Guhagar District, Ratnagiri (M. S.) 415703, India

P. B. Gawali
Indian Institute of Geomagnetism, Navi Mumbai (M. S.) 410218, India

magnetic enhancement is an indication of presence of ferromagnetic minerals in agricultural soil from the studied area. Heavy metals in soil samples were analyzed by using double beam atomic absorption spectrophotometer. The mean concentration of Mn (6.760 mg/kg) followed by Fe (3.929 mg/kg), Cu (2.284 mg/kg), Pb (1.328 mg/kg), Zn (0.936 mg/kg), Cd (0.682 mg/kg) and Ni (0.595 mg/kg) was observed. The evaluation of anthropogenic influence and contamination with trace elements in soil from study area was carried out using geoaccumulation index. Soil geoaccumulation index (Igeo) shows that maximum values of Fe (5.599) and least value of Cd (−0.976) were observed. The geoaccumulation class (Igeo class) sequence was observed to be Cd > Pb > Ni > Zn > Cu > Mn > Fe. The interpretation of the obtained field measurements and the laboratory analyzes indicates that Cd, Pb and Ni provide the potential risk, whilst the other heavy metals are in the safe limits.

Keywords Soil · Magnetic susceptibility · Heavy metals contamination · Geoaccumulation index

21.1 Introduction

Magnetic susceptibility depends on composition and grain size of magnetic minerals presents in soil, sediments and rocks (Venkatachalapathy et al. 2011). Magnetic particles generated by industrial processes which have a diameter >2 μm (Flanders 1994); Matzka and Maher (1990) suggest that those particles are smaller than <2.5 μm which comes from the vehicular emission. Pedogenic ferromagnetic minerals are predominantly in super paramagnetic <0.02 μm; stable single domain which has 0.02–0.04 μm grain size (Venkatachalapathy et al. 2011), whereas anthropogenic magnetic particles have been multi-domain size >105 μm (Hay et al. 1997). The complexity of relations between pollutants and magnetic particles, and the fact that they have a common sources at are usually separate particles make it impossible to derive a single function to calculate pollutants concentrations from susceptibility measurements. For each new area of investigation, new ways of interpretation have to be found. It is difficult to assess how far magnetic susceptibility on its own is providing us with information on pollution (Hanesch and Scholger 2002; Xue and Yong 2005).

Since anthropogenic pollution usually has strong magnetic signature, this non-destructive magnetic technique looks promising in monitoring soil pollution (Mohamed et al. 2011). For the present, similar technique has been employed by ourselves to successfully characterize and quantify the degree of pollution in agricultural soil of Kopargaon area, Ahmadnagar District, Maharashtra, India.

21.2 Study Area

The present study area is located in the northern part of Ahmadnagar district of Maharashtra, India and lies at the bank of River Godavari which is major river flowing in the Southern India. The area lies in between 19°53'N to 19°88'N and 74°29'E to 74°48'E. There are around 79 villages in Kopargaon taluka amongst nearly 18 villages located on the bank of River Godavari (Fig. 21.1). The economy of the Kopargaon Tehsil is driven mainly by agriculture with sugarcane being the major cash-crop cultivated in studied region. The groundwater quality is also getting affected due to industrial pollution from sugar and allied industries like distillery and dairy. The effluent generated from these sugar and allied industries are causing environmental problems in the studied area.

Geologically, study area covers Deccan volcanic province (DVP) which is consisting of a various types of basaltic rock layers (Fig. 21.2). The basalts from the area are dark to grey in colour and fine to medium-grained in texture. They show typical spheroidal weathering that gives rise to large rounded boulders on the outcrops. The weathering starts along the well-developed joints, first rounding off the angles and the corners and then producing thin concentric shells or layers, which become soft and fall off gradually. Basalts display two sets of prominent vertical as well as horizontal joints, and the flows are highly jointed and fractured all over the

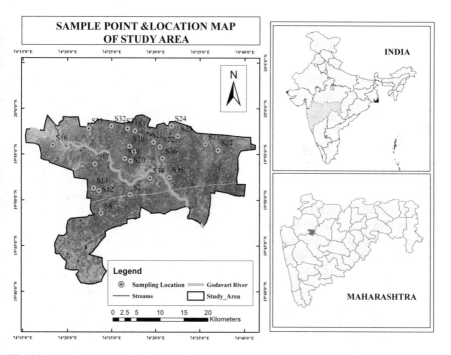

Fig. 21.1 Location map of the soil sampling from the study area

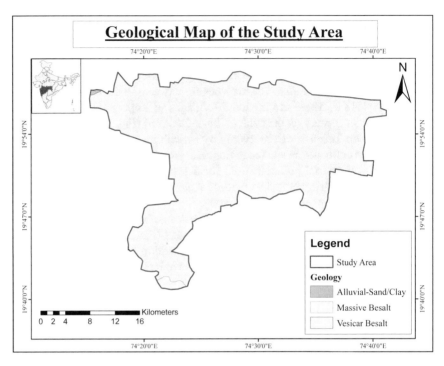

Fig. 21.2 Geological map of the study area

basin. The study area is having black cotton soil, originated mainly due to in situ decomposition of the basalts. The thickness of soil layer is highly variable, from 30 cm to 15 m. The soils of this region are highly saline (Krishna and Govil 2005). They are calcareous, neutral to alkaline (pH 7.5–8.5) and clays with a high amount of bases and have water holding capacity (Shankar et al. 1994).

The delineation of the geomorphic unit is based on interpretation from remote sensing data as well as observations from topography, relief and slope. The geomorphic unit in the study area is depicted in Fig. 21.3. A lineament is a linear feature in a landscape which is an expression of an underlying geological structure such as a fault, fracture zones and shear zones. Origins may be radically different from those of terrestrial lineaments due to the different tectonic processes involved. The lineament map is useful in evaluating the tectonic history of the continental area. Geomorphology and lineament pattern show the tectonic activity in the area. Many streams are characterized by straight segment with straps angles. The lineaments map prepared (Fig. 21.4) with the help of topographical maps. Total of 10 lineaments were observed in the present study area, out 01 lineament is structural, and other 09 lineaments are become geomorphic. The River Godavari is flow parallel to the structural lineaments. The lineaments are identified and mapped from MRSAC data as shown in (Fig. 21.4). It is clearly observed that major river channels flow along the structural lineaments.

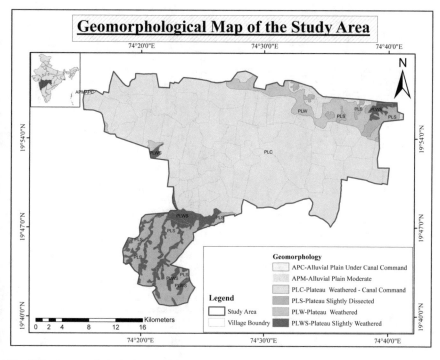

Fig. 21.3 Geomorphological map of the study area

In the study area, most of the urbanizations is on the bank of River Godavari channel which is seen in the red colour in the land use map of the study area (Fig. 21.5). Whereas as the agricultural and thin forest area marked by yellow colour in the land use map, whilst water bodies were marked in blue tinge. The agricultural practices in the area under investigation mainly sugarcane, pulses and wheat, but main attention is focussed on cash crops like sugarcane.

21.3   Materials and Method

The selection criteria for soil sampling locations were considered on the basis of human activities and waste water stream surrounding the agricultural field. Due to the use of waste water for irrigating, the agricultural land soil of this area is contaminated by heavy metals. The present study area shows black cotton soil which is derived from weathering of pre-existing host rock, i.e. basalts. Along the bank of River Godavari, the nature of soil is alluvial; it is derived from weathering products of basalts. The nature of the soil in the study area shows silt, loamy and clay texture and high water holding capacity.

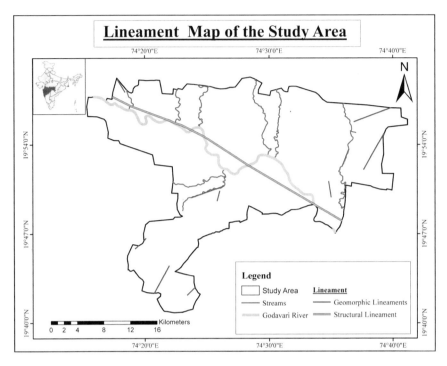

Fig. 21.4 Lineament map of the study area

A total of 23 top soil samples were collected in the month of May 2013 and their locations marked by global position system (GPS) make GARMIN. The soil samples were collected from the agriculture land that is under the influence of nearby sugar factory (Table 21.1). The method of collection soil samples was followed United Stated Environmental Protection Agency (USEPA 1986).

The samples were taken after clearing the area to be sampled of any surface debris (e.g. twigs, rocks, litter). Then, the samples were collected within 0–20 cm depth by using hand auger. The samples sealed in the polythene bags. Then, the samples mixed thoroughly in homogenization container to obtain homogeneous sample.

21.3.1 Environmental Magnetic Analysis

Magnetic susceptibility measurement is a powerful tool for the assessment of the heavy metal contamination. Distributions of magnetic susceptibility minerals in the soil profile are often similar to the distribution of heavy metals (Hanesch and Scholger 2002). Commonly, maximum magnetic susceptibility in the soil profile is observed at the same depth as highest concentration of heavy metal. Magnetic susceptibility measurements were carried out on soil sample which was dried at 40 °C and

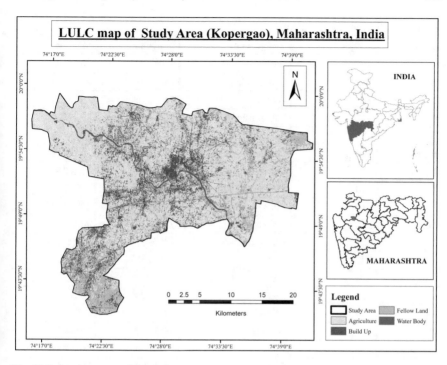

Fig. 21.5 Land use map of the study area

disaggregated. Samples were packed into 10 ml plastic container using cling-film to immobilize the soil sample. To ensure that the variable sample volume did not influence result, containers were filled to at least half of their capacity. Environmental magnetic analysis of soil samples was carried out at Indian Institute of Geomagnetism (IIG, Mumbai). Magnetic susceptibility was conducted using AGICO-MFK1-FA multifunction frequency Kappabridge KLY4S with low frequency susceptibility (F1) 976 Hz and high frequency susceptibility (F3) 15,616 Hz.

For all the measurements, the sensitivity was set at 1.0. Measurements were carried out three times; first air reading, sample reading and a second air reading before and after each series for drift correction (Kanu et al. 2013). The sensor is a handy laboratory sensor which makes use of 10 ml samples in plastic containers. It has the ability of taking measurements at two different frequencies, i.e. at 976 Hz (low frequency) and 15,616 Hz (high frequency). When the 10 ml cylindrical plastic bottles are in use, the accuracy of the instrument is 1% (Dearing 1999). The susceptibility measurements were done at both low (976 Hz) and high (15,616 Hz) frequencies which were further used to compute the frequency dependent susceptibility (χ FD). The percentage frequency depended susceptibility χ FD (%) was calculated from the expression,

Table 21.1 Soil sampling locations

S. No	Sample ID	Village name	Longitude	Latitude
1	S1	Dauch	74°42′23″	20°02′26″
2	S2	Deuce (Khurd)	74°27′11″	19°55′30″
3	S8	Pohegaon	74°32′06″	20°00′08″
4	S12	Madhi Budruk	74°34′58″	19°58′00″
5	S13	Velapur	74°33′54″	20°04′23″
6	S16	Manjur	74°23′31″	20°03′43″
7	S17	Mahegaon	74°33′21″	19°54′54″
8	S18	Malegaon Thadi	74°35′04″	19°57′41″
9	S19	Dharangaon	74°41′09″	20°08′54″
10	S20	Murshetpur	74°37′31″	20°04′24″
11	S21	Khirdi Ganesh	74°40′53″	20°06′42″
12	S22	Bolki	74°44′18″	20°01′18″
13	S23	Karanji	74°40′37″	20°00′26″
14	S24	Ogdi	74°38′24″	19°59′18″
15	S25	Shirasgaon	74°38′27″	19°58′58″
16	S27	Ukkadgaon	74°54′16″	20°04′35″
17	S30	Takli	74°41′53″	20°11′37″
18	S31	Brahmangaon	74°33′09″	20°05′40″
19	S32	Rawande	74°34′16″	20°10′18″
20	S33	Sangvi Bhusar	74°23′07″	20°07′57″
21	S34	Bet	74°31′25″	20°02′45″
22	S35	Savatsar	74°35′01″	20°07′18″
23	S36	Shingnapur	74°45′37″	20°02′37″

$$\chi FD(\%) = \frac{(\chi LF - \chi HF)}{\chi LF} \times 100 \tag{21.1}$$

where

χ LF = magnetic susceptibility at low frequency.

χ HF = magnetic susceptibility at high frequency.

21.3.2 Heavy Metal Analysis

The heavy metal analysis was carried out at School of Environmental Earth Science, North Maharashtra University Jalgaon by using double beam atomic absorption spectrophotometer (AAS) as per the standard procedures. The heavy metals such as Cd,

Cu, Fe, Mn, Ni, Pb and Zn were analyzed. Before the analysis of the soil, samples were digested. The digestion is necessary for destroying the organic matter and it preventing from interfering with the specific test. The 1 gm of soil is mixed with nitric acid for digestion in beaker on the hot plate at 150 °C (USEPA 1986). Beakers are cover by glass and digested the solution. Wash the beaker with distilled water and filter the solution using Whatman filter paper number 42. Make the volume up to 50 ml by using de-ionized water. Then, these samples run on atomic absorption spectrophotometer. Measure the concentration of different elements on atomic absorption spectrophotometer (AAS) directly at specific wavelength in nm Cd^{2+} (228.3), Cu^{2+} (324.8), Fe^{2+} (248.3), Mn^{2+} (279.5), Ni^{2+} (232), Pb^{2+} (283.3) and Zn^{2+} (213. 9).

21.3.2.1 Preparation of Spatial Variation Maps

Iso-concentration/contour maps were prepared with the help of Surfer 7.0 geoscientific software to delineate spatial variation regarding heavy metals from the study area's soil samples.

21.3.2.2 Assessment of Metal Contamination

The evaluation of anthropogenic influence and contamination with trace elements in soil from study area was carried out using geoaccumulation index. Igeo value is used that permit the assessment of degree of soil contamination with respect to global standards. Igeo is calculated by using the following mathematical equation (Muller 1969).

$$\text{Igeo} = \log 2\frac{Cn}{1.5} \times Bn \qquad (21.2)$$

where

Cn = mean measured concentration of the examined element in the studied soil

Bn = average (crustal) geochemical surrounding value for concentration of the element 'n' in basalts (average basalts) and

1.5 = the factor to compensate the surrounding data (correction factor) due to lithogenic effect

This method has been broadly employed in European trace element studies since the late 1960s. The parent rock in study area is basalts, hence the average concentrations of concern heavy metals in basalts were used for calculation of Igeo values. The average concentration of heavy metals in basalt has presented in Table 21.2. According to Muller (1969), the Igeo for each trace element is calculated and classified as the following orders.

i. Uncontaminated (Igeo ≤ 0)

Table 21.2 Average composition of heavy metals volcanic rocks (basalts)

heavy metals	(values in mg/kg)
Cd	0.15
Cu	110
Fe	77,600
Mn	1280
Ni	76
Pb	7.8
Zn	86

All parameters are expressed in mg/kg
Source (Mason and Moore 1982)

ii.   Uncontaminated to moderately contaminated (0 < Igeo ≤ 1)
iii. Moderately contaminated (1 < Igeo ≤ 2)
iv.   Moderately to heavily contaminated (2 < Igeo ≤ 3)
v.   Heavily contaminated (3 < Igeo ≤ 4)
vi.   Heavily to extremely contaminated (4 < Igeo ≤ 5)
vii. Extremely contaminated (Igeo ≥ 5).

21.4 Results and Discussion

21.4.1 Magnetic Susceptibility

The mass specific magnetic susceptibility values of the soil samples collected from study area were given in Table 21.3. The values of mass magnetic susceptibility at low frequency were observed in ranged from 16.83×10^{-7} to 59.38×10^{-7} m^3/kg^{-1} with an average value of 34.25×10^{-7} m^3/kg^{-1}. This significant magnetic enhancement is indication of ferromagnetic minerals present in the analyzed soil samples and thus increased pollution. Whereas the values of mass magnetic susceptibility at high frequency are found between 16.18×10^{-7} and 56.36×10^{-7} m^3/kg^{-1} with an average value of 32.74×10^{-7} m^3/kg^{-1}. The frequency dependent mass magnetic susceptibility was observed in ranged from 3.08 to 5.93% with an average value of 4.40%. Earlier researcher's soil has been classified based on values of frequency dependence susceptibility (Dearing 1999). According to Dearing (1999), if χ FD % values > than 14%, which indicates presence of coarse multi domains grains or superparamagnetic grains (<0.05 μm), if χFD% values between 12 and 14% called as virtually superparamagnetic grains and the values ranged between 2 and 10% indicating the presence of a mixture of superparamagnetic (SP) and multi domain (MD). The results of the percentage frequency dependence susceptibility showed that all soil samples have a mixture of superparamagnetic (SP) grains and multi domain (MD) magnetic grains because all samples dependent mass magnetic susceptibility

Table 21.3 Magnetic susceptibility of soil samples from the study area

S. No	Sample ID	mass (g)	χ LF (10^{-5} SI)	χ HF (10^{-5} SI)	χ LF (10^{-7} m^{3}kg^{-1})	χ HF (10^{-7} m^{3}kg^{-1})	χ FD (%)
1	S1	8.49	250.67	237.83	29.53	28.01	5.12
2	S2	8.1	177.7	170.12	21.94	21.00	4.27
3	S8	8.49	318.96	303.7	37.57	35.77	4.78
4	S13	10.03	168.78	162.23	16.83	16.17	3.88
5	S15	8.19	486.46	461.72	59.40	56.38	5.09
6	S16	8.69	148.68	140.74	17.11	16.20	5.34
7	S17	9.46	351.93	339.02	37.20	35.84	3.67
8	S18	8.93	300.35	287.02	33.63	32.14	4.44
9	S19	8.54	179.91	172.75	21.07	20.23	3.98
10	S20	8.9	418.68	400.24	47.04	44.97	4.40
11	S21	9.43	345.6	329.02	36.65	34.89	4.80
12	S22	8.83	344.38	323.94	39.00	36.69	5.94
13	S23	9.21	491.04	469.71	53.32	51.00	4.34
14	S24	8.62	361.29	342.07	41.91	39.68	5.32
15	S25	9.08	253.83	240.04	27.95	26.44	5.43
16	S26	9.4	215.82	207.21	22.96	22.04	3.99
17	S28	9.68	490.36	471.86	50.66	48.75	3.77
18	S29	8.97	283.47	274.75	31.60	30.63	3.08
19	S30	8.79	229.97	218.49	26.16	24.86	4.99
20	S31	10.06	210.68	200.28	20.94	19.91	4.94
21	S32	9.45	484.11	468.57	51.23	49.58	3.21
22	S33	8.83	262.42	251.5	29.72	28.48	4.16
23	S34	8.46	227.41	219.88	26.88	25.99	3.31
24	S35	8.89	334.72	319.67	37.65	35.96	4.50
25	S36	8.71	332.48	321.64	38.17	36.93	3.26
Maximum		10.06	491.04	471.86	59.40	56.38	5.94
Minimum		8.10	148.68	140.74	16.83	16.17	3.08
Mean		8.97	306.79	293.36	34.24	32.74	4.40

Where *LF*—low frequency, *HF*—high frquency; χLF ($10-5$ SI)—it is the low frequency bulk susceptibility, χHF ($10-5$ SI)—it is the high frequency bulk susceptibility; χLF ($10-7$ m^{3}kg^{-1})—it is the mass specific susceptibility obtained by dividing the bulk susceptibility with sample density at low frequency; χHF ($10-7$ m^3 kg^{-1})—it is the mass specific susceptibility obtained by dividing the bulk susceptibility with sample density at high frequency; sample volume was used is 10 cc values were observed between 2 and 10%. The frequency dependent mass magnetic susceptibility ranges between 3.08 and 5.93% with an average value of 4.40%. These values are very low, it implies that the magnetic enhancement at the station is more of anthropogenic origin than lithogenic and geogenic (Dearing 1999).

21.4.1.1 Vertical Distribution of Magnetic Susceptibility

The vertical distribution of κ value in soil which shows that presence of top soil magnetic susceptibility enhancement as showed as vertical distribution of depth and magnetic susceptibility (Fig. 21.6). The enhanced magnetic susceptibility levels are recorded in the uppermost horizons of the soil profile. Uppermost horizon is recipient of atmospheric pollution deposited on the soil surface including magnetic particles and related heavy metals. The anthropogenic peak of κ value is characteristics of areas with strong anthropogenic pressure (Strzyszcz and Magiera 1998; Hanesch and Scholger 2002).

In the present study, maximum magnetic level enhancement was found at top layer to 6 cm depth, also at the depth of 18 cm, the κ value increases. Whereas at the depth

Fig. 21.6 Vertical distribution of magnetic susceptibility in soil profile

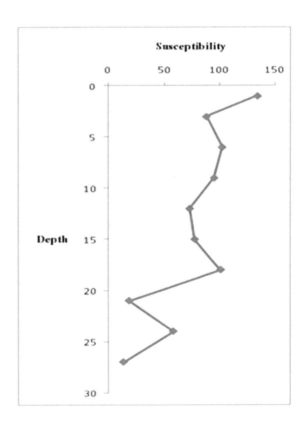

of 9, 12 and 15 cm, the magnetic susceptibility becomes stable and increases at 18 cm depth. In the deeper part of the soil profile, the magnetic susceptibility was constantly decreasing and some of the profile reached constant value. The maximum value of magnetic susceptibility was observed in most of the samples in different depth of the soil profile. In this soil horizon, the majority of anthropogenic contaminants and heavy metals was accumulated due to use of untreated wastewater from sugar factory for irrigation in the study area. Hence, it can be deducted that the magnetic susceptibility and heavy metal concentrations are both high. It indicates that the increasing of magnetic susceptibility is due to the anthropogenic activities.

21.4.2 Heavy Metals

The analytical results heavy metals in agricultural soil from study area were presented in Table 21.4. The Cd concentartion in soil samples was found ranged from 0.021 to 0.963 mg/kg with an average value of 0.682 mg/kg. The average abundance of Cd in parent rock of the studied area, i.e. basalts is about 0.15 mg/kg (Mason and Moore 1982). According to Ellis and Revitt (1982), Cd is due to the influence of fossil fuel burning. The geo-spatial distribution of cadmium in the soil samples from the study areas is higher in alluvium is than the hard rock terrain zone (Fig. 21.7). The higher concentration is found in central portion of the study area, and this contaminant area is an urban settlement area (Fig. 21.7).

Cu was observed is varies from 1.52 to 3.30 mg/kg with an average value of 2.284 mg/kg. The average abundance of Cu in background rock of basalts is about 110 mg/kg (Mason and Moore 1982). The geo-spatial distribution of Cu in soil shows higher concentration at north central and south central part (Fig. 21.8).

Fe concentration is varied from 2.13 to 5.28 mg/kg with an average value of 3.929 mg/kg. The average abundance of Fe in background rock of basalts is about 77,600 mg/kg (Mason and Moore 1982). The geo-spatial distribution of Fe is observed to be more in the central part of the study area (Fig. 21.9).

Mn concentration varies from 3.200 to 8.63 mg/kg with an average value of 6.760 mg/kg. The average abundance of Mn in basalts is about 1280 mg/kg (Mason and Moore 1982). The geo-spatial distribution of Mn in soil showed that all area has high concentration except few patches in the central portion of the study area (Fig. 21.10).

Ni concentration is ranged from 0.270 to 1.190 mg/kg with an average value of 0.595 mg/kg. The average abundance of Ni in background rock of basalts is about 76 mg/kg (Mason and Moore 1982). The geo-spatial distribution of Ni in soil shows that few patches in the south, north and west have low concentration compared to other part of the study area (Fig. 21.11). The main source of nickel is industries and the use of liquid type of manure for agricultural practices which is composted materials and agrochemicals such as fertilizers and pesticides (Gowd et al. 2010).

Pb concentration is varied from 0.27 to 2.35 mg/kg with an average value of 1.328 mg/kg. The average abundance of Pb in background rock of basalts is about

Table 21.4 Heavy metals concentration in soil samples from the study area

Sr. no	Sample ID	Cu	Fe	Mn	Zn	Pb	Cd	Ni
1	S1	1.670	4.570	5.690	1.260	0.890	0.190	0.990
2	S2	1.640	4.650	5.470	0.599	0.670	0.021	0.510
3	S8	2.710	3.270	5.380	0.667	0.310	0.510	0.480
4	S12	2.950	4.030	7.540	0.578	0.360	0.570	0.360
5	S13	2.600	4.170	7.680	0.566	0.270	0.690	0.440
6	S16	2.230	2.770	4.300	0.452	1.770	0.310	1.170
7	S17	2.390	2.810	3.200	0.778	1.370	0.770	0.350
8	S18	2.310	3.390	5.980	0.889	1.230	0.890	0.840
9	S19	2.240	4.210	6.470	1.270	1.410	0.500	0.290
10	S20	2.290	4.430	8.260	1.360	1.170	0.917	0.380
11	S21	2.360	4.590	6.750	1.210	0.980	0.923	0.760
12	S22	3.300	4.610	6.120	1.440	1.940	0.944	0.820
13	S23	2.910	4.470	6.370	1.120	1.330	0.963	0.390
14	S24	1.890	4.790	8.630	0.772	1.630	0.941	0.360
15	S25	1.970	5.110	7.540	1.560	0.800	0.672	0.470
16	S27	2.460	2.130	7.470	1.340	0.690	0.756	0.510
17	S30	2.880	4.510	8.410	0.673	1.180	0.781	0.330
18	S31	2.810	4.660	7.390	1.350	1.370	0.872	0.770
19	S32	1.520	3.010	8.190	0.894	1.970	0.853	1.130
20	S33	2.110	3.190	7.660	0.674	2.090	0.787	0.270
21	S34	1.670	3.310	6.890	0.661	2.350	0.038	1.190
22	S35	1.740	3.400	6.750	0.778	2.110	0.876	0.490
23	S36	1.880	4.280	7.330	0.642	2.660	0.903	0.390
Maximum		3.300	5.110	8.630	1.560	2.660	0.963	1.190
Minimum		1.520	2.130	3.200	0.452	0.270	0.021	0.270
Average		2.284	3.929	6.760	0.936	1.328	0.682	0.595

Heavy metals concentration in soil samples were expressed in mg/kg

7.8 mg/kg (Mason and Moore 1982). Presence of Pb in soils could be due to the burning of fossils fuels (Garty et al. 1996) or due to excessive use of poly vinyl chloride (PVC) pipes for irrigation and their degradation even in small quantities. The geo-spatial distribution of Pb in soil showed that higher concentration was found in central and northwest part of the study area (Fig. 21.12).

Zinc concentration in soil varies from 0.452 to 1.560 mg/kg with an average value is 0.936 mg/kg. The average abundance of Zn in basalts is about 86 mg/kg (Mason and Moore 1982). The geo-spatial distribution of Zn in soil shows that few patches of the north part have high concentration than the other part of the study area (Fig. 21.13).

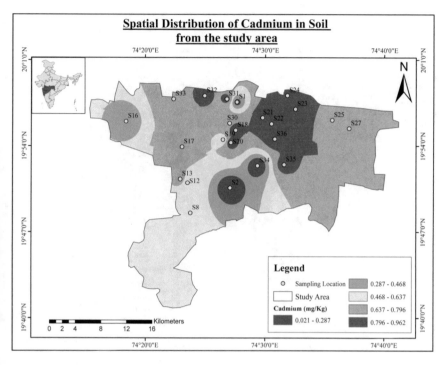

Fig. 21.7 Spatial variation of cadmium in soil from the study area

21.4.2.1 Assessment of Metal Contamination

The Igeo values of trace elements in soils for each site are given in Table 21.5. According to the geoaccumulation indices which are calculated by using Eq. (21.1) and listed in Table 21.5, the analyzed soil samples are extremely contaminated with Fe; heavily to extremely contaminated with manganese and moderately to heavily contaminate with zinc and cooper. Whereas the soils samples are uncontaminated with cadmium and uncontaminated to moderately contaminated with nickel and lead. The results of the geoaccumulation index also confirmed that the iron, manganese, zinc and copper have the highest contamination effects, whilst cadmium and nickel have no or nearly contamination effects, respectively. The geoaccumulation class (Igeo class) sequence is observed to be Cd > Pb > Ni > Zn > Cu > Mn > Fe. The data suggest that geoaccumulation with trace element proves the degree of obtainable trace element concentration in soil of the area is exceed toxic limits except the metal cadmium. This data also showed that the heavy metal contamination due to anthropogenic sources because he geoaccumulation index.

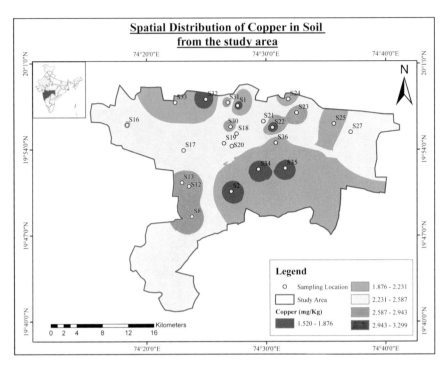

Fig. 21.8 Spatial variation of copper in soil from the study area

21.4.2.2 Correlation Between Magnetic Susceptibility and Heavy Metals

The Pearson correlation coefficients between the heavy metals concentrations and magnetic susceptibility values were observed and presented in Table 21.6. The results were classified according to the correlation coefficient R (Yu and Hu 2005), as the following orders.

  i. R value ranged from 0.8 to 1.00 suggests a strong correlation;
  ii. R value ranged from 0.5 to 0.8 suggests a significant correlation;
  iii. R value ranged from 0.3 to 0.5 suggests a weak correlation;
  iv. R value is less than 0.3 suggests an insignificant correlation.

The resulting correlation analysis demonstrated that concentrations of Fe ($R = 0.46$) and Cu ($R = 37$) were weakly correlated with MS, whereas Zn ($R = 0.29$), Cd ($R = 0.10$) and Ni ($R = 0.04$) show insignificant correlation. A weakly positive and insignificant correlation between the heavy metals with magnetic susceptibility were observed for most of the selected elements, indicating a pollution source is not common.

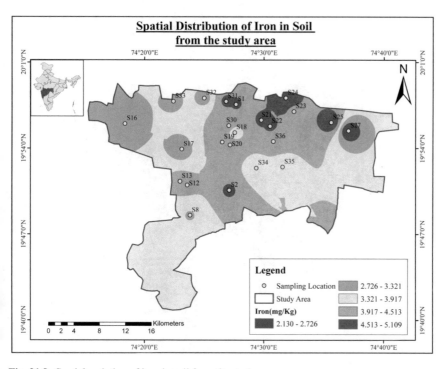

Fig. 21.9 Spatial variation of iron in soil from the study area

21.5 Conclusion

The frequency dependent mass magnetic susceptibility of soil samples from the study areas is observed between 3.08 and 5.93% which is indication of anthropogenic contamination, and in addition, it is seen that magnetic particles have been superparamagnetic and multi-domain size > 105 μm. Concentration of Cd in all soil samples shows non contamination. However, other heavy metals like Cu, Fe, Pb, Mn, Ni and Zn in soils from all sites were observed higher concentration. The Igeo values suggest that contamination of all the analyzed heavy metals from soils is higher in all soils samples from study area except the metal cadmium. Agricultural soils are also significantly influenced by heavy metals which are derived from anthropogenic activities. The sources of heavy metals in the studied soils are mainly derived from traffic sources due to burning of fossil fuels. However, the sources of heavy metals element in agricultural soils are mainly influenced by parent materials (parent rock), extensive use of agrochemical fertilizer and pesticides which contain heavy metals. It is highly recommended carrying out a remediation action in the studied region as well as surroundings using suitable techniques in order to remove the contamination effects and to safe the human health.

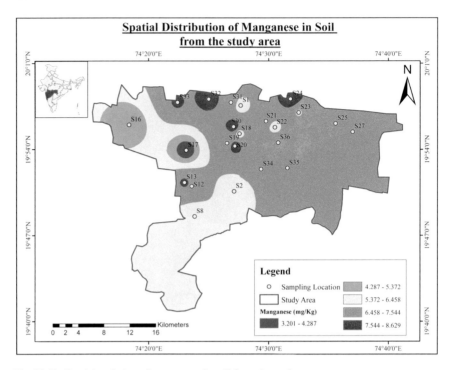

Fig. 21.10 Spatial variation of manganese in soil from the study area

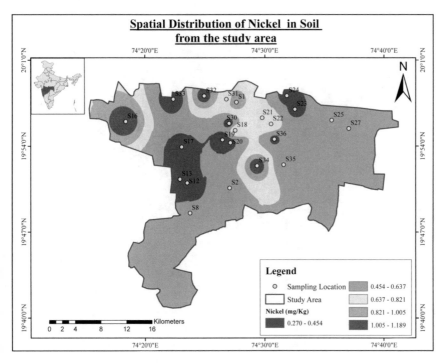

Fig. 21.11 Spatial variation of nickel in soil from the study area

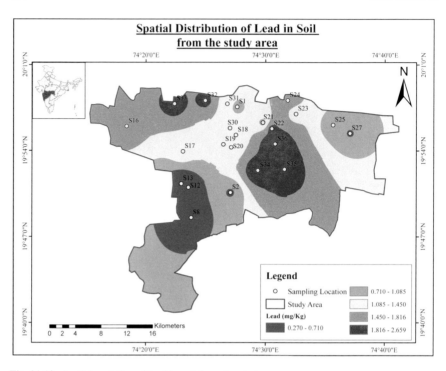

Fig. 21.12 Spatial variation of lead in soil from the study area

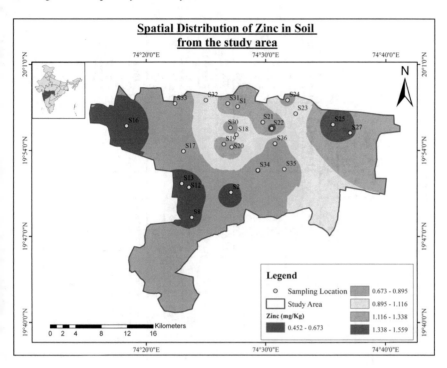

Fig. 21.13 Spatial variation of zinc in soil from the study area

Table 21.5 Geoaccumulation index of soil samples from the study area

Sample ID	Element	Cu	Fe	Mn	Zn	Pb	Cd	Ni
S1	Igeo	2.389	5.675	3.987	2.160	0.966	−1.420	2.001
	Igeo class	2 < Igeo ≤ 3	Igeo ≥ 5	3 < Igeo ≤ 4	2 < Igeo ≤ 3	0 < Igeo ≤ 1	Igeo ≤ 0	2 < Igeo ≤ 3
S2	Igeo	2.381	5.682	3.970	1.837	0.843	−2.377	1.713
	Igeo class	2 < Igeo ≤ 3	Igeo ≥ 5	3 < Igeo ≤ 4	1 < Igeo ≤ 2	0 < Igeo ≤ 1	Igeo ≤ 0	1 < Igeo ≤ 2
S8	Igeo	2.599	5.529	3.963	1.884	0.508	−0.991	1.687
	Igeo class	2 < Igeo ≤ 3	Igeo ≥ 5	3 < Igeo ≤ 4	1 < Igeo ≤ 2	0 < Igeo ≤ 1	Igeo ≤ 0	1 < Igeo ≤ 2
S12	Igeo	2.636	5.620	4.110	1.821	0.573	−0.943	1.562
	Igeo class	2 < Igeo ≤ 3	Igeo ≥ 5	3 < Igeo ≤ 4	1 < Igeo ≤ 2	0 < Igeo ≤ 1	Igeo ≤ 0	1 < Igeo ≤ 2
S13	Igeo	2.581	5.635	4.118	1.812	0.448	−0.860	1.649
	Igeo class	2 < Igeo ≤ 3	Igeo ≥ 5	3 < Igeo ≤ 4	1 < Igeo ≤ 2	0 < Igeo ≤ 1	Igeo ≤ 0	1 < Igeo ≤ 2
S16	Igeo	2.515	5.457	3.866	1.715	1.265	−1.208	2.074
	Igeo class	2 < Igeo ≤ 3	Igeo ≥ 5	4 < Igeo ≤ 5	1 < Igeo ≤ 2	0 < Igeo ≤ 1	Igeo ≤ 0	2 < Igeo ≤ 3
S17	Igeo	2.545	5.464	3.737	1.950	1.154	−0.812	1.550
	Igeo class	2 < Igeo ≤ 3	Igeo ≥ 5	3 < Igeo ≤ 4	1 < Igeo ≤ 2	0 < Igeo ≤ 1	Igeo ≤ 0	1 < Igeo ≤ 2
S18	Igeo	2.530	5.545	4.009	2.008	1.107	−0.750	1.930
	Igeo class	2 < Igeo ≤ 3	Igeo ≥ 5	3 < Igeo ≤ 4	2 < Igeo ≤ 3	0 < Igeo ≤ 1	Igeo ≤ 0	1 < Igeo ≤ 2
S19	Igeo	2.517	5.639	4.043	2.163	1.166	−1.000	1.468
	Igeo class	2 < Igeo ≤ 3	Igeo ≥ 5	4 < Igeo ≤ 5	2 < Igeo ≤ 3	1 < Igeo ≤ 2	Igeo ≤ 0	1 < Igeo ≤ 2
S20	Igeo	2.526	5.661	4.149	2.193	1.085	−0.737	1.586
	Igeo class	2 < Igeo ≤ 3	Igeo ≥ 5	4 < Igeo ≤ 5	2 < Igeo ≤ 3	1 < Igeo ≤ 2	Igeo ≤ 0	1 < Igeo ≤ 2
S21	Igeo	2.539	5.677	4.061	2.142	1.008	−0.734	1.887
	Igeo class	2 < Igeo ≤ 3	Igeo ≥ 5	4 < Igeo ≤ 5	2 < Igeo ≤ 3	1 < Igeo ≤ 2	Igeo ≤ 0	1 < Igeo ≤ 2

(continued)

Table 21.5 (continued)

Sample ID	Element	Cu	Fe	Mn	Zn	Pb	Cd	Ni
S22	Igeo class	2 < Igeo ≤ 3	Igeo ≥ 5	4 < Igeo ≤ 5	2 < Igeo ≤ 3	1 < Igeo ≤ 2	Igeo ≤ 0	1 < Igeo ≤ 2
	Igeo	2.685	5.679	4.019	2.218	1.305	−0.724	1.920
S23	Igeo class	2 < Igeo ≤ 3	Igeo ≥ 5	4 < Igeo ≤ 5	2 < Igeo ≤ 3	1 < Igeo ≤ 2	Igeo ≤ 0	1 < Igeo ≤ 2
	Igeo	2.630	5.665	4.036	2.109	1.141	−0.715	1.597
S24	Igeo class	2 < Igeo ≤ 3	Igeo ≥ 5	4 < Igeo ≤ 5	2 < Igeo ≤ 3	1 < Igeo ≤ 2	Igeo ≤ 0	1 < Igeo ≤ 2
	Igeo	2.443	5.695	4.168	1.947	1.229	−0.725	1.562
S25	Igeo class	2 < Igeo ≤ 3	Igeo ≥ 5	4 < Igeo ≤ 5	1 < Igeo ≤ 2	1 < Igeo ≤ 2	Igeo ≤ 0	1 < Igeo ≤ 2
	Igeo	2.461	5.723	4.110	2.253	0.920	−0.872	1.678
S27	Igeo class	2 < Igeo ≤ 3	Igeo ≥ 5	4 < Igeo ≤ 5	2 < Igeo ≤ 3	0 < Igeo ≤ 1	Igeo ≤ 0	1 < Igeo ≤ 2
	Igeo	2.557	5.343	4.105	2.187	0.856	−0.820	1.713
S30	Igeo class	2 < Igeo ≤ 3	Igeo ≥ 5	4 < Igeo ≤ 5	2 < Igeo ≤ 3	0 < Igeo ≤ 1	Igeo ≤ 0	1 < Igeo ≤ 2
	Igeo	2.626	5.669	4.157	1.887	1.089	−0.806	1.524
S31	Igeo class	2 < Igeo ≤ 3	Igeo ≥ 5	4 < Igeo ≤ 5	1 < Igeo ≤ 2	1 < Igeo ≤ 2	Igeo ≤ 0	1 < Igeo ≤ 2
	Igeo	2.615	5.683	4.101	2.190	1.154	−0.758	1.892
S32	Igeo class	2 < Igeo ≤ 3	Igeo ≥ 5	4 < Igeo ≤ 5	2 < Igeo ≤ 3	1 < Igeo ≤ 2	Igeo ≤ 0	1 < Igeo ≤ 2
	Igeo	2.348	5.493	4.145	2.011	1.311	−0.768	2.059
S33	Igeo class	2 < Igeo ≤ 3	Igeo ≥ 5	4 < Igeo ≤ 5	2 < Igeo ≤ 3	1 < Igeo ≤ 2	Igeo ≤ 0	2 < Igeo ≤ 3
	Igeo	2.491	5.519	4.116	1.888	1.337	−0.803	1.437
S34	Igeo class	2 < Igeo ≤ 3	Igeo ≥ 5	4 < Igeo ≤ 5	1 < Igeo ≤ 2	1 < Igeo ≤ 2	Igeo ≤ 0	1 < Igeo ≤ 2
	Igeo	2.389	5.535	4.070	1.880	1.388	−2.119	2.081
	Igeo class	2 < Igeo ≤ 3	Igeo ≥ 5	4 < Igeo ≤ 5	1 < Igeo ≤ 2	1 < Igeo ≤ 2	Igeo ≤ 0	2 < Igeo ≤ 3

(continued)

Table 21.5 (continued)

Sample ID	Element	Cu	Fe	Mn	Zn	Pb	Cd	Ni
S35	Igeo	2.407	5.546	4.061	1.950	1.341	−0.756	1.696
	Igeo class	$2 < \mathrm{Igeo} \leq 3$	$\mathrm{Igeo} \geq 5$	$4 < \mathrm{Igeo} \leq 5$	$1 < \mathrm{Igeo} \leq 2$	$1 < \mathrm{Igeo} \leq 2$	$\mathrm{Igeo} \leq 0$	$1 < \mathrm{Igeo} \leq 2$
S36	Igeo	2.440	5.646	4.097	1.867	1.442	−0.743	1.597
	Igeo class	$2 < \mathrm{Igeo} \leq 3$	$\mathrm{Igeo} \geq 5$	$4 < \mathrm{Igeo} \leq 5$	$1 < \mathrm{Igeo} \leq 2$	$1 < \mathrm{Igeo} \leq 2$	$\mathrm{Igeo} \leq 0$	$1 < \mathrm{Igeo} \leq 2$
Average	Igeo	2.515	5.599	4.052	2.003	1.071	−0.976	1.733
	Igeo class	$2 < \mathrm{Igeo} \leq 3$	$\mathrm{Igeo} \geq 5$	$4 < \mathrm{Igeo} \leq 5$	$2 < \mathrm{Igeo} \leq 3$	$1 < \mathrm{Igeo} \leq 2$	$\mathrm{Igeo} \leq 0$	$1 < \mathrm{Igeo} \leq 2$

Table 21.6 Pearson correlation coefficient (R) matrix between heavy metals and magnetic susceptibility

	Cu	Fe	Mn	Zn	Pb	Cd	Ni	χ FD (%)
Cu	1.00	0.12	−0.04	0.17	−0.35	0.38	−0.28	0.38
Fe		1.00	0.32	0.35	−0.15	0.10	−0.23	0.46
Mn			1.00	0.16	0.06	0.38	−0.23	−0.03
Zn				1.00	−0.11	0.31	0.02	0.29
Pb					1.00	0.18	0.26	−0.30
Cd						1.00	−0.36	0.10
Ni							1.00	0.04
χ FD (%)								1.00

Acknowledgements The authors are thankful to the Department of Applied Geology, School of Environmental and Earth Sciences, Kavayitri Bahinabai Chaudhari North Maharashtra University Jalgaon, India for providing the necessary resources. Authors also thankful to Indian Institute of Geomagnetism Kalamboli, Navi Mumbai for testing of soil samples for magnetic susceptibility. Author (RBG) officially acknowledge to the Principal, G.B. Tatha Tatyasaheb Khare Commerce, Parvatibai Gurupad Dhere Art's and Shri. Mahesh Janardan Bhosale Science College, Guhagar District Ratnagiri (M.S.), India.

References

Dearing JA (1999) Environmental magnetic susceptibility, using the Bartington MS2 system, 2nd edn. Chi Publishing, England

Ellis JB, Revitt DM (1982) Incidence of heavy metals in street surface sediments: solubility and grain size studies. Water Air Soil Pollut 17:87–100

Flanders PJ (1994) Collection, measurement, and analysis of airborne magnetic particulates from pollution in the environment. J Appl Phys 75:5931–5936

Garty J, Kauppi M, Kauppi A (1996) Accumulation of airborne elements from vehicles in transplanted lichens in urban sites. J Environ Qual 25:265–272

Gowd SS, Reddy RM, Govil PK (2010) Assessment of heavily metal contamination in soils at Jajmau (Kanpur) and Unnao industrial areas of the Ganga Plain, Uttar Pradesh, India. J Hazard Mater 174:113–121

Hanesch M, Scholger R (2002) Mapping of heavy metal loadings in soils by means of magnetic susceptibility measurements. Environ Geol 42:857–870

Hay KL, Dearing JA, Baban SM, Lovel J (1997) A preliminary attempt to identify atmospherically-derived pollution particles in English topsoil from magnetic susceptibility measurements. Phys Chem Earth 22:207–210

Kanu MO, Meludu OC, Oniku SA (2013) Measurement of magnetic susceptibility of soils in jalingo, N-E Nigeria: a case study of the jalingo mechanic villag. World Appl Sci J 24:178–187

Krishna AK, Govil PK (2005) Heavy metal distribution and contamination in soils of Thane-Belapur industrial development area, Mumbai Western India. Environ Geol 47(8):1054–1061

Mason B, Moore CB (1982) Principle of geochemistry, 4th edn. Wliely, New York

Matzka J, Maher BA (1990) Magnetic bio-monitoring of roadside tree leaves: identification of spatial and temporal variations in vehicle-derived particulates. Amos Environ 33:4565–4569

Mohamed EB, Mohamed S, Ahmed B, Samir N (2011) Heavy metal pollution and soil of Beni Mellal City (Morocco). Environ Earth Sci. Springer, Berlin. https://doi.org/10.1007/s12665-011-1215-5

Muller G (1969) Index of geo-accumulation in sediments of the Rhine River. Geo Jour 2:108–118

Shankar V, Martin M, Bhatt A, Erkman S (1994) Draft on status of implementation of the hazardous waste Rules 1989. Ministry of Environment and Forests, New Delhi

Strzyszcz Z, Magiera T (1998) Magnetic susceptibility and heavy metals contamination in soils of southern Poland. Phys Chem Earth 23:1127–1131

United Stated Environmental Protection Agency (1986) Acid digestion of sediment, sludge and soils' USEPA test methods for evaluating soil waste SW-846. USEPA, Cincinnati, OH, United States of America

Venkatachalapathy R, Veerasingam S, Basavaiah N, Ramkumar T, Deenadayalan K (2011) Environmental magnetic and geochemical characteristics of Chennai coastal sediments, Bay of Bengal India. J Earth Syst Sci 120(5):885–895

Xue SW, Yong Q (2005) Correlation between magnetic susceptibility and heavy metals in urban topsoil: a case study from the city of Xuzhou China. J Enviro Geol 49:10–18

Yu J, Hu X (2005) Application of data statistical analysis with SPSS. Post and Telecom Press, Beijing, pp 163–173

Printed in the United States
by Baker & Taylor Publisher Services